The Great Civil Engineering Overhaul

Other Titles of Interest

Standard Practice for Sustainable Infrastructure, ASCE/COS 73-23, by the American Society of Civil Engineers (ASCE/COS 2023). Standard Practice for Sustainable Infrastructure, ASCE/COS 73-23, addresses development and implementation of sustainable infrastructure solutions through the entire infrastructure life-cycle process. (ISBN 978-0-7844-1573-3)

Adapting Infrastructure and Civil Engineering Practice to a Changing Climate, by the Committee on Adaptation to a Changing Climate; edited by J. Rolf Olsen (ASCE/CTA 2015). Adapting Infrastructure and Civil Engineering Practice to a Changing Climate presents an accurate discussion of the potential significance of climate change to engineering practice. (ISBN 978-0-7844-7919-3)

Climate-Resilient Infrastructure: Adaptive Design and Risk Management, by the Committee on Adaptation to a Changing Climate; edited by Bilal M. Ayyub (ASCE/CACC 2018). MOP 140 provides guidance for developing or enhancing of methods for infrastructure analysis and design to achieve infrastructure resilience targets while minimizing life-cycle costs in a changing climate. (ISBN 978-0-7844-1519-1)

Climate Change Modeling, Mitigation, and Adaptation, edited by Rao Y. Surampalli, Tian C. Zhang, C. S. P. Ojha, B. R. Gurjar, R. D. Tyagi, and C. M. Kao (ASCE/EWRI 2013). This title contains 25 invited chapters that present the most current thinking on the environmental mechanisms contributing to global climate change and explore scientifically grounded steps to reduce the buildup of greenhouse gases in the atmosphere. (ISBN 978-0-7844-1271-8)

Impacts of Future Weather and Climate Extremes on United States Infrastructure: Assessing and Prioritizing Adaptation Actions, by Mari R. Tye and Jason P. Giovannettone (ASCE/CACC 2021). This report provides prioritization frameworks in accommodating projected future weather and climate extremes for policy makers and engineers involved in infrastructure planning and design. (ISBN 978-0-7844-1586-3)

The Great Civil Engineering Overhaul

Bill Wallace

ASCE PRESS

Published by the American Society of Civil Engineers

Cataloging-in-Publication Data on file with the Library of Congress

LCCN 2024002995

Published by American Society of Civil Engineers
1801 Alexander Bell Drive
Reston, Virginia 20191-4382
www.asce.org/bookstore | ascelibrary.org

***Errata:* Errata, if any, can be found at https://dx.doi.org/10.1061/9780784485446.**

ISBN 978-0-7844-1622-8 (print)
ISBN 978-0-7844-8544-6 (PDF)
ISBN 978-0-7844-8545-3 (ePub)

Manufactured in the United States of America.

28 27 26 25 24 1 2 3 4 5

Book cover credit: Graph courtesy of NASA. The graph shows the seasonal cycle of temperature variation on the Earth's surface and how those temperatures deviate from the average from 1951 to 1980. The global mean surface air temperature for that period was 14 °C (57 °F), with an uncertainty of several tenths of a degree.

Contents

Preface

Here is my B.H.A.G. My Big, Hairy, Audacious Goal (Collins and Porras 1994) in writing this book is to instigate a major overhaul in civil engineering standards and practices, an overhaul made necessary by the consequences of a warming and increasingly harsh climate. I intend to provide students, engineers, and other practitioners with the concepts, methods, and tools for handling the greatest professional challenge and the greatest opportunity they will ever have in their careers. For others, I offer a more straightforward explanation of sustainability and its relationship to the natural environment, climate, and human well-being. I thought it would be helpful to know how we got to this point in our economic development and why extensive changes are necessary if we want our progress to continue.

Why is this overhaul necessary? Increasing atmospheric concentrations of greenhouse gases[1] (GHGs, mostly carbon dioxide), produced primarily from burning fossil fuels for heat, mobility, and electrical power, have overwhelmed the Earth's climate-regulating systems. The result is global warming. Increased thermal energy retained in the atmosphere is disrupting a once stable climate. The Earth's changing climate is having disastrous effects, disrupting economies and societies worldwide. Sea levels are rising and are projected to inundate many coastal cities. Rising seas and storm surges are overtopping barriers engineered for flood protection. Extreme temperatures and heat waves are taxing power systems. Extended droughts are creating water scarcities that were never anticipated. More frequent and intense rainstorms and melting snows are flooding cities, towns, and agricultural lands. Six of the most expensive disaster years have occurred in the United States in the last 11 years. At US$155 billion, 2021 was the third most costly year since 1980. Then 2022 came in at US$165 billion, knocking 2021 out of third place (NOAA NCEI 2024,

[1]Greenhouse gases (GHGs) are carbon dioxide (76%), methane (16%), nitrous oxide (6%), and fluorinated gases (2%). The warming potential of GHGs is measured in carbon dioxide equivalents (CO_2eq). Since carbon dioxide is the primary greenhouse gas, GHG emissions reduction efforts generally focus on reducing carbon dioxide emissions, referring to carbon dioxide and other greenhouse gases as simply carbon.

Munich 2021). Since 1980, extreme weather events have cost this country over US$2.6 trillion (NOAA NCEI 2024).

Climate change is overturning a foundational civil engineering assumption: past conditions are good predictors of future conditions. That assumption, known as stationarity, has allowed engineers to design long-lived infrastructure projects with confidence, knowing that the completed works will perform safely and reliably throughout their design life. But climate change is creating new and harsh conditions markedly different from what was universally understood, accepted, and codified into engineering standards and practices. The costly disasters are primarily the result of civil infrastructure assets being overwhelmed by weather events reaching levels of intensity, duration, and frequency that were never accounted for in infrastructure designs. Existing infrastructure assets need to be strengthened or relocated. New infrastructure assets must avoid climate hazards and be made more resilient and adaptive to changing climate conditions. At the same time, GHG emissions from these assets must be reduced dramatically, not just emissions from operations but emissions released throughout the entire life cycle of the infrastructure asset.

Climate change is an urgent problem. The world's climate scientists tell us that unless GHG emissions are dramatically reduced, climate conditions will become markedly worse. Some conditions may become irreversible. At the end of 2022, global temperatures had risen 1.2 °C (2.2 °F) above preindustrial levels. According to the Intergovernmental Panel on Climate Change (IPCC), unless global temperature increases are kept below 1.5 °C (2.7 °F), climate-related risks to health, livelihoods, food security, water supply, human security, and economic growth will increase substantially (IPCC 2018). Staying below 1.5 °C (2.7 °F) means the world's nations must reach net-zero carbon emissions by 2050. Getting on that emissions reduction track means cutting the world's carbon emissions in half by 2030. The nations' pledges and targets as of December 2023 will only limit warming to 2.1 °C (3.8 °F) (Climate Analytics 2023).

As I stated in a 2018 article for *Engineering News-Record* (Wallace 2018), we engineers are part of the problem. The primary sources of GHG emissions come from the transportation, power, and industrial sectors, sectors for which we plan, design, build, and operate the facilities and systems that produce those emissions. Now, we can and should be part of the solution: changing out fossil energy systems and replacing them with low- or no-carbon-emitting systems, using materials with low embodied carbon, and designing and building infrastructure projects that are resilient and adaptive to changing climate conditions. We know these industry sectors. We know what works and doesn't work. We can use our knowledge and experience to handle the problems of a changing climate.

So far, the civil engineering community has been slow to respond. This is understandable. Civil engineers, as the saying goes, are between a rock and a hard place. Ethically and morally, civil engineers are responsible for protecting public health, safety, and welfare. Accordingly, they are obligated to deviate from many accepted climate-related standards and practices, recognizing that they are no longer reliable for handling changing climate conditions. At the same time, a civil engineer's license to practice is predicated on their knowledge and competency in applying currently accepted standards and practices. The engineer's defense against claims of negligence on a project is to demonstrate their adherence to the standard of care (i.e., accepted standards and practices were applied over the course of the project). Unfortunately, the currently accepted climate-related standards and practices have yet to be revised to account for a changing climate.

Adhering to unreliable standards and practices can increase the engineer's exposure to claims and litigation. In their book *Building a Resilient Tomorrow,* Hill and Martinez-Diaz (2022) issued this warning:

> At the greatest legal risk will be professionals who rely only on historical experience to make decisions even when credible authorities—such as the authors of the US Fourth National Climate Assessment, a congressionally-mandated analysis issued in 2018 identifying climate risks to the nation—assert that it is no longer valid to assume that future climate conditions will resemble the past. In such cases, judges and juries may begin to find that architects, planners, and engineers acted unreasonably when they made consequential decisions based on outdated information. For example, using building codes and standards that do not reflect the current and future risk of climate-exacerbated extreme weather may be viewed as falling below the standard of care required by professionals.

This standard of care interpretation is already working its way through the courts. In a recent case, a District Court in Louisiana held that engineering companies involved in designing a highway barrier could be held liable for flood damage from a 1,000-year flood event even though the Louisiana State standards for highway barriers call for protection from a 50-year flood event (GovInfo 2017). In 2016, a highway barrier built on I-12 in accordance with state standards acted as a dam and flooded more than 100,000 homes and businesses. Many were not covered by flood insurance because they were outside what were designated as high-risk flood zones in which insurance was required. Damage costs are in the neighborhood of US$20 billion (Jones 2016).

Unfortunately, the assumption of stationarity is baked into the civil engineering discipline. Standards developed for handling a stable climate are presumed to be effective: products of a long history of satisfactory results under historical climate conditions. Changing accepted engineering standards is made difficult by design. Any changes must undergo a rigorous vetting process to protect public health, safety, and welfare from unproven modifications. In addition, standards have become increasingly prescriptive over time as project owners and engineers seek to reduce project costs, banking on both a stable climate and past project successes.

Civil engineers have become indoctrinated, accepting stationarity as an operating principle. They are educated, tested, and licensed to practice based on their proficiency in applying accepted standards and practices based on historic climate stability. Workflow in the engineering industry is preordained: respond to project proposal requests, compete for and win projects, follow a defined scope of work using accepted standards and practices, complete the project, and position for the next. New science is interesting but irrelevant. Working outside the box is risky as it exposes engineers to claims and litigation. Innovation is confined to current methods, just doing them faster, better, or cheaper.

Despite these challenges, civil engineers must start the overhaul process. Standards must be revised to handle climate uncertainties and drastically reduce GHG emissions. And while these revisions are underway, interim guidance must be developed and broadly disseminated. This guidance must state that project owners and engineers are responsible for incorporating climate change considerations into project planning, design, construction, and operation and recommend how such provisions can be made. General guidance is available from the US Department of Transportation, US Army Corps of Engineers, and other government agencies. Whereas deviating from accepted standards is risky, equally risky is not seeking to incorporate climate considerations into the infrastructure project delivery process. This guidance should show that project delivery methodologies for handling climate uncertainties exist and are being applied successfully. This book summarizes five of these methodologies and details two that seem most practical. Example applications are provided.

As of this writing, funds are becoming available to make this happen. In 2022, the 117th Congress passed two landmark pieces of legislation: the Infrastructure Investment and Jobs Act (IIJA, also known as the Bipartisan Infrastructure Law or BIL) and the Inflation Reduction Act (IRA). Combined, they represent an investment of US$1.25 trillion in carbon emissions reduction and resilient, adaptive infrastructure (Tomer 2022).

To engage, engineers need to understand why an overhaul is necessary, what needs to be done, and how to do it. Here is how this book is organized.

Chapter 1 starts with the *why*. Civil engineering practices need to change in response to a changing climate. Damages are mounting. Today's infrastructure was designed assuming a stable climate, but the climate is no longer stable. The cause of this instability is global warming. Burning fossil fuels releases carbon dioxide and other so-called greenhouse gases. Increasing atmospheric concentrations of these gases have disrupted the Earth's climate-regulating systems. The additional thermal energy retained in the atmosphere brings about higher temperatures and more extreme weather events. Scientists are predicting that if society does not reduce carbon emissions to net zero by 2050, harsh climate and weather events will become markedly worse (IPCC 2018). The world's nations have committed to achieving net-zero carbon emissions by 2050, but current efforts are falling woefully short of what is needed.

Chapter 2 goes back to basics, connecting today's civil infrastructure with the assumption of stationarity and the work of civil engineers. As it turns out, civil engineers have always considered weather and climate conditions in their designs, but they assumed that those conditions were statistically constant, an assumption called *stationarity*. Civil engineering standards and practices handled all but the most extreme climate conditions. Climate and weather extremes were accounted for up to the point at which their chances of happening were too remote to worry about.

Chapter 3 discusses the relationships among civil infrastructure, natural capital, and sustainability. It also deals with the question, "Are we sustainable?" In 1987, the United Nations Brundtland Commission defined sustainable development as development that "meets the needs of the present without compromising the ability of future generations to meet their own needs." This chapter answers many questions. What are those needs, and what are we doing to meet those needs? Is society's approach to development sustainable? Clearly, human society has made great progress. How did we get to where we are today, and at what cost?

Chapter 4 describes the consequences of unsustainable development. Much of what human society gained in development has been at the expense of Earth's resources and services. As one author describes it, human progress can be tracked as a climb up the heat ladder, moving from one energy source to the next: one that is more convenient, potent, or economical. Humans are stuck on the rung of fossil fuels, and the consequences are enormous. Burning fossil fuels has released vast amounts of greenhouse gases, which have degraded the Earth's climate-regulating services. Consequently, there has been a general warming of the Earth's atmosphere, increasing extreme and damaging weather events, products of a changing climate. This damage has placed additional stresses on society, environmentally, economically, and socially. Civil engineers are in uncharted territory, seeing extreme values of climate and weather variables for which they have little or no experience.

Chapter 5 describes the *how*, engineering responses to a changing climate. According to former White House Science Advisor John Holdren (2007),

> We basically have three choices – mitigation, adaptation, and suffering. We're going to do some of each. The question is what the mix is going to be. The more mitigation we do, the less adaptation will be required, and the less suffering there will be.

Designing new or refurbishing existing infrastructure will involve mitigation and adaptation: mitigation to address the causes of a changing climate and adaptation to address the consequences. Both need to be done in ways that are fair and equitable. The scale of the response must be comprehensive, involving both new and existing infrastructure assets. Mitigation must focus on greenhouse gas emissions reduction, changing out fossil energy sources for low- or no-carbon sources, and using materials with low- or no-embodied carbon. Adaptation must cover new and existing infrastructure assets, adding robustness and resilience across the board. The chapter also discusses Plan B, "last-resort" engineering of the planet to reduce solar radiation input and remove greenhouse gases from the atmosphere.

Chapter 6 extends the *how*. It describes the concept of deep uncertainty as it applies to global climate change and civil infrastructure design. Here, reference is made to a book by Marchau et al., *Decision Making Under Deep Uncertainty* (2019), and summarizes five methodologies for handling climate and other uncertainties along with example applications. Of the five methodologies, dynamic adaptive planning (DAP) and dynamic adaptive policy pathways (DAPP) stand out as the most understandable and helpful in handling the risks from changing climate conditions as well as the other conventional infrastructure project risks. The DAP methodology applies to individual infrastructure projects, covering a broad range of uncertainties, including climate and weather. The DAPP methodology is more suitable for handling one or more climate-related objectives within a defined area such as a city, region, or watershed.

Chapter 7 offers a derivation of DAP, adding resiliency to the original process. The process enables engineers to (1) meet project owner requirements and specifications while (2) being sufficiently durable, adaptable, and resilient to handle changing climate conditions and (3) improving sustainable performance, mainly by reducing the use of fossil fuels. As it turns out, engineers and project owners have been using a form of DAP all along to handle typical infrastructure project uncertainties (e.g., the loss of a key person on a project, changes in demand for services, and reductions in funding for operations). In this derivation, climate and weather uncertainties are also identified and accounted for.

Chapter 8 offers a derivation of dynamic adaptive policy pathways (DAPP), a methodology for reducing exposures to a limited set of climate-related risks across a defined area. This is accomplished through a series of actions (called pathways), each designed to handle a defined range of conditions. The pathways are sequenced so that as conditions change, a new pathway is taken when the previous one is no longer effective. Pathways can be physical (constructed works), operational (adjustments to existing operations), or policy (guidance, regulations, laws) intended to achieve some overall objective. In DAPP, the pathways are designed to get and keep the plan on track to achieve success. Each pathway (or group of pathways) has its own set of conditions for success. Keeping the plan on track requires an application of DAP, detecting and responding to deviations from those conditions.

Chapter 9 concludes with a summary of actions being taken globally, along with guidance on how engineers are expected to support these actions. Today, there is broad acknowledgment that the Earth has warmed significantly, and the warming is human-caused. Virtually all of the world's nations agree that carbon emissions must be reduced to net zero by 2050, but achieving that goal is another matter. As of this writing, warming has reached 1.3 °C (2.3 °F) above preindustrial levels and continues to rise. Nations have committed to reducing their carbon emissions, but many of the individual national ambitions are limited. Because of new climate legislation, the United States can now reach its net-zero commitment. The United Kingdom's Institution of Civil Engineers offers six ways for civil engineers to take action on climate change.

Will this overhaul be done? The answer is yes, but when and by whom? Some engineers are working on it, reviewing existing standards and practices, locating climate-related content, and starting the updating process. Some are talking to their bosses or clients, warning them of the climate-related risks they face and changing how projects are planned, designed, and delivered. But for most others, climate change has yet to register as an urgent issue. When asked in a survey what they are doing about climate change, most architecture and engineering professionals said they are waiting for mandatory codes and standards (Laxo 2023). In effect, they want someone to tell them to do it.

Civil engineers: If that is what you want, so be it. If not, it is time to lead. You are the best equipped to respond to a changing climate. You know what works and doesn't work. Your project delivery experience is all about managing change. If not you, then who?

References

Climate Analytics. 2023. Carbon Action Tracker: *The CAT thermometer.* December 2023 Update. Accessed March 14, 2024. https://climateactiontracker.org/global/cat-thermometer/.

Collins, J. C., and J. I. Porras. 1994. *Built to last: Successful habits of visionary companies.* New York: HarperCollins.

GovInfo. 2017. City of Walker et al v. State of Louisiana through the Department of Transportation and Development et al, Docket No. 654,278; § 26. 19th Judicial District Court, Parish of East Baton Rouge, State of Louisiana. Accessed March 2, 2023. https://www.govinfo.gov/app/details/USCOURTS-lamd-3_17-cv-00055/summary.

Hill, A. C., and L. Martinez-Diaz. 2022. *Building a resilient tomorrow.* Oxford, UK: Oxford University Press. Kindle ed., 46–47.

Holdren, J. 2007. Quote from "World Scientists Near Consensus on Warming." *New York Times,* January 30. Accessed February 25, 2024. https://www.nytimes.com/2007/01/30/world/30climate.html.

IPCC (Intergovernmental Panel on Climate Change). 2018. "Summary for policymakers." *Global warming of 1.5°C,* edited by V. Masson-Delmotte, P. Zhai, H.-O. Pörtner, D. Roberts, et al. New York: Cambridge University Press, 3–24.

Jones. T. L. 2016. "Walker mayor says DOTD, barrier wall responsible for town's flooding, threatens lawsuit," *The Advocate,* August 25. Accessed August 14, 2017. https://www.theadvocate.com/louisiana_flood_2016/article_59c52500-6a28-11e6-af6e-877659a40ce5.html.

Laxo, A. 2023. "Climate forward? How architects and engineers are(n't) using climate projections to inform design." HGA and the University of Minnesota Climate Adaptation Partnership. https://hga.com/climate-forward.

Marchau, V. A. W. J., W. E. Walker, P. J. T. M. Bloemen, and S. W. Popper (eds.). 2019. *Decision making under deep uncertainty.* https://doi.org/10.1007/978-3-030-05252-2.

Munich, R. E. 2021. "Record hurricane season and major wildfires – The natural disaster figures for 2020." Accessed May 17, 2021. https://www.munichre.com/en/company/media-relations/media-information-and-corporate-news/media-information/2021/2020-natural-disasters-balance.html.

NOAA NCEI (NOAA National Centers for Environmental Information). 2024. "U.S. billion-dollar weather and climate disasters." Accessed February 24, 2024. https://www.ncei.-noaa.gov/access/billions/. DOI: 10.25921/stkw-7w73.

Tomer, A. 2022. "Will the infrastructure law and Inflation Reduction Act transform American transportation? It's complicated." Accessed January 11, 2024. https://www.brookings.edu/research/will-the-infrastructure-law-and-inflation-reduction-act-transform-american-transportation-its-complicated/.

Wallace, W. A. 2018. "Viewpoint: Lead on climate change solutions before it's too late." *Eng. News-Record,* October 25. https://www.enr.com/authors/1182-william-wallace.

Acknowledgments

For writing this book, I place the blame on dozens of my engineering friends and colleagues. I wrote this book largely out of frustration. Some of you have told me climate change is not happening even after I showed you the work and conclusions of climate experts at NOAA, NASA, and the National Academies. Folks, these are federal agencies and organizations whose mission success depends on understanding the threats posed by a changing climate. Why don't you believe them? Some of you have suggested that sustainability is just a buzzword and not something real engineers need to worry about. Others are confused about sustainability and resiliency. I have been told that infrastructure must be more resilient; sustainability is just about conserving energy and resources. Huh? Infrastructure needs to be more resilient because your infrastructure projects are not sustainable. Your projects have and continue to destabilize the Earth's climate. They burn fossil fuels to create energy for heat, power, and mobility. Burning fossil fuels releases huge quantities of carbon dioxide, a greenhouse gas that has significantly damaged the Earth's once stable climate-regulating systems. Higher temperatures, extended heat waves, rising sea levels, more frequent and intense storms: these conditions, products of a changing climate you helped create, are overwhelming the very infrastructure assets you designed. Now, you say these assets need to be resilient. What a surprise. You had a hand in changing the climate. Now, it's your turn. You have to change.

Change is good. For decades, the civil engineering industry has been whining that they are not getting paid for what they are worth. Ladies and gentlemen, your work has become commoditized. Your value has been eroded by standards and practices that are becoming increasingly prescriptive. Today, in this era of climate change, most climate-related standards and practices no longer apply. Your client and bosses now want to know what you think. What are the risks of a changing climate? How could it affect my project? What do I need to do to protect it from damage? You civil engineers are best equipped to respond. You know what works and doesn't work. Your project delivery experience is all about managing change. If not you, then who?

I also wrote this book thanks to the encouragement and support of many people. My colleagues, members, and staff of the ASCE Committee on Sustainability (COS) always provided sound feedback and put up with my occasional rants. Dave Ellison and Ryan Dougherty provided me with much-needed guidance on project delivery methods. Larry Roth gave me extremely helpful critiques of the manuscript and shared his project experience. David Lincoln provided outstanding advice and introduced me to the subject of deep uncertainty. Last but certainly not least, I want to acknowledge my wife, Diane, my best friend, companion, and counsel. Her ability to read through my all too frequent gobbledygook, edit, and come up with excellent suggestions is unsurpassed.

CHAPTER 1

What's Past Is No Longer Prologue

"What's past is prologue."

from William Shakespeare's The Tempest, Act 2, Scene I.

Engraved on a stone panel at the entrance to the National Archives Building in Washington, DC

"For engineers working in the built environment, what's past is no longer prologue."

Bill Wallace, Presentation to the Atlantic Canada Water and Wastewater Association, October 2012

Global Warming and Its Consequences

Ready or not, civil engineering is about to get a major overhaul, one that will visit every aspect of this engineering discipline related to climate. Other engineering disciplines, chemical, mechanical, electrical, or any that work in the built environment, are not exempt. The driver is global warming. Increasing atmospheric concentrations of greenhouse gases (GHGs) are trapping more thermal energy, raising the Earth's temperature, and significantly disrupting the Earth's climate-regulating systems. As a result, climate conditions are changing in ways and at a pace that was never contemplated nor incorporated into civil infrastructure design.

Infrastructure assets were designed to handle all but the most unlikely weather events, extreme events that were possible but highly improbable. Relying on historical climate data and past project performance, engineers had judged such instances too rare to worry about. Recent experience shows otherwise. Now and well into the future, severe weather events and climate conditions that were once considered improbable are becoming more probable and, by implication, more dangerous.

Today, the United States and the rest of the world are experiencing the consequences of this warming phenomenon: a new and harsh climate marked by frequent and devastating weather extremes. Importantly (and thankfully), the public is beginning to realize that these extremes are not one-off incidents but rather a harbinger of things to come. The public is also suffering the consequences: extensive damage, prolonged disruptions, huge dollar losses, frequent harm, and loss of lives. In 2020, the United States suffered through 22 separate billion-dollar weather and climate disasters, costing the nation $95 billion in damages (Figure 1-1). The year 2020 shattered the previous annual event record, 16 for 2011 and 2017.

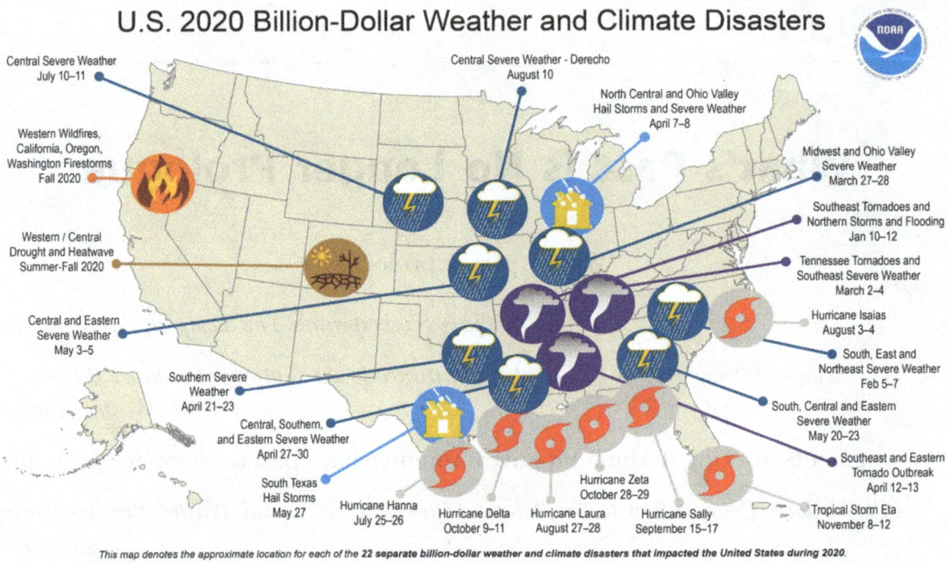

Figure 1-1. 22 billion-dollar weather and climate disasters in the United States in 2020. Each disaster resulted in at least $1 billion in damages.
Source: Smith (2020).

Over the last 11 years, 90% of US counties have experienced a climate-related disaster (Rebuild by Design 2022), a weather event with destructive effects beyond the response capabilities of state and local governments. In January 2024, the National Oceanic and Atmospheric Administration's (NOAA) National Centers for Environmental Information (NCEI) reported since 1980, there have been 376 weather and climate disasters, each costing $1 billion or more. The accumulated cost since 1980, adjusted for inflation, is more than $2.6 trillion. Moreover, the number of billion-dollar events per year is increasing. As shown in Figure 1-2 (NOAA NCEI 2024a), the years with the greatest number of billion-dollar events occurred in the last decade. The years 2011 to 2020 were the warmest on record (WMO 2020). The year 2021 had 20 separate billion-dollar disasters (Smith 2022), the second-highest number and the third-highest yearly cost on record.

Climate-Related Assumptions Are No Longer Reliable

Why are these disasters happening? The answer lies in a fundamental assumption of civil infrastructure design. Society's civil works are being built assuming a stable climate, a climate in which past climate conditions were good predictors of future climate conditions. This assumption, known as *stationarity*, allowed civil engineers and others to build long-lived infrastructure projects, confident that their projects

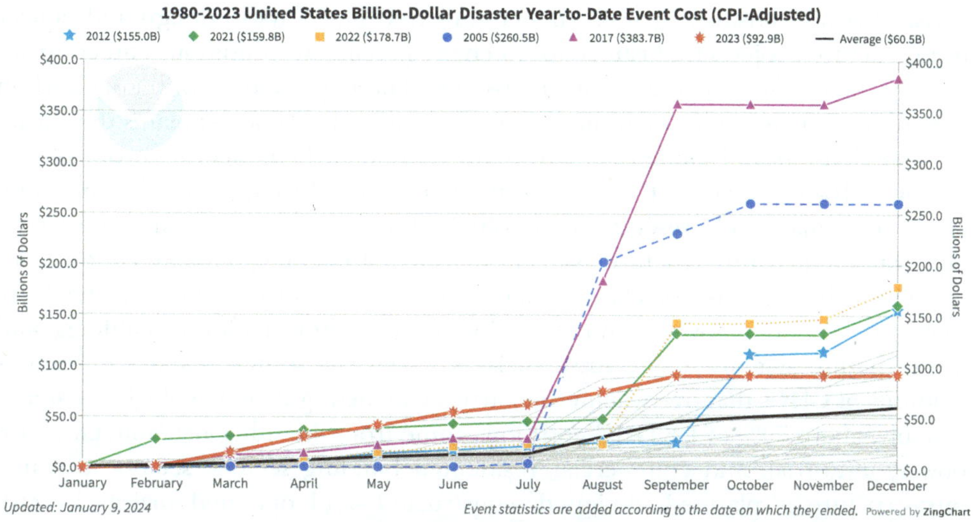

Figure 1-2. Month-by-month accumulation of billion-dollar weather and climate disasters: 1980–2023 CPI adjusted.
Source: NOAA Climate.gov, NOAA image by NOAA NCEI (2024a).

would be safe, reliable, and economical throughout their design life. Today, that assumption is no longer valid. As a result, infrastructure projects built according to current climate-related standards and practices are less safe and less reliable, subject to the vagaries of a warming and harsher planet.

Unfortunately, current engineering standards and practices have not kept up with changing climate conditions, and practitioners are generally at a loss on how to proceed. The US Department of Transportation's Federal Highway Administration (2017a) makes this observation.

> To date, few transportation agencies in the United States have explicitly required design changes in anticipation of future climate change due, in part, to a lack of understanding of how best to incorporate projections into designs. This is particularly true in terms of developing projected values for extreme conditions.

Laxo et al. (2023) surveyed current architecture and engineering (A&E) practices. They found that the industry, for the most part, still applies historical climate information for making civil infrastructure project design decisions. For those A&E firms that offer services in sustainable design, climate change adaptation services are generally not included. The reasons offered included few client requests for incorporating climate projections into designs, no codes and standards for using climate projection data, and the need for internal capacity to apply climate projection data to projects.

Engineering standards and practices must change. Continuing to absorb the increasing economic, environmental, and social costs of weather and climate disasters is not sensible. Civil engineers must find new ways to design and deliver infrastructure projects that can adapt to changing climate conditions and confront their primary cause: excess GHG emissions. These emissions, primarily carbon dioxide (CO_2), are more than the Earth can handle. This book explains the causes of these disasters and the reasons why many current civil engineering protocols are deficient. It also describes the engineering fixes needed to meet the demands of a changing climate and presents methodologies for designing, delivering, operating, and maintaining infrastructure assets in the face of changing climate conditions.

The primary cause of climate change is society's reliance on burning fossil fuels for heating, moving, and powering most of this nation's buildings and infrastructure systems. Using fossil fuels as an energy source releases enormous amounts of GHGs that overwhelm the Earth's climate-regulating ecological systems. Notably, about 70% of US GHG emissions come from the transportation, electrical power, industrial, and commercial/residential building sectors, sectors in which infrastructure is planned, designed, constructed, and operated mainly by civil engineers.

Many of these infrastructure assets already have or will fail in the future because infrastructure asset owners and engineers have not taken changing climate conditions into account. These failures take many forms, ranging from annoying disruptions to early deterioration to serious and expensive damage. In the worst cases, the effects are severe, causing significant losses in property and loss of life. Unless changes are made, these failures will continue, becoming more frequent and more severe. Some notable examples of recent weather and climate disasters are described as follows.

Hurricane Sandy (October 2012). Cost: $65 billion, 159 deaths

Continuous, heavy rains and storm surges flooded New York City (Figure 1-3), submerging areas well outside the designated 500-year flood zones (Colvin 2020). The wind, rain, and storm surges ruined entire communities, shut down power, leveled buildings, and started fires. The level of damage from "Superstorm Sandy," as it was called, was incredible. Over 650,000 houses were damaged or destroyed, and 8.5 million people lost power. Some areas were without power for weeks or months. One hundred fifty-nine deaths were attributed directly or indirectly to this storm (Blake et al. 2013, USACE 2015).

Storm surges combined with rising sea levels overwhelmed the barriers designed to prevent flooding and pushed millions of gallons of seawater into the subway tunnels. This deluge flooded the underground platforms and train tracks and destroyed electrical and mechanical equipment, which, because of the urban setting, were located below ground level (FEMA 2013). Scientists studying the relationship between climate change and damage costs from this storm determined that $8 billion in losses was attributable to the rise in sea level caused by climate change (Strauss et al. 2021).

Figure 1-3. Flooded road in downtown Manhattan from Hurricane Sandy on October 30, 2012, in New York.
Source: Kobby Dagan/Shutterstock.com.

Camp Fire, Northern California (November 2018). Cost: $16.5 billion, 85 deaths

Caused by a faulty electrical transmission line, the Camp Fire in northern California (Figure 1-4) burned more than 153,000 ac and almost destroyed the towns of Paradise and Concow. Losses totaled $16.5 billion (Associated Press 2019), $4 billion of which were not insured, making it the most expensive natural disaster of 2018. In the past, faulty transmission lines have caused wildfires. However, in the Camp Fire case, climate was a significant contributor. A combination of high temperatures and a long, severe drought created the conditions for intense and rapidly spreading fires. This fire spread faster than officials realized (Tribune Media Wire 2019). Slow or no action on improving evacuation routes, creating fuel reduction zones, and enforcing home-building moratoriums made this disaster worse (Gafni 2018).

Since 2000, US wildfires have become larger, more frequent, and covered more land area. To the north, Canada was not immune. Canadian wildfires in 2023 burned 45 million ac of forest, more than 6 times the national average (NYT 2023). These fires not only endanger communities but also change the structure and composition of vegetation, altering snowpack and reducing water supplies (ScienceDaily 2022). Fires can be so hot as to turn soil into impermeable surfaces, preventing infiltration and increasing the chances of flooding (NYT 2023).

Figure 1-4. An image from a NASA Earth satellite showing the extent of the Camp Fire in California. The photo uses shortwave infrared light to highlight the active fire.
Source: NASA, Joshua Stevens.

Midwestern US Floods (January–March 2019). Cost: $10.8 billion, 3 deaths

In the spring of 2019, multiple extreme weather events in the Midwestern United States combined to create a major flood disaster. From January to early March, many areas in the Midwest experienced low temperatures accompanied by record snowfalls. In late March, temperatures in parts of Nebraska reached 60 °F (18.3 °C) over 3 days and experienced 1.5 in. of rain. With the ground frozen from previous low temperatures, melting snow and rainwater flowed to lakes and rivers, raising their levels to record heights. Strong water flows broke upriver ice, sending large pieces of ice downstream. Later that year, heavy rains—four times what is considered normal—fell in Montana, the Dakotas, and Nebraska. The damage was extensive. Council Bluffs, Iowa, had 30 levee failures. The Spencer Dam in Nebraska collapsed, sending an 11 ft wall of water downstream that destroyed three bridges. Communities in Iowa, Missouri, and Nebraska (Figure 1-5) had to be evacuated. Farmers in these states lost hundreds of millions of dollars in crops and livestock.

Aging levees, inadequate maintenance, and inconsistent regulations were part of the problem. However, the primary cause was insufficient flood control capacity. Record-breaking precipitation produced record runoffs that overwhelmed flood defense systems, systems designed for much lower floodwater levels. The US Army Corps of Engineers operates many dams on the Missouri River, balancing river stakeholder concerns of flood control, barge shipping, environmental protection, and recreation. The Corps' priorities are safety, flood mitigation, and repairing

Figure 1-5. A levee break in the Midwestern United States flooded the entire town of Pacific Junction, Iowa, and its residents.
Source: William Blake/Shutterstock.com.

damages from earlier events. Often, these priorities conflict with many stakeholders' priorities (Smith and Schwartz 2019).

At the time of writing, heavy rains caused the Edenville and Sanford dams in central Michigan to overflow, forcing the evacuation of around 10,000 people. Making matters worse, the Dow Chemical complex in Midland, Michigan, and a Superfund hazardous waste cleanup project are in the path of floodwaters. If floodwaters reach the Superfund project, they could mobilize contaminated sediments and send them downstream, ruining many years of cleanup work (NYT 2020a; Flesher 2020).

As of this writing, Boyce Hydro, the owner of the Edenville Dam, is in the Michigan government's crosshairs. The license for this former hydroelectric dam was revoked in 2018 because of, among other issues, a "longstanding failure to increase the project's spillway capacity to safely pass flood flows." The Federal Energy Regulatory Commission (FERC) that oversees hydroelectric dams noted that to comply with agency requirements, the dam must upgrade its spillway or otherwise protect against damage from overtopping (NYT 2020b). According to FERC, "Boyce Hydro argued that the chances of a 'probable maximum flood' event in the next five to ten years [are] five to ten in one million" (Yoders 2020).

Similar Problems, Other Places

Not surprisingly, the United States is not the only country experiencing climate-related problems. Figure 1-6 shows the scale of global economic losses between 1970

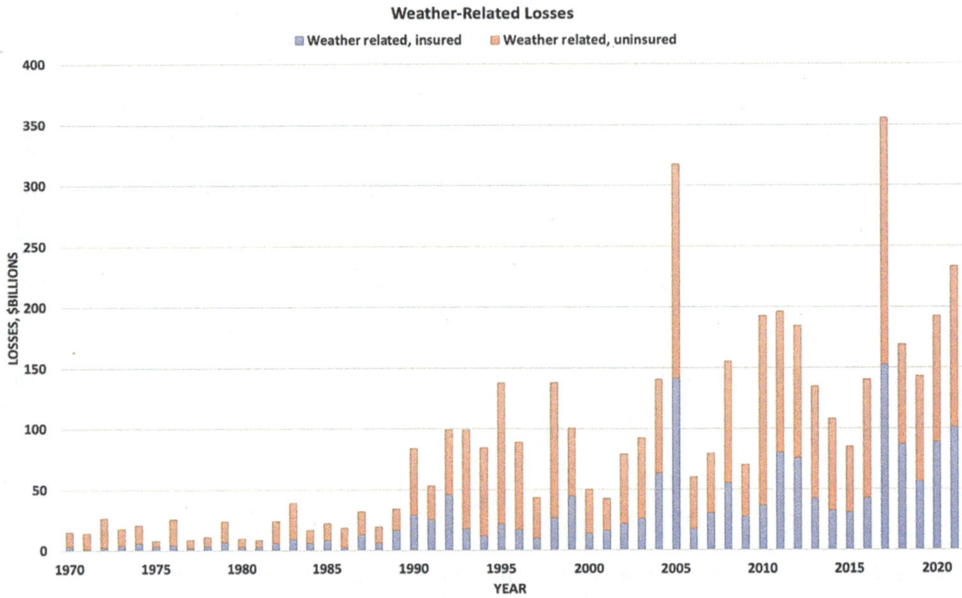

Figure 1-6. Global insured and uninsured economic losses 1970 to 2021 in constant US dollars. *Source*: Data from Swiss RE Institute (2022).

and 2020 attributable to natural disasters. Figure 1-7 presents a sample of global extreme events in 2022 attributed to climate change. These events are significant both in terms of cost and potency. However, they do not illustrate the huge associated social costs: people displaced and stressed, personal items and property lost, damaged or destroyed, income reduced, and more. Furthermore, they do not reflect the additional vulnerabilities of people living in less developed nations. For example, frequent flooding and drought are especially damaging to communities dependent on rain-fed agriculture.

The European Heat Wave of 2003

A tragic example of unexpected weather extremes happened in Europe (Figure 1-8) not too long ago. In 2003, an extended heat wave killed more than 70,000 people. France was hit the hardest, reporting over 14,000 fatalities (BBC News 2003), primarily older people living alone and unable to care for themselves. The houses were designed to complement France's typical climate: mild daytime temperatures followed by cool nights. Taking advantage of these temperature cycles and local materials, designers constructed buildings with stone, brick, or masonry, giving each a thermal mass that would absorb and store heat during the day and release heat during the evening hours. If a building's thermal mass is tuned to local climate conditions, air conditioning becomes less necessary. But in 2003, extended periods

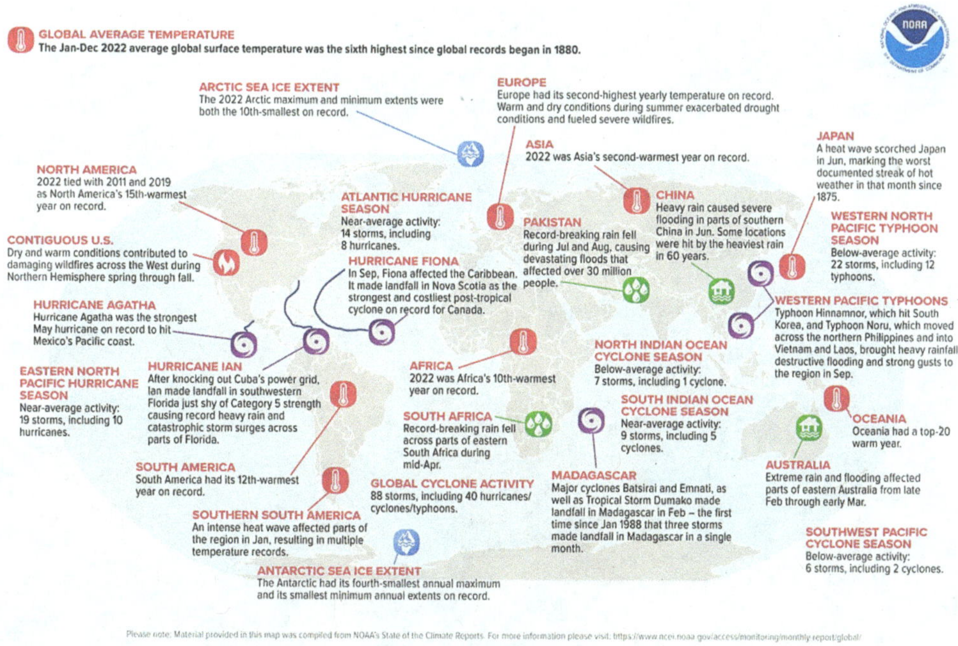

Figure 1-7. Selected significant climate anomalies and events in 2022.
Source: NOAA NCEI (2024b).

of high temperatures built up heat in the thermal mass of buildings, making occupancy at best uncomfortable, at worst, dangerous.

For France, social norms made this heat wave a disaster. The heat wave occurred in August, when a large portion of the French population was away on holiday, leaving older people to care for themselves as typically is the case. Regrettably, many older people had not experienced this level of extreme heat and could not or did not know how to respond. Many of the elderly died in their homes, unable to cope with the heat. Some bodies were not found for weeks after the heat wave subsided.

The French government contributed to this disaster. Not only did it fail to recognize the threat posed by the heat wave, but it also dismissed the early warnings from medical professionals. Instead, it initially declared that everything was under control.

The government's actions closely followed the governance model of "Deny, Deflect, Defend," as described by Lagadec (2004).

1. **Deny**. Early reports from doctors regarding stressed medical services were dismissed as just grousing over hospital budget cuts. Media reports were dismissed as typical media hype. The French government assumed that less information was better to avoid public panic. They stated that everything was under control. The situation is normal. When it gets hot, the death rate for older people increases.

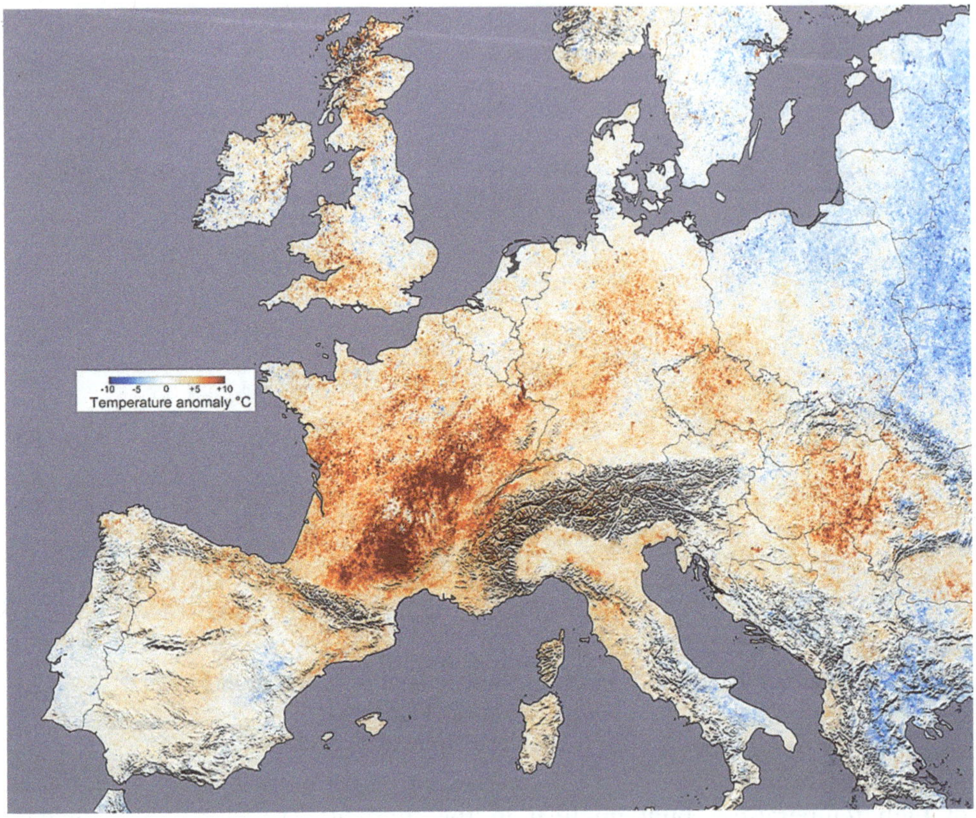

Figure 1-8. 2003 heat wave temperature variations relative to July 2001 temperatures in Europe. *Source*: Courtesy of Reto Stockli and Robert Simmon, based on data provided by the MODIS Land Science Team, Public domain, via Wikimedia Commons.

2. **Deflect.** When the scale of deaths became known, the government then stated that it was the fault of the families, as they tend to abandon their elderly parents. Another statement claimed that it was the doctors' fault for going on holiday.
3. **Defend.** When people questioned why the government did not do something early on, the government's statements changed again. Government officials now declared that they had not been informed. If they had, the outcome would have been different.

Thankfully, the French government heeded the lessons from the 2003 heat wave. During a recent series of heat waves in France (June 24 to July 7 and July 21 to 27, 2019), officials took multiple steps to limit heat-related casualties, including issuing alerts, establishing emergency hotlines, opening parks and swimming pools, closing schools, and canceling public events (Figure 1-9). During those periods, the death toll exceeded 1,400 (BBC News 2019), half of whom were over 75 years of age. However, the number was ten times less than the death toll of 2003.

Figure 1-9. June 2019, in Paris: Tourists and locals cool down in the Trocadero fountains. Temperatures rose to 40 °C (104 °F). The normal June high temperature in Paris is 23 °C (74 °F). *Source*: Spech/Shutterstock.com.

When Cape Town Almost Hit Day Zero

For some cities, the lessons learned after a climate-induced crisis can bring the necessary structural and operational changes. On February 5, 2018, the City of Cape Town, South Africa, announced that it was about to become the first major city in the world to run out of water. The city had been in an extended drought, the worst in over 100 years (Figure 1-10). Day Zero was set for May 11, the day that water levels in dams were projected to be so low that water taps needed to be turned off. Once turned off, people would have to bring containers to pick up their ration of water of 25 L per person per day at one of 200 distribution points, all guarded by security forces. This drastic measure would give the city 3 months of supply at highly restricted flow rates, enough to keep the reservoir levels above 13.5% of capacity until the fall rainy season (City of Cape Town 2018).

Luckily, the fall rains came as hoped, and the severe "Day Zero" preparation restrictions were lifted. But before then, city officials set limits on water use and imposed fines for violations. Residents responded by collecting wash water and using it to flush toilets. Showering was postponed or time-limited. Even the radio stations responded by setting up playlists of songs that lasted 2 min, timed for a quick shower.

Figure 1-10. January 25, 2018: Lines of people waiting to collect natural spring water for drinking during the drought in Cape Town, South Africa.
Source: Mark Fisher/Shutterstock.com.

A "No More!" Moment?

Near disasters like Cape Town's Day Zero close call may be what cities need to make the necessary changes. As one Cape Town resident noted (Mahr 2018), "We'll never, ever take water for granted again." Authors Hill and Martinez-Diaz (2020) suggest that these extreme events cause people and governments to declare, "No more!" and begin to make changes to their water infrastructure that should have been made years, if not decades, before. In the case of Cape Town, there was little resilience in their water supply system. The city relied heavily on surface water sources but less on groundwater, water conservation, or desalination (Naidoo 2018). Today, the City of Cape Town has maintained its water restrictions. The city is fast-tracking projects to improve and diversify its water supplies, making them more resilient in the face of a drier climate. The city has also initiated an extensive communications campaign designed to build public awareness about the current state of its water resources and how they need to be managed. Water usage, dam storage, and water production rates are now tracked and published by the City of Cape Town (2020) in their "Weekly Water Dashboard."

The water restrictions put in place in 2018 still apply. Cape Town residents are limited to 105 liters of drinking water per day. Washing vehicles with their drinking water allotment can only be done if residents use a bucket. Using non-potable water or other vehicle-cleaning processes is encouraged.

Figure 1-11. Chennai metro water distribution in Korattur, India, November 19, 2019. *Source*: FocalPlane Photography/Shutterstock.com.

"Day Zero" was not unique to Cape Town. In the summer of 2019, Chennai, India, a city of 8 million people, began distributing drinking water by railcars and trucks (Figure 1-11). That summer, the levels in Chennai's reservoirs had dropped to 1% capacity. Water distribution by rail and truck continued into the fall. Finally, in late November, the water shortage was declared officially over, at least for that year. Like Cape Town, Chennai is diversifying its water supplies by building desalination plants and improving how the city handles its wastewater (Russell 2020).

The Past Is No Longer Prologue

The common theme in these examples is the inability of society's current civil infrastructure to handle the additional stresses caused by a changing climate. Throughout history, civil engineers assumed that local climate conditions, although fluctuating across seasons, were statistically constant. Regardless of fluctuations in climate variable values, the probability distribution of these values would remain the same.

Statisticians call this assumption stationarity: the statistical properties of engineering design parameters will be the same in the future as they have been in the past. That assumption enables engineers to design civil infrastructure projects that function reliably and safely throughout their design life. It also allows engineers to create climate-related design codes and standards, which, if followed, would consistently produce satisfactory results.

Unfortunately, these codes and standards are based on a level of climate stability that no longer exists. Nonstationarity is now the new normal. *Climate stressors*,[1] climate variables (temperature, precipitation, wind) at extreme values and in combinations of intensity, duration, and frequency, can degrade infrastructure function or performance. Climate stressors differ from other physical stressors in infrastructure design and operation. Stressors such as vehicle weight on a bridge or floor loading in a building can be controlled. Climate stressors cannot.

The Earth's climate has been relatively stable for the last 10,000 years (Hansen and Sato 2011). And in this formerly stable climate, engineers had climate stressors figured out. Using historical climate data, engineers developed predictive climate stressor models. Climate conditions were handled statistically as probability distributions of climate stressor variables. By knowing the probability distributions of these variables, engineers could make reasonable estimates of future climate conditions assuming climate stationarity (i.e., past climate conditions are good predictors of future climate conditions).

Global warming changed all that. Today, the Earth's ambient temperature is rising rapidly, bringing new climate stressor extremes: record temperatures, extended heat waves and droughts, extreme storms, and sea surges. Above normal temperatures are melting land ice and expanding seawater as it warms, causing sea levels to rise (NASA 2023). These extremes are no longer manageable using conventional climate-related civil engineering standards and practices. Figure 1-12 shows the extent to which global average land surface temperatures have risen in the northern and southern hemispheres over the last 70 years. Since 1950, the summer months (June–July–August) in the northern hemisphere have warmed significantly, showing a change in both the mean and extreme values. The winter months showed a similar shift but with fewer extreme values.

The net result is that the range and frequency distribution of climate stressor variables are no longer predictable to the degree of accuracy needed for creating reliable and climate-resilient infrastructure designs. Examples of climate stressors and the affected infrastructure assets are presented in Table 1-1. Consequently, the codes and standards that assumed sufficient climate predictability are no longer reliable and require a major rework. Indeed, the civil engineering profession itself requires a course correction, tackling the causes and making civil infrastructure assets more durable, adaptable, and resilient in the face of a changing climate.

This change in the statistical properties of climate stressors and loss of predictability can be traced back to society's unsustainable use of renewable and

[1]Climate stressors have been loosely defined by NOAA as a "condition, event, or trend related to climate variability and change that exacerbate hazards." FHWA defines them as a "[v]ariation in a climate variable that may lead to a climate impact." See US Climate Resistance Toolkit. https://toolkit.climate.gov/content/glossary. Also, FHWA, Synthesis of Approaches for Addressing Resilience in Project Development, FHWA-HEP-17-082, p. D-5.

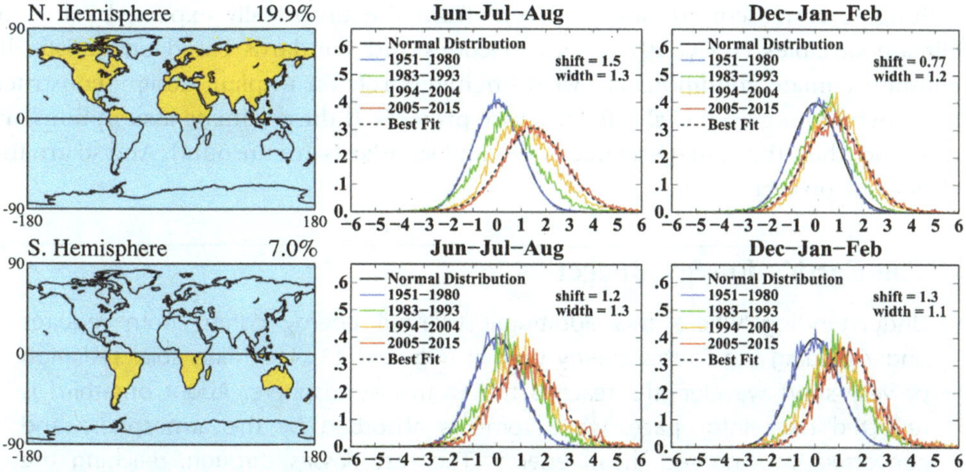

Figure 1-12. Frequency of occurrence of local temperature anomalies (relative to 1951 to 1980 mean) divided by local standard deviation (horizontal axis) for land areas shown on the map. *Source*: Hansen and Sato, Figure 1 (2011). Creative Commons Attribution 3.0.

Table 1-1. US Department of Transportation List of Climate Stressors that Could Affect Infrastructure Assets.

Infrastructure Asset	Climate Stressor: Extreme Temperature	Flooding/ Precipitation	Storm Surge	Wind	Drought	Freeze/ Thaw	Permafrost Thaw
Pavements	✓	✓	✓		✓	✓	✓
Bridges	✓	✓	✓	✓		✓	✓
Culverts		✓	✓				
Slopes and Soils	✓	✓	✓		✓		
Mechanical/ Electrical Equipment	✓	✓	✓	✓			✓

Source: Adapted from FHWA (2017b), Table 5.

nonrenewable resources, notably the burning of fossil fuels. The use of these fuels has and continues to degrade the Earth's climate-regulating system. Enormous quantities of CO_2 and other so-called GHGs are being released, increasing atmospheric GHG concentrations to levels not seen in millions of years (Zhang et al. 2013). At these concentrations, the additional thermal energy retained by the Earth's oceans and atmosphere overwhelms the Earth's climate-regulating systems.

This excess thermal energy is significantly raising the Earth's average temperature and disrupting climate equilibrium. Climate statistics are being reshaped, shifting to forms considerably different from the historically expected values of climate variables (IPCC 2013). Design codes and standards based on statistically constant climate conditions are what engineers rely on to plan, design, construct, and operate long-lived civil infrastructure projects. If these climate assumptions are unsound, then the climate-related codes and standards are unsound. And so are the associated projects.

The Earth's Energy Budget

Understanding where this additional thermal energy comes from requires understanding the Earth's energy budget (Figure 1-13). Incoming solar radiation of very short wavelengths reaches the Earth's atmosphere. About one-third is reflected back into space, while some is absorbed by the atmosphere and converted into infrared (heat) energy. The rest passes through, reaching the Earth's surface, where it is absorbed. The energy absorbed is radiated back into space as infrared energy. However, because of low but significant concentrations of carbon dioxide and other greenhouse gases in the atmosphere, some of this energy is retained (IPCC 2007). How much is retained depends mainly on the atmospheric concentrations of greenhouse gases. Without the greenhouse effect, the Earth's average temperature would be about −18 °C (0 °F) (University of Calgary Energy Education 2024).

A Matter of Ethical and Moral Responsibility ... and Liability

Engineering standards and practices have not kept pace with changing climate conditions. Consequently, civil engineers face a serious dilemma and professional risk. Ethically and morally, civil engineers are responsible for protecting public health, safety, and welfare. Accordingly, they are obligated to deviate from many accepted codes and standards, recognizing that they are no longer applicable in a changing climate. At the same time, a civil engineer's license to practice is predicated on their knowledge and competency in applying currently accepted codes and standards. Indeed, their legal defense against negligence claims is proof that those codes and standards were followed.

This risk was further underscored by authors Hill and Martinez-Diaz (2020):

> At the greatest legal risk will be professionals who rely only on historical experience to make decisions even when credible authorities—such as the authors of the US Fourth National Climate Assessment, a congressionally-mandated analysis issued in 2018 identifying climate risks to the nation—assert that it is no longer valid to assume that future climate conditions will resemble the past. In such cases, judges and juries may begin to find that architects, planners, and engineers acted unreasonably when they made consequential decisions based on outdated information. For example, using

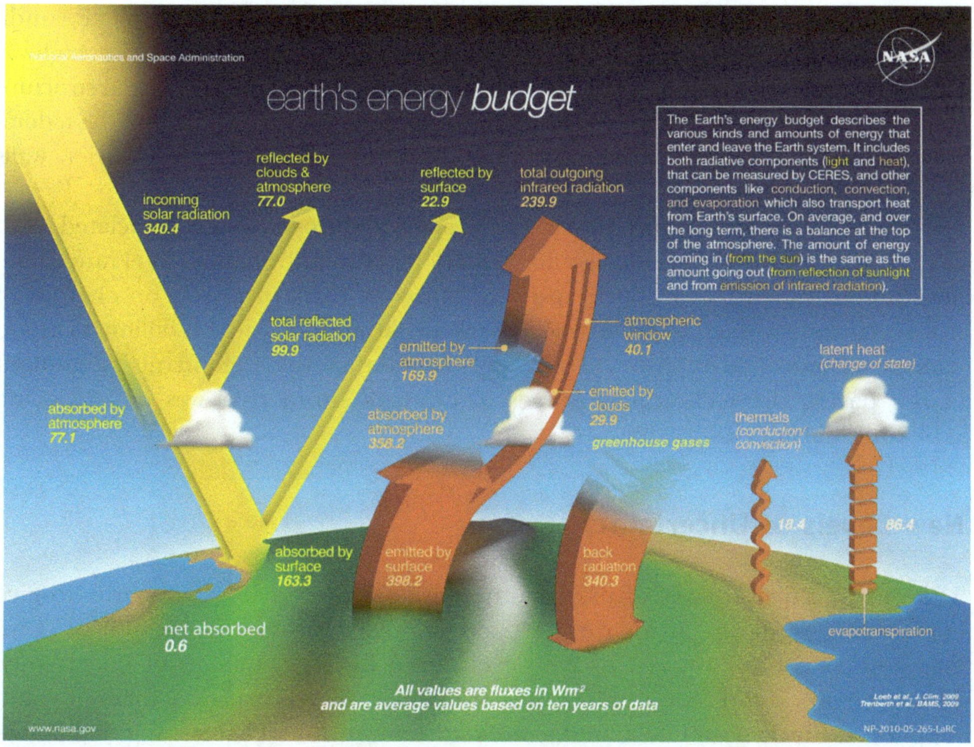

Figure 1-13. The Earth's energy budget: the balance between the radiant energy that reaches Earth from the sun and the energy that flows from Earth back to space. Energy flux units are watts per square meter (W/m^2).
Source: Gillard 2017.

> building codes and standards that do not reflect the current and future risk of climate-exacerbated extreme weather may be viewed as falling below the standard of care required by professionals.

Unfortunately, litigation along these lines is already working its way into the courts. In a recent case, a District Court in Louisiana decided that engineering companies involved in designing a highway barrier could be held liable for flood damage from a 1,000-year flood event even though the engineering companies involved followed the Louisiana State standards for highway barriers: protection from a 50-year flood event.[2] The plaintiffs are telling the court that the designers

---

[2]City of Walker et al. versus State of Louisiana through the Department of Transportation and Development et al., Docket No. 654,278; Sec. 26. 19th Judicial District Court, Parish of East Baton Rouge, State of Louisiana. In 2016, a highway barrier built on I-12 in accordance with state standards acted as a dam and flooded more than 100,000 homes and businesses. Many were not covered by flood insurance because they were not in high-risk flood zones. Damage costs were around $20 billion.

should have known that the standard was not adequate for today's climate conditions. State standards aside, what was evident to observers was that the highway barrier not only acted like a dam but was also designed like a dam (i.e., a solid structure without draining holes to release excess stormwater). Since this flood event, the federal government now requires draining holes to be placed in future barrier walls (Chawla 2020).

Given the scale of the physical damages and loss of life associated with climate disasters, civil liability could extend to criminal liability. In France, the government sentenced a mayor and his deputy to years in prison on charges of manslaughter for their roles in the deaths of 50 people and €1 billion in flood damages from a 2010 storm. The judge ruled that the defendants' argument of not having the technical resources to evaluate storm risks was unconvincing (Hill and Martinez-Diaz 2020).

New Levels of Uncertainty

The relationship of levels of uncertainty, stationarity, and nonstationarity to civil engineering is shown in Figure 1-14 and discussed further in Chapter 6. Civil engineers previously worked in the Level 1 and Level 2 domains of uncertainty. The strength, composition, and other properties of construction materials and supplies reside in Level 1, manufactured to known and reliable ranges. Climate conditions, properties of soils, and other environmental variables reside in Level 2. Soils could be sampled and tested, from which a model of the subsurface and its properties can be developed. Historical climate data could be collected and analyzed to produce probability distributions of climate variable values. Tracking and analyzing climate variable probability distributions over time would confirm the statistical stability of climate variables.

With a changing climate, climate variability has slipped into Level 3 and Level 4 uncertainty. Climate system models have sought to bracket the range of outcomes based on mathematical representations of the multiple climate factors: surface albedo (a measure of how much energy is reflected back toward space), ocean effects, aerosols in the atmosphere, the carbon cycle, cloud cover, soil moisture, and more. Possible human interventions, their timing, and their effectiveness expand the number of alternative futures. In addition, there are knock-on effects (Level 4b uncertainty), unknown effects as climate variables reach new extremes, potentially creating new tipping points that produce other stressors or reinforce existing stressors.

Making decisions regarding how to address climate change has been called "super wicked" problems (Bhave et al. 2016) "because of [the] characteristics of severe uncertainty, divergent stakeholder understanding and interests, dynamic socioenvironmental interactions and limited understanding of natural and societal responses to future climatic changes."

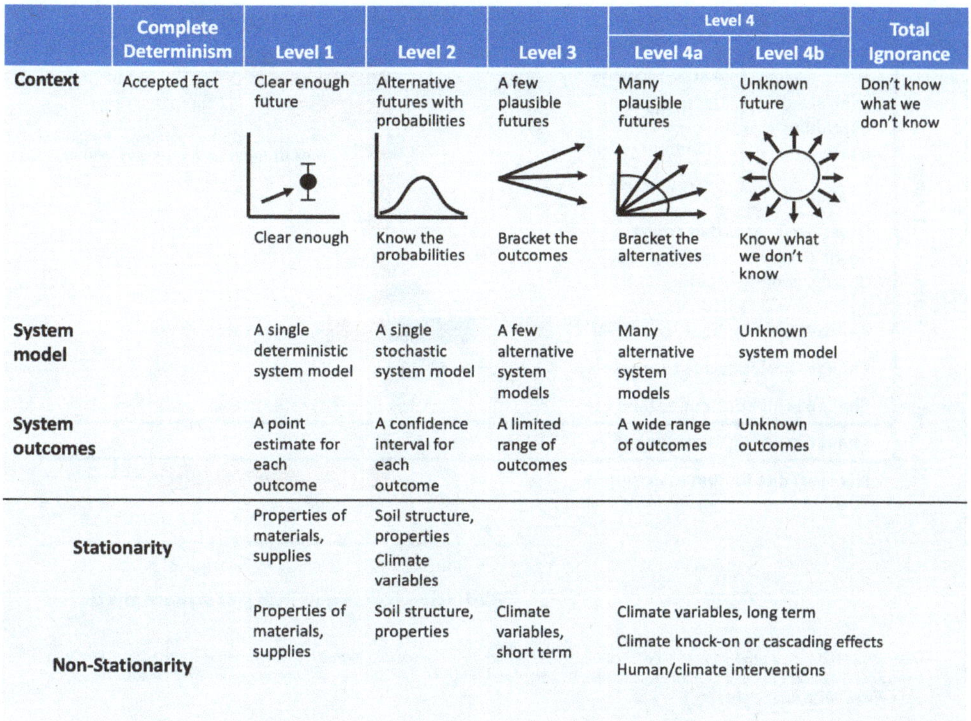

Figure 1-14. Matching stationarity and non-stationarity in climate and other design variables with levels of uncertainty.

Source: Adapted from Marchau et al. (2019). Creative Commons Attribution 4.0 International License.

Climate Change: The State of Play

As this book is being prepared, quite a few civil engineers, many of whom hold professional engineering licenses and currently practice, have yet to accept the findings of respected US and international institutions regarding climate change, its causes, or its threat to the built environment and the public. Those engineers and others are urged to read the current peer-reviewed reports produced by the Intergovernmental Panel on Climate Change (IPCC), the National Academies, NASA, NOAA, EPA, and other respected scientific and engineering institutions for which mission success depends on a sound understanding of climate science.[3]

[3]For further information, see https://www.climate.gov/, https://www.ipcc.ch/, https://climate.nasa.gov/, https://sites.nationalacademies.org/sites/climate/index.htm, and https://www.epa.gov/climate-indicators.

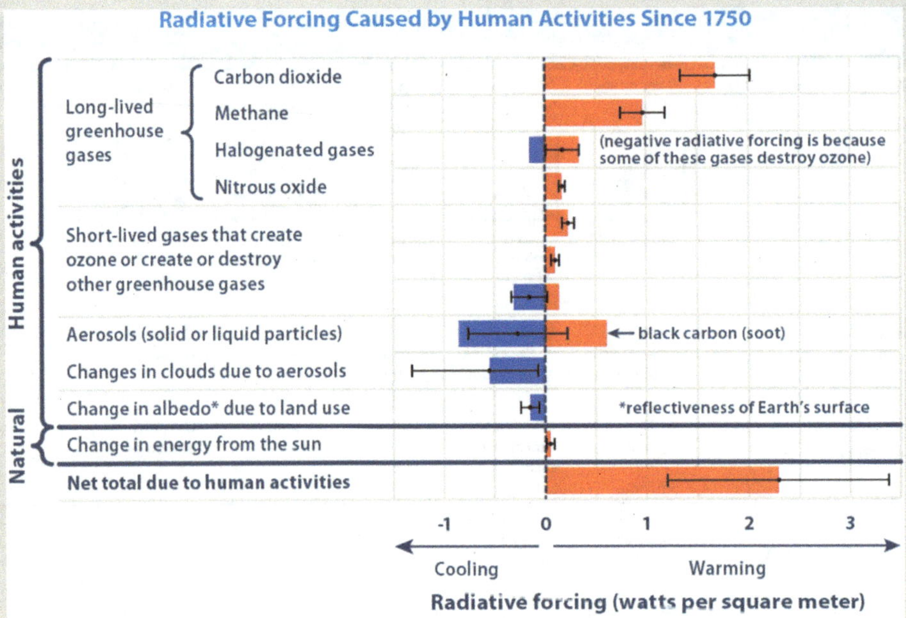

Data source: IPCC (Intergovernmental Panel onClimate Change). 2013. Climate change 2013: The physical science basis. Working Group I contribution to the IPCC Fifth Assessment Report. Cambridge, United Kingdom: Cambridge University Press. www.ipcc.ch/report/ar5/wg1.

For more information, visit U.S. EPA's "Climate Change Indicators in the United States" at www.epa.gov/climate-indicators.

Figure 1-15. Radiative forcing caused by human activities since 1750.
Source: EPA Climate change indicators: Climate forcing (2013). Data: IPCC (2013).

In August 2021, the IPCC released its sixth assessment report (IPCC 2021). The report leaves no doubt that human activities caused the climate to change, and there is no basis for denial. In its "Summary for Policy Makers", the IPCC states, "It is unequivocal that human influence has warmed the atmosphere, ocean and land. Widespread and rapid changes in the atmosphere, ocean, cryosphere, and biosphere have occurred."

Here are the specifics.

1. The Earth's climate has changed significantly and will continue to change well into the future. As of this writing, the Earth's ambient temperature has risen approximately 1.3 °C (2.3°F) over pre-industrial levels.
2. The predominant cause is the increased atmospheric concentrations of GHGs, primarily carbon dioxide produced by burning fossil fuels for heat, mobility, and electric power generation (Figure 1-15). These human-produced gases have changed the climate significantly. NOAA's Annual Greenhouse Gas Index compares the warming influence of GHGs today to their influence in 1990. In 1990, the signers of the UN Kyoto Protocol agreed to use it as a benchmark for measuring the results of efforts to reduce emissions. Figure 1-16 shows that human influence on global warming has increased by 49% since that benchmark year.

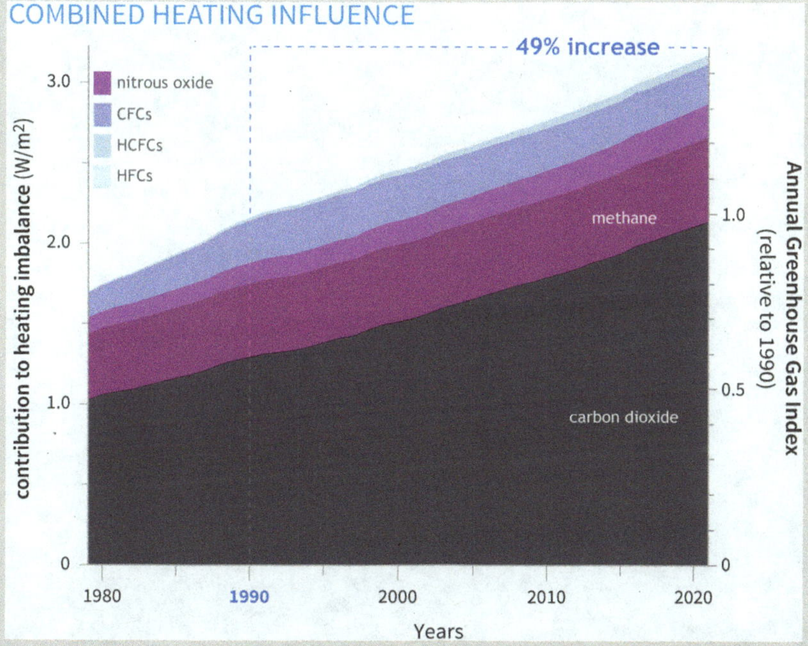

Figure 1-16. This graph shows the heating imbalance caused by the significant human-produced greenhouse gases: carbon dioxide (gray), methane (dark purple), nitrous oxide (medium purple), chlorofluorocarbons (CFCs, lavender), hydrochlorofluorocarbons (HCFCs, blue), and hydrofluorocarbons (HFCs, light blue). Relative to conditions in 1750, today's atmosphere absorbs more than three excess watts of energy per square meter of Earth's surface.
Source: NOAA Climate.gov.

3. People have argued that this warming is natural, a product of changes in the sun's radiation and variations in our Earth's orbit. These factors can indeed cause the Earth's climate to change naturally. However, the Earth's temperature increase is happening too fast for orbital changes and is too large to be attributed solely to solar cycles, the natural changes in solar radiation. Furthermore, in the current solar cycle, solar irradiance has been trending downward, meaning that natural warming has been less of a contributor (Figure 1-17). The most recent solar minimum occurred in 2019. The next solar cycle will push solar irradiance toward a maximum, with the effects felt in mid-decade 2025 (National Weather Service 2023).

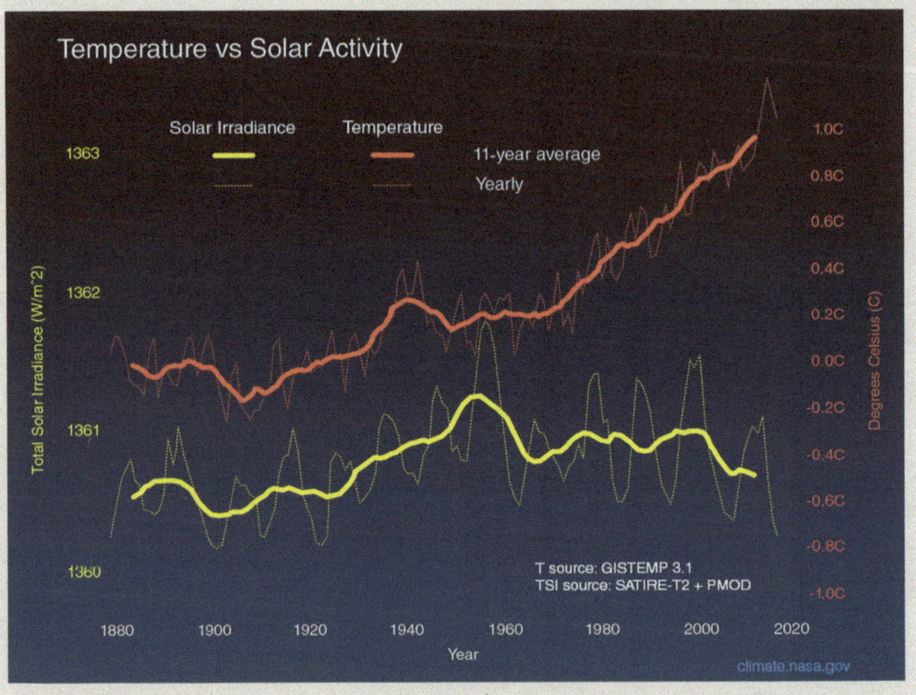

Figure 1-17. Comparison of global surface temperatures and the energy from the sun received by Earth.
Source: NASA (2019).

4. Climate scientists have measured the three other natural causes of climate change: incoming energy from the sun, the Earth's albedo, and changes in atmospheric gases or particles. They determined that the only explanation for the temperature changes since the Industrial Revolution is the increase in atmospheric concentrations of GHGs.
5. Even before the IPCC's AR6 report (IPCC 2021), this finding has been confirmed to a five-sigma level of statistical significance. In other words, the probability of the observed warming being from natural causes is one chance in 3.5 million (McKinsey Global Institute 2020). Five-sigma is the gold standard for scientists, the level of certainty used to confirm scientific discoveries such as the existence of the Higgs boson subatomic particle (Doyle 2019).
6. Increasing atmospheric concentrations of GHGs are causing the Earth's atmosphere to retain thermal energy that, in turn, is warming the planet. This increase in thermal energy is disrupting the equilibrium of the climate system, resulting in extreme storms, extended heat waves, droughts, sea level rise, and more.

7. These new and changing weather and climate conditions are exceeding levels that were never accounted for in civil infrastructure designs. As a result, civil infrastructure assets are being severely damaged, disrupting services and harming the public.
8. Greenhouse gases are still being released into the atmosphere. Therefore, global warming and climate change will continue well into the future. Actions must be taken now to reduce GHG emissions and protect society and the built environment from the effects of a changing climate.

Here is the bottom line: climate change is happening, is human-caused, and has devastating consequences. The cause can be traced back to the burning of fossil fuels to produce heat, provide mobility, and generate electrical power. About 70% of fossil fuel use can be traced back to civil infrastructure: transportation, electric power generation, residential and commercial heating and cooling, and industrial facilities. Unless the emissions from the burning of fossil fuels are radically reduced, extreme temperature and precipitation events will continue to increase across the globe. Unless civil infrastructure is reinforced, refurbished, or redesigned to withstand and adapt to changing climate conditions, the devastation will become worse.

Reduce Carbon Emissions to Net Zero By 2050

Taking meaningful action to reduce carbon emissions is now extremely urgent. A recent special report issued by the IPCC (2018) states that unless global warming is limited to 1.5 °C (2.7 °F) above pre-industrial levels [as of this writing in 2024, it is currently about 1.3 °C (2.3 °F)], the potential for devastation will increase dramatically.

The report points out that limiting global warming to 1.5 °C (2.7 °F) requires a dramatic reduction in carbon emissions plus atmospheric carbon removal, cutting emissions by 45% by 2030 and achieving net-zero emissions by 2050. Accordingly, achieving net zero requires a rapid and wholesale withdrawal from fossil energy sources, replacing them with low- or no-carbon energy sources. It also requires radical reductions in energy use, including the reduction of the embodied energy in construction materials such as steel and concrete. Last, any remaining carbon emissions must be captured, stored, or absorbed.

In the business-as-usual scenario, RCP 8.5,[4] that is, no meaningful reductions in GHG emissions, many formerly habitable places on the Earth will become uninhabitable in 2100. In this scenario, the IPCC projects a 100% chance that the global annual mean temperature will rise 3.0 °C (5.4 °F) above pre-industrial levels (IPCC 2013, p. 1056).

[4]RCP 8.5 is the Representative Concentration Pathway scenario for the worst case: radiative forcing values reach 8.5 watts per square meter. For this scenario, carbon dioxide equivalent concentrations surpass 1,200 ppm and temperatures reach 3.7 °C (6.7 °F) above pre-industrial levels (range 2.6 °C (4.7 °F) to 4.8 °C (8.6 °F)).

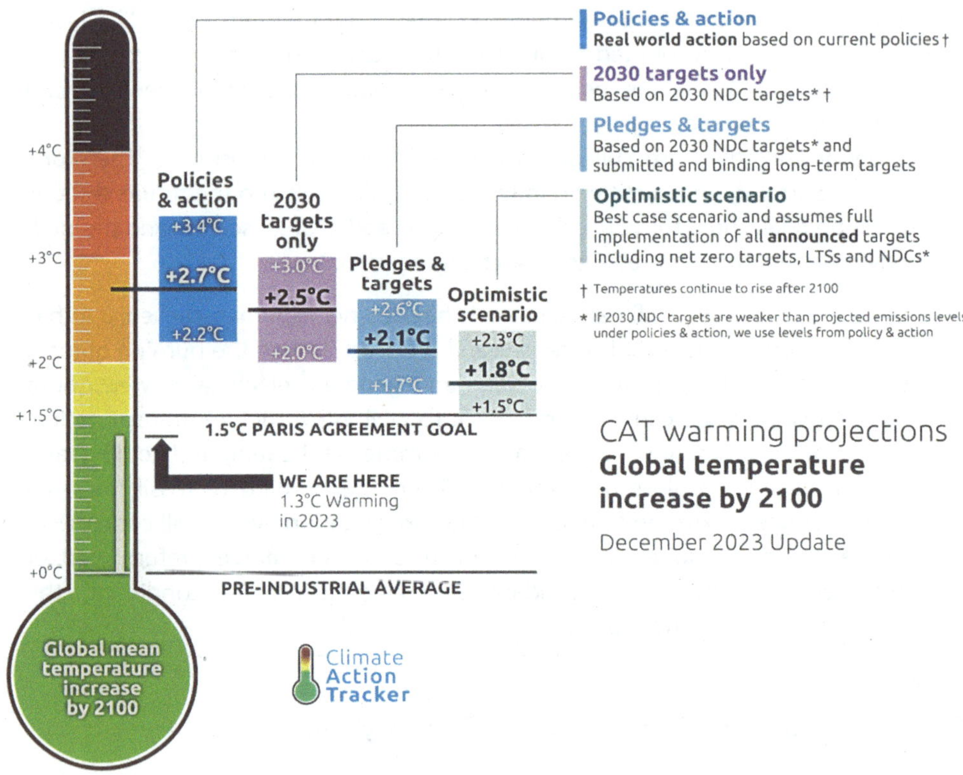

Figure 1-18. The CAT thermometer as of December 2023.
Source: © 2023 by Climate Analytics and NewClimate Institute. All rights reserved. Reprinted with permission.

Regrettably, the world's nations do not have the ambition to do what is necessary (Climate Action Tracker 2023). As of this writing, current national policies and actions will result in an increase of 2.7 °C (4.9 °F) by the year 2100. Even in the best case, assuming all national promises are kept, the projected temperature will still surpass 1.5 °C (2.7 °F) (Figure 1-18).

The need for this overhaul in civil engineering standards and practices is urgent. Absent new guidance and collaboration with regulators and project owners, engineers will continue to plan, design, build, and operate civil infrastructure assets using currently accepted civil engineering codes and standards. With rare exceptions, these codes and standards do not account for changing climate conditions.

The Hazards and Risks of Business as Usual

In a study on the risks and economics of climate change, the McKinsey Global Institute (2020) points out that without meaningful efforts to reduce GHG emissions, the frequency and severity of acute climate hazards will increase worldwide, including lethal heat waves and intense storms. Chronic hazards such as drought, heat stress, and rising sea levels will also increase. The socioeconomic impacts will likely be nonlinear as climate stressor tipping points are reached, causing secondary and tertiary effects.

If warming continues unabated, the effects and impacts of a changing climate will change the world economically. The McKinsey study (2020) looked at five sectors applying a business-as-usual scenario: negligible GHG emissions reduction.

1. **Livability and workability**. Parts of the world will become excessively hot and humid, affecting not only the ability to work outdoors but also survival. Not only lost work hours, but the probability of lethal heat waves will likely increase in parts of India. Agriculture and tourism will likely decline in the Mediterranean region as the chances of drought and extreme heat increase. In addition, the number of climate refugees will grow, comprised of people whose livelihoods have been severely degraded because of harsh climate conditions. They will seek refuge and livelihoods in other countries with more hospitable climates.
2. **Food systems**. The chance of a 15% or greater decrease in global agricultural yield is expected to double by 2030 and double again in 2050.
3. **Physical assets**. Storm surges and flooding will increase in coastal areas. Damage to homes, businesses, and infrastructure assets will have compounding effects on local economies, real estate values, and tax revenues.
4. **Infrastructure services**. Investment in climate adaptation will be required as individual assets and the economic activity they support are increasingly at risk from climate hazards. Twenty-five percent of the world's top 100 airports are 10 m or less above sea level. Power grids are also vulnerable to extreme heat as electricity demands increase as the weather becomes hotter.
5. **Natural capital**. Forests, fish stocks, coral reefs, land ice, freshwater, and permafrost are at increasing risk of deterioration. Glaciers throughout the world are expected to shrink. Glacier-fed rivers supply fresh water to one-sixth of the world's population.

Changes in Scale

Importantly, the scale of these disasters is increasing, raising the question of whether they can be handled using existing tools and resources. For example, the western states in the United States are now experiencing wildfires covering 1 million acres. Fighting fires of this scale requires the mobilization of firefighters and equipment from multiple states and takes months to contain. The Dixie Fire in northern California started on July 13, 2021, and was not 100% contained until October 30, 2021 (CAL FIRE 2022). "Fifteen years ago, a 100,000-acre fire would be the largest fire of your career. Now, we have one-million-acre fires," said a 25-year veteran firefighter (McDonald et al. 2021). "We're the size of Rhode Island, working toward

Delaware." In Canada, firefighters talk not only about fire scale but about a new kind of fire that produces new weather: fire wind, fire whirls, and fire tornados. There are also fire thunderstorms created by pyrocumulonimbus clouds that can stretch 200 mi wide and produce pyrogenic lighting (NYT 2023).

In October 2021, another record was broken, this time in Northern Italy. The commune of Rossiglione in Genoa, Liguria, received 29.2 in. of rain in 12 h, causing flooding and landslides. This happened in a region averaging 50 in. of rain annually (Sky News 2021).

Is Achieving Net-Zero Carbon by 2050 Possible?

In a related article (Henderson et al. 2020), McKinsey researchers described what it would take to achieve net-zero carbon releases by 2050. While this destination is technically possible, the pathways to get there seem impossible to navigate, given the world's current state of affairs. The article describes three possible pathways distinguished by the speed of decarbonization and their associated tradeoffs. The shifts needed in five business, economic, and societal areas are described.

1. **Food and forestry**. Changes are required in food supplies, moving away from animal protein, and changing current practices for rice cultivation. Growing rice in flood conditions contributes 12% of global methane emissions, a potent GHG (Lamb 2019). Reductions in food waste and deforestation are also required.
2. **Electrification**. Transportation and buildings decarbonize by changing from fossil energy to renewable energy sources.
3. **Industry**. Power for low- to medium-temperature heating is supplied by renewable energy sources. Energy efficiency is increased through the use of recycled materials.
4. **Power and fuel.** Massive reduction in electricity generation from coal- and gas-fired plants is achieved. Hydrogen and bioenergy play an increasing role.
5. **Carbon capture and sequestration**. Carbon emissions are captured and used as feedstocks or placed in storage. Large-scale reforestation, at its peak, reforests a land area the size of Iceland each year.

The article (Henderson et al. 2020) points out that all parts of the economy must participate, as the failure of one must be taken on by the others, or the goal will be missed. However, if the world's collective response to the COVID-19 pandemic is any example, exceeding 1.5 °C (2.7 °F) is likely. Even with the clarity of this deadly disease spreading rapidly worldwide, the leaders of many nations, particularly the United States, responded erratically, following more or less Lagadec's thesis of Deny, Deflect, Defend. Given this present-day example, it is hard to imagine that the industries and governments of the world could institute the level of collaboration needed to achieve net-zero carbon by 2050.

Is There Hope?

Yet, there is some hope. At this writing, more than 4,400 companies worldwide (SBTi 2023) have set targets for achieving significant carbon emissions [i.e., below 2 °C (3.6 °F), well below 2 °C (3.6 °F), or 1.5 °C (2.7 °F)]. As of this writing, the US Congress has passed and President Biden has signed legislation that will repair and upgrade US infrastructure, reducing its carbon footprint and making it stronger and more resilient to a changing climate. This legislation also improves what the federal administration calls *human infrastructure*: environment, housing, education, child-care, and health care.

In the fall of 2019, leaders from governments and the private sector met at the United Nations Climate Action Summit to discuss the next steps in the battle against climate change. At the end of the meeting, UN Secretary-General António Guterres announced that 77 countries, 10 regions, and more than 100 cities have committed to net-zero carbon emissions by 2050. Actions by the private sector included one-third of the world's banks committing to the goals of the Paris Agreement on climate change, investors holding $1 trillion in investments making commitments to make their portfolios carbon neutral by 2050, and 87 major companies (net capitalization over $2.3 trillion) committing to work toward the goal of limiting warming to 1.5 °C (2.7 °F). A report and implementation plan for these commitments was presented at the twenty-fifth meeting of the Conference of the Parties (COP 25) in Madrid, Spain, in December 2019 (IISD 2019).

Sadly, COP 26 in Glasgow adjourned at the end of 2021, producing the Glasgow Pact, which, as some analysts concluded (O'Grady 2021), has left the goal of limiting warming to 1.5 °C (2.7 °F) on life support. COP 27, held in Sharm El Sheikh, Egypt, in 2022, sought to build momentum for global decarbonization but had to contend with an energy crisis, supply-chain issues, and trade tensions (Wallace-Wells 2022). In November 2023, COP 28 presented the results of the first global stocktake, an assessment of what the world's nations have done or not done to meet the goals of the Paris Agreement. Since COP 27, not much has changed (Climate Action Tracker 2023).

What Needs to Be Done

Given the current situation and prospects for the future, industry leaders, government officials, infrastructure project owners, and engineers at multiple levels must coordinate and accomplish the following: (1) enhance infrastructure resilience, (2) decrease exposure to climate stressor extremes, and (3) reduce the source of climate stressor extremes: GHG emissions.

Enhance Infrastructure Resilience

The essential task for infrastructure project owners—federal agencies, states, cities, and counties—will be to add durability and resilience to their infrastructure assets and associated systems. Existing assets need to be strengthened or refurbished to adapt to changing conditions. New infrastructure projects must be designed to handle expanded ranges of climate stressor variable values. Project owners must consider strategies that avoid exposure to climate hazards. They must also develop projects that can accommodate or adapt to changing conditions. This can be accomplished by employing new project delivery methodologies that can handle the uncertainties associated with climate change. This book proposes the application of modified dynamic adaptive planning and dynamic adaptive policy pathways methodologies to make this happen. These are discussed in Chapters 7 and 8.

Decrease Exposure to Climate Stressor Hazards

Stresses on infrastructure asset functionality can be moderated through broad-scale reductions in climate stressor exposures. For example, flood exposure to infrastructure assets located within a watershed or coastal area may be reduced through flood risk reduction projects implemented at predetermined flood water levels. Water stress can be reduced through a series of projects for managing water supplies and use, carried out in order of cost-effectiveness.

Responsibilities for managing such climate stressors vary based on the area (e.g., region, city, watershed, coastal communities) under consideration. Responsibilities and authorities may be complicated because of the size of the area and the existence of multiple jurisdictions.

Reduce the Source of Climate Stressor Extremes: GHG Emissions

Unless GHG emissions are reduced substantially, the Earth's climate will continue to change, increasing the frequency, intensity, and duration of weather extremes. This means making immediate and drastic cuts in carbon emissions. Federal, state, and local officials must call for and facilitate these emissions reductions, replacing fossil energy sources with low- or no-carbon sources. Because about 70% of carbon emissions come from infrastructure assets, the task of replacement falls on the project owners and the engineers who design, build, and operate these assets. The requirements include

- Making radical and across-the-board improvements in energy conservation;
- Reducing substantially embodied carbon in infrastructure construction materials and processes;
- Designing new and retrofitting existing assets to be compatible with low- or no-carbon energy sources; and
- Redesigning manufacturing, transportation, operational, and business processes to reduce energy consumption and GHG emissions.

Critical milestones for the path to net zero are presented in Figure 1-19. It should be noted that this pathway has the developed nations (advanced economies) achieving overall net zero emissions for electricity in 2035. Electrical power generation using unabated coal (i.e., coal combustion without carbon capture and storage) is to be phased out by 2030.

How the Civil Engineering Industry Should Respond

If governments and industries follow up on their commitments, there will be a strong demand for functioning and efficient infrastructure projects that can handle changing climate conditions and their effects. Civil engineers must be equipped

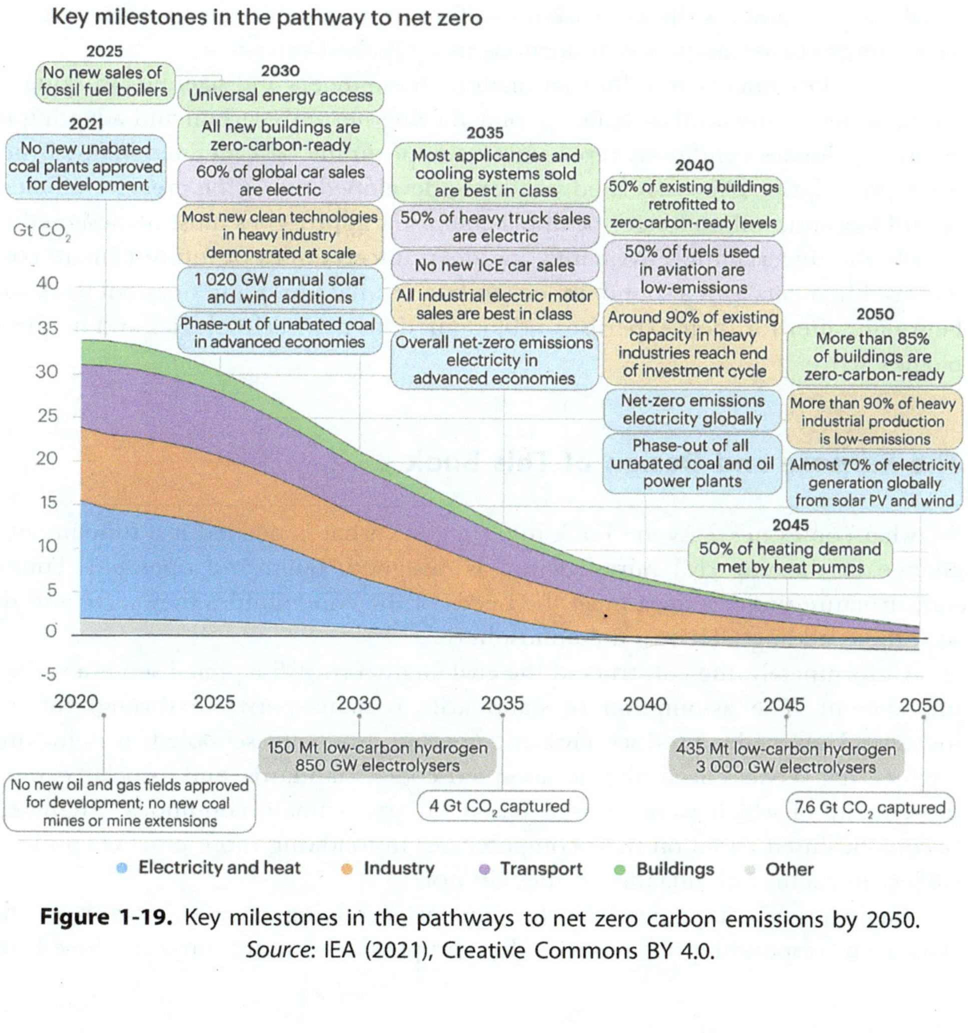

Figure 1-19. Key milestones in the pathways to net zero carbon emissions by 2050. *Source*: IEA (2021), Creative Commons BY 4.0.

to meet that demand even in the face of climate uncertainties. New infrastructure design and delivery methods are needed—standards and practices that enable engineers to produce projects that will meet owner requirements while adapting to changing climate conditions and severely limiting or eliminating carbon emissions.

However, it is unlikely that governments and industries will act in unison and implement the steps required to limit global warming to 1.5 °C (2.7 °F) above pre-industrial levels. Near-term demands for continued local, regional, and national economic development will likely supersede many efforts to reduce carbon emissions. As a result, planning for a business-as-usual course (RCP 8.5) for climate change remains a reasonable strategy.

For engineers, the recommendation for infrastructure project delivery is to design for increased robustness and resiliency, along with maximum flexibility for adaptation. For coastal areas, managed retreat (i.e., moving infrastructure assets out of harm's way from sea rise and storm surges) may be the most cost-effective option. In all cases, engineers should work on shifting to renewable energy sources along with energy conservation and reductions in embodied carbon.

For civil engineers and, for that matter, all engineers and scientists working in the built environment, the challenge of achieving net zero carbon and adapting to changing climate conditions represents an opportunity of a lifetime. Many, if not most, prescriptive, climate-related standards, developed out of the press for efficiency and low cost, will no longer be applicable. New approaches must be designed to handle the uncertainties, frequently the deep uncertainties of future climate conditions. Engineers with a sound understanding of what works and does not work will be sought after for their expertise and judgment as new technologies and methodologies arrive.

The Purpose and Design of This Book

So, what comes next? As the book title suggests, what is needed is a fundamental change in the way civil infrastructure is designed, built, and operated. Future infrastructure projects must meet the needs of the communities they serve and do so without adding GHGs to the atmosphere.

Unfortunately, the construct of the civil engineering discipline itself is a serious impediment. The assumption of stationarity remains pervasive throughout the industry. Universities produce civil engineering graduates schooled in delivering engineering services according to accepted codes, standards, and associated practices, many of which were devised based on past climate conditions. Engineers become licensed based on their competencies in following those practices and are subject to claims and litigation if they do not.

Engineering companies have become project-doing engines, organized and skilled in responding to requests for proposals, winning projects based on demonstrated successes on past projects, designing and delivering projects in accordance with applicable standards and practices, and moving on to the next. Innovation is limited to doing the same things, only faster, better, or cheaper. Doing anything outside established boundaries is risky. If anything goes wrong, their only defense is this: "We completed the project just like any other local engineering firm would have done. We followed accepted standards and practices."

Not surprisingly, civil engineering is becoming commoditized. If climate conditions are predictable, then as application experience increases, standards and practices can be made more prescriptive, repetitions of what others have already done successfully, adjusted for location and owner prerogatives. As standards and practices become more prescriptive, the need for experienced engineers diminishes. Criteria for being selected for project work, stated or unstated, shifts from qualifications to cost.

Another purpose of this book is to help guide the industry out of this commoditization trap. But to respond effectively to a changing climate, civil engineers must not only convince others that a corresponding industry change is warranted, they must also convince themselves. Having been immersed in "stationarity world" throughout their careers, the notion of a changing climate is still difficult for civil engineers to grasp. It alters one of the fundamental assumptions of their engineering discipline. However, it also opens up an abundance of opportunities for engineers to use their knowledge and experience and move away from commodity engineering. Ready or not, they will be called on to figure out how to make civil infrastructure climate-resilient under changing climate conditions.

References

Associated Press. 2019. "California's deadly camp fire was the costliest natural disaster of 2018, says insurer." Accessed August 27, 2020. https://web.archive.org/web/20190109033048/http://time.com/5497613/california-camp-fire-worlds-costliest-natural-disaster/.

BBC News. 2003. "France buries its unclaimed dead." Accessed January 11, 2023. http://news.bbc.co.uk/1/hi/world/europe/3077464.stm.

BBC News. 2019. "Summer heat killed nearly 1,500 in France, officials say." Accessed January 11, 2023. https://www.bbc.com/news/world-europe-49628275.

Bhave, A. G., D. Conway, S. Dessai, and D. A. Stainforth. 2016. "Barriers and opportunities for robust decision-making approaches to support climate change adaptation in the developing world." *Clim. Risk Manage.* **14** (2026): 1–10. https://doi.org/10.1016/j.crm.2016.09.004.

Blake, E. S., T. B. Kimberlain, R. J. Berg, J. P. Cangialosi, and J. L. Beven II. 2013. "Tropical cyclone report: Hurricane Sandy (AL182012)." Accessed October 22–29, 2012. https://www.nhc.noaa.gov/data/tcr/AL182012_Sandy.pdf.

CAL FIRE. 2022. "Dixie fire incident." Accessed January 11, 2023. https://www.fire.ca.gov/incidents/2021/7/13/dixie-fire/.

Chawla, K. 2020. "A win for Walker: Judge denies dismissing parts of I-12 barrier wall lawsuit." Accessed August 16, 2021. https://www.wafb.com/2020/06/08/win-walker-judge-denies-dismissing-parts-i-barrier-wall-lawsuit/.

City of Cape Town. 2018. "Water outlook 2018 report." Accessed January 23, 2024. https://resource.capetown.gov.za/documentcentre/Documents/City%20research%20reports%20and%20review/Water%20Outlook%202018%20-%20Summary.pdf.

City of Cape Town. 2020. "Weekly water dashboard, 01 June 2020." Accessed June 7, 2020. https://www.google.com/search?hl=en&q=City%20of%20Cape%20Town%20Weekly%20Water%20Dashboard.

Climate Action Tracker. 2023. "The CAT thermometer, December 2023." Accessed January 23, 2024. https://climateactiontracker.org/global/cat-thermometer/.

Colvin, J. 2020. "FEMA redrawing city's flood zone after Superstorm Sandy." Accessed August 27, 2020. https://www.dnainfo.com/new-york/20121206/new-york-city/fema-redrawing-citys-flood-zone-after-superstorm-sandy.

Doyle, A. 2019. "Evidence for man-made global warming hits 'gold standard': Scientists." Accessed January 23, 2024. https://www.reuters.com/article/idUSKCN1QE1ZV/.

EPA (US Environmental Protection Agency). 2013. "Climate change indicators: Climate forcing." Accessed February 26, 2024. https://www.epa.gov/climate-indicators/climate-change-indicators-climate-forcing.

FEMA. 2013. "Mitigation assessment team report: Hurricane Sandy in New Jersey and New York, building performance, observations, and technical guidance." Accessed January 23, 2024. https://rucore.libraries.rutgers.edu/rutgers-lib/44511/.

FHWA (Federal Highway Administration). 2017a. *Vulnerability assessment and adaptation framework*. FHWA-HEP-18-020. Washington, DC: FHWA. https://www.fhwa.dot.gov/environment/sustainability/resilience/adaptation_framework.

FHWA 2017b. *Synthesis of approaches for addressing resilience in project development*. FHWA-HEP-17-082. Washington, DC: FHWA.

Flesher, J. 2020. "Were years of cleanup at Michigan toxic waste site undone after dams failed?" Accessed August 27, 2020. https://globalnews.ca/news/6973450/michigan-flood-dow-chemical-dioxin/.

Gafni, M. 2018. "Rebuild Paradise? Since 1999, 13 large wildfires burned in the footprint of the camp fire." Accessed August 27, 2020. https://www.mercurynews.com/2018/12/02/rebuild-paradise-since-1999-13-large-wildfires-burned-in-the-footprint-of-the-camp-fire/.

Gillard, E. 2017. "Keeping an eye on the Earth's energy budget." Accessed February 24, 2024. https://climate.nasa.gov/news/2641/keeping-an-eye-on-earths-energy-budget/.

Hansen, J. E., and M. Sato. 2011. "Earth's climate history: Implications for tomorrow." Accessed January 11, 2023. https://www.giss.nasa.gov/research/briefs/2011_hansen_15/PaleoImplications.pdf.

Henderson, K., D. Pinner, M. Rogers, B. Smeets, C. Tryggestad, and D. Vargas. 2020. "Climate math: What a 1.5-degree pathway would take." Accessed January 23, 2024. https://www.mckinsey.com/~/media/mckinsey/business%20functions/sustainability/our%20insights/climate%20math%20what%20a%201%20point%205%20degree%20pathway%20would%20take/climate-math-what-a-1-point-5-degree-pathway-would-take-final.pdf.

Hill, A. C., and L. Martinez-Diaz. 2020. *Building a resilient tomorrow*, 9. New York: Oxford University Press.

IEA (International Energy Agency). 2021. "Net zero by 2050: A roadmap for the global energy sector." Accessed January 23, 2024. https://www.iea.org/reports/net-zero-by-2050.

IISD, SDG Knowledge Hub. 2019. "77 countries, 100+ cities commit to net zero carbon emissions by 2050 at climate summit." Accessed January 12, 2023. https://sdg.iisd.org/news/77-countries-100-cities-commit-to-net-zero-carbon-emissions-by-2050-at-climate-summit/.

IPCC (Intergovernmental Panel on Climate Change). 2007. "The physical science basis." In *Fourth assessment report: Climate change 2007: Working group I.* Geneva: IPCC.

IPCC. 2013. "Climate change 2013: The physical science basis." In *Working group I contribution to the fifth assessment report of the intergovernmental panel on climate change.* Geneva: IPCC.

IPCC. 2018. "Summary for policymakers." In *Global warming of 1.5°C. IPCC special report on the impacts of global warming of 1.5°C above pre-industrial levels and related global greenhouse gas emission pathways, in the context of strengthening the global response to the threat of climate change, sustainable development, and efforts to eradicate poverty.* Geneva: IPCC.

IPCC. 2021. "Summary for policymakers." In *Sixth assessment report: Working group I: The physical science basis.* Geneva: IPCC.

Lagadec, P. 2004. "Understanding the French 2003 heat wave experience: Beyond the heat, a multi-layered challenge." *J. Contingencies Crisis Manage.* **12** (4): 160–169. https://doi.org/10.1111/j.0966-0879.2004.00446.x.

Lamb, E. 2019. "Should we eat less rice?" Accessed March 20, 2023. https://blogs.scientificamerican.com/roots-of-unity/should-we-eat-less-rice/.

Laxo, A., B. Hoppe, H. Roop, and P. Cipriano. 2023. "Climate forward? How architects and engineers are(n't) using climate projections to inform design." Accessed January 23, 2024. https://hga.com/climate-forward.

Lindsey, R. 2022. "Climate change: Annual greenhouse gas index." Accessed February 26, 2024. https://www.climate.gov/news-features/understanding-climate/climate-change-annual-greenhouse-gas-index.

Mahr, K. 2018. "How Cape Town was saved from running out of water." Accessed January 23, 2024. https://www.theguardian.com/world/2018/may/04/back-from-the-brink-how-cape-town-cracked-its-water-crisis.

Marchau, V. A., W. E. Walker, T. M. Bloemen, and S. W. Popper. 2019. *Decision making under deep uncertainty.* Cham, Switzerland: Springer. https://doi.org/10.1007/978-3-030-05252-2_1.

McDonald, B., S. Burrous, E. Weingart, and M. Felling. 2021. "Inside the massive and costly fight to contain the Dixie Fire." Accessed December 21, 2021. https://www.nytimes.com/interactive/2021/10/11/us/california-wildfires-dixie.html.

McKinsey Global Institute. 2020. "Climate risk and response: Physical hazards and socio-economic impacts." Accessed January 12, 2023. https://www.mckinsey.com/capabilities/sustainability/our-insights/climate-risk-and-response-physical-hazards-and-socioeconomic-impacts.

Naidoo, D. 2018. "Day zero on the back of drought in Southern Africa: Lessons for the future." Accessed January 11, 2023. https://www.brookings.edu/blog/africa-in-focus/2018/02/23/day-zero-on-the-back-of-drought-in-southern-africa-lessons-for-the-future/.

NASA (National Aeronautics and Space Administration). 2019. "What is the sun's role in climate change?" Accessed February 26, 2024. https://climate.nasa.gov/explore/ask-nasa-climate/2910/what-is-the-suns-role-in-climate-change/.

NASA. 2023. "Sea level change: Observations from space, understanding sea level." Accessed March 20, 2023. https://sealevel.nasa.gov/understanding-sea-level/overview.

National Weather Service. 2023. "Hello solar cycle 25." Accessed January 18, 2023. https://www.weather.gov/news/201509-solar-cycle.

NOAA Climate.gov. 2024. U.S. billion-dollar weather and climate disasters in historical context. Accessed February 26, 2024. https://www.climate.gov/disasters2020.

NOAA National Centers for Environmental Information (NCEI). 2024a. "U.S. billion-dollar weather and climate disasters." https://www.ncei.noaa.gov/access/billions/.

NOAA NCEI. 2024b. "Monthly global climate report for annual 2022." Accessed February 26, 2024. https://www.ncei.noaa.gov/access/monitoring/monthly-report/global/202213.

NYT (New York Times). 2020a. "Floodwaters breach Michigan dams, forcing evacuations." Accessed August 27, 2020. https://www.nytimes.com/2020/05/20/us/michigan-dams.html.

NYT. 2020b. "Michigan dams breached, forcing evacuations in midland: Live updates." Accessed January 11, 2023. https://www.nytimes.com/2020/05/20/us/michigan-dam-flooding.html.

NYT. 2023. "'It's like our country exploded': Canada's year of fire." Accessed October 24, 2023. https://www.nytimes.com/2023/10/24/magazine/canada-wildfires.html.

O'Grady, C. 2021. "Glasgow pact leaves 1.5°C goal on life support." *Science* **374** (6570): 920–921. https://doi.org/10.1126/science.acx9654.

Rebuild by Design. 2022. "Atlas of disaster: Facts + figures for US climate disasters 2011–2021." Accessed November 17, 2022. https://rebuildbydesign.org/atlas-of-disaster/.

Russell, P. R. 2020. "Threats of 'day zero' scarcity multiply." Accessed January 23, 2024. https://www.enr.com/articles/49714-threats-of-day-zero-water-scarcity-multiply.

SBTi (Science Based Targets initiative). 2023. "Companies taking action." Accessed January 12, 2023. https://sciencebasedtargets.org/companies-taking-action/.

ScienceDaily. 2022. "University of Colorado at Boulder: US fires four times larger, three times more frequent since 2000: New analysis confirms a palpable change in fire dynamics already suspected by many." Accessed January 10, 2023. https://www.sciencedaily.com/releases/2022/03/220316145723.htm.

Sky News. 2021. "Italy: 29 inches of rain in 12 hours sets new European record as extreme weather lashes country." Accessed October 9, 2021. https://news.sky.com/story/italy-29-inches-of-rain-in-12-hours-sets-new-european-record-as-extreme-weather-lashes-country-12428005.

Smith, A. B. 2020. "2010–2019: A landmark decade of US billion-dollar weather and climate disasters." Accessed August 22, 2020. https://www.climate.gov/news-features/blogs/beyond-data/2010-2019-landmark-decade-us-billion-dollar-weather-and-climate.

Smith, A. B. 2022. "2021 US billion-dollar weather and climate disasters in historical context." Accessed September 13, 2022. https://www.climate.gov/news-features/blogs/beyond-data/2021-us-billion-dollar-weather-and-climate-disasters-historical.

Smith, M., and J. Schwartz. 2019. "'Breaches everywhere': Flooding bursts Midwest levees, and tough questions follow." Accessed August 27, 2020. https://www.nytimes.com/2019/03/31/us/midwest-floods-levees.html.

Strauss, B. H., P. M. Orton, K. Bittermann, M. K. Buchanan, D. M. Gilford, and R. E. Kopp, et al. 2021. "Economic damages from Hurricane Sandy attributable to sea level rise caused by anthropogenic climate change." Accessed January 23, 2024. https://doi.org/10.1038/s41467-021-22838-1.

Swiss RE Institute. 2022 "Sigma explorer." Accessed October 31, 2022. https://www.swissre.com/institute/research/sigma-research/data-explorer.html.

Tribune Media Wire. 2019. "First Camp Fire 911 calls reveal what Paradise dispatchers told residents: 'We're not in any danger'." Accessed August 27, 2020. https://ktla.com/news/local-news/first-camp-fire-911-calls-reveal-what-paradise-dispatchers-told-concerned-residents-were-not-in-any-danger/.

University of Calgary Energy Education. 2024. "Earth temperatures without GHGs." Accessed February 26, 2024. https://energyeducation.ca/encyclopedia/Earth_Temperature_without_GHGs

USACE. 2015. "North Atlantic coast comprehensive study: Resilient adaptation to increasing risk: Main report, 2." Accessed January 23, 2024. https://usace.contentdm.oclc.org/digital/collection/p266001coll1/id/2793/.

Wallace-Wells, D. 2022. "The world took a bold, toothless step forward on climate justice." Accessed December 3, 2022. https://www.nytimes.com/2022/11/23/opinion/environment/cop27-climate-decarbonization.html.

WMO (World Meteorological Organization). 2020. "State of the global climate 2020: Provisional report." Accessed May 15, 2021. https://library.wmo.int/index.php?lvl=notice_display&id=21804.

Yoders, J. 2020. "Federal regulator cited failed Michigan dam multiple times." Accessed January 23, 2024. https://www.enr.com/articles/49422-federal-regulator-cited-failed-michigan-dam-multiple-times.

Zhang, Y. G., M. Pagani, Z. Liu, S. M. Bohaty, and R. DeConto. 2013. "A 40-million-year history of atmospheric CO_2." Accessed January 23, 2024. https://royalsocietypublishing.org/doi/10.1098/rsta.2013.0096.

Additional References (Not Cited)

Barros, V. R., D. J. Dokken, K. J. Mach, M. D. Mastrandrea, T. E. Bilir, and M. Chatterjee, et al. 2014. "Contribution of working group II to the fifth assessment report of the intergovernmental panel on climate change." In *Climate change 2014: Impacts, adaptation, and vulnerability. Part A: Global and sectoral aspects.* Cambridge, UK: Cambridge University Press.

Kane, D. 2018. "Running dry in Cape Town." Accessed January 23, 2024. https://www.nytimes.com/2018/02/01/opinion/cape-town-drought-day-zero.html.

Keys, B. 2023. "Your homeowners' insurance bill is the canary in the climate coal mine." *New York Times.* May 7, 2023. Accessed May 7, 2023. https://www.nytimes.com/2023/05/07/opinion/climate-change-homeowners-insurance-housing-market.html.

Mail & Guardian. 2018. "Here are the city of Cape Town's guidelines for level 3 water usage." Accessed January 11, 2023. https://mg.co.za/article/2018-11-29-here-are-the-city-of-cape-towns-guidelines-for-level-3-water-usage/.

NRC (National Research Council). 2020. *Climate change: Evidence and causes: update 2020.* Washington, DC: The National Academies Press. https://doi.org/10.17226/25733.

NOAA (National Oceanic and Atmospheric Administration). 2023. "Topping the charts: September 2023 was Earth's warmest September in 174-year record." Accessed October 24, 2023. https://www.noaa.gov/news/topping-charts-september-2023-was-earths-warmest-september-in-174-year-record.

NOAA. 2024. "Assessing the global climate in 2023." Accessed February 29, 2024. https://www.ncei.noaa.gov/news/global-climate-202312.

Summers, J., L. Harwell, K. Buck, L. Smith, et al. 2020. "Development of a cumulative resilience screening index (CRSI) for natural hazards: An assessment of resilience to acute meteorological events and selected natural hazards." EPA/600/R-20/274. Washington, DC: USEPA. https://cfpub.epa.gov/si/si_public_file_download.cfm?p_download_id=543576&Lab=CEMM.

Winter, K. 2023. "Cape Town's water: Heading in the wrong direction." Accessed October 24, 2023. https://www.news.uct.ac.za/article/-2023-02-13-cape-towns-water-heading-in-the-wrong-direction.

CHAPTER 2

Civil Infrastructure and the Assumption of Stationarity

"Engineers are probably the single most indispensable group needed for maintaining and expanding the world's economic well-being and its standard of living..."

Richard Weingardt (1998)

"Architects are selected based on their successful design innovations; engineers are selected based on how many times they've delivered the same design successfully."

Rolf Saegesser (2006)

Civil engineering needs to transform itself, moving from an engineering discipline based on static climate conditions to one that can accommodate a changing climate. This transformation will be difficult. Until now, engineers and other practitioners have operated successfully under the assumption of stationarity. That assumption has been the basis for the design and delivery of infrastructure projects since the practice of civil engineering began.

To reject that assumption and to understand why this rework is needed, it is necessary to start at a basic level. What is civil infrastructure? Why is it important? What do civil engineers do? What design assumptions are being used, particularly regarding climate? How do these assumptions relate to infrastructure reliability, public health, safety, and welfare? Why has this approach worked so far?

What is Civil Infrastructure?

Civil infrastructure encompasses buildings, roads, schools, power plants, water treatment systems, waste disposal systems, communication systems, and more. It is the underlying physical framework of societies. It constitutes the facilities and systems on which the growth, development, and day-to-day operation of a community, city, state, or country depends.

Civil infrastructure projects range in complexity, from simple road repair to large and intricate constructed works. Some are ingenious devices for moving vessels through waterways, as in the case of the Falkirk Wheel shown in Figure 2-1. Others, such as the Brooklyn Bridge (Figure 2-2), have become historic landmarks, internationally recognized symbols of American engineering excellence.

Figure 2-1. The Falkirk Wheel is an unusual civil infrastructure project in central Scotland. This rotating boat lift connects two canals as part of the UK inland waterways infrastructure. *Source*: TreasureGalore/Shutterstock.com.

Typically, civil infrastructure assets are long-lived, community-approved structures and systems that provide an efficient and equitable flow of resources and services to populations in cities, towns, and rural communities. These assets are continually maintained, expanded, and upgraded, serving growing populations with evolving needs and taking advantage of technological improvements. Maintenance and upgrades are financed primarily through taxes and fees.

In its Report Card for America's Infrastructure (ASCE 2021), ASCE groups infrastructure assets into 17 categories:

1. **Aviation:** Airports handling both commercial and general aviation services.
2. **Bridges:** Structures that convey motorized and nonmotorized traffic over obstacles such as rivers or valleys.
3. **Dams:** Structures to hold back water. They provide water storage, irrigation, hydropower, flood control, and recreation.
4. **Drinking Water:** Water treatment systems and underground piping operated by utilities that deliver treated water of drinking water quality.
5. **Energy:** Electrical power facilities and transmission lines that deliver electrical energy.
6. **Hazardous Waste:** Waste disposal and storage facilities for managing dangerous wastes. This category includes uncontrolled hazardous waste sites, i.e., land containing significant quantities of dangerous compounds. This category also includes brownfield sites, properties for which surface or subsurface contamination complicates redevelopment or reuse.

Figure 2-2. The Brooklyn Bridge spans the East River and connects Brooklyn to Manhattan. It was designed by John A. Roebling and built by his son, Washington Roebling, with the help of Washington's wife, Emily Roebling. The bridge opened in 1883. In 1964, it was designated a National Historic Landmark, and a National Historic Civil Engineering Landmark in 1972 (NYC DOT 2023).

Source: IM photo/Shutterstock.com.

7. **Inland Waterways:** Networks of inland navigation channels and intercoastal waterways that move materials, goods, and agricultural products.
8. **Levees:** Structures, mainly consisting of earthen embankments, that control the flow of water and reduce flood risk.
9. **Public Parks:** Areas of land that are set aside for public recreational use. Parks include natural and artificial open spaces, landscaping, facilities, playing fields, and buildings.
10. **Ports:** Maritime facilities with one or more harbors for loading and unloading ship cargo.
11. **Rail:** Freight and passenger rail line networks, including the structures, buildings, land, and equipment necessary for support.
12. **Roads:** Interstate, state, and county smoothed or paved surfaces carrying vehicle traffic.
13. **Schools:** The kindergarten through senior year school building infrastructure used for learning, libraries, sports, and feeding children. School buildings are also used for community meetings, emergency centers, and shelters.

14. **Solid Waste:** Facilities and systems for collecting, storing, transferring, and processing municipal solid waste. Solid waste includes product packaging, yard trimmings, furniture, clothing, bottles, cans, food, newspapers, appliances, electronics, and batteries discarded from residential, commercial, or institutional sources. This waste does not include industrial, hazardous, construction, and demolition waste.
15. **Stormwater:** Systems for collecting and transferring stormwater, including sewers, ditches, flood control reservoirs, rain gardens, and natural riverine systems.
16. **Transit:** Public transportation systems for moving people, goods, and services. These systems provide access to jobs, schools, shopping, healthcare, and other services while enabling equitable access and sustainable mobility options.
17. **Wastewater:** Facilities and systems designed to collect, transfer, and treat wastewater from households, businesses, and industries. This category includes sanitary sewer pipes and wastewater treatment facilities.

Why Is Civil Infrastructure Important?

A sound and working set of infrastructure assets and systems is essential for community growth and economic development. Infrastructure assets and systems enable society to

- Extract and harvest raw materials efficiently.
- Convert materials to valuable finished products through efficient manufacturing facilities at home and abroad.
- Move people, materials, products, and services efficiently through multiple interconnected modes of transportation and communication.
- Live, work, and play safely, comfortably, and productively.
- Process for reuse or disposal of industrial and municipal discards safely and efficiently and with minimal environmental degradation.
- Cope with and recover from extreme climate and weather events quickly, safely, and economically.

Regrettably, infrastructure maintenance and upgrading in the United States have not kept up with overall needs. Every 4 years, ASCE rates this nation's infrastructure in terms of its physical conditions and investment needs. US infrastructure's most recent grade reported in 2021 is C minus. The estimated cost to bring US infrastructure up to a good state of repair and earn a B grade is $2.59 trillion invested over 10 years (ASCE 2021).

Civil Infrastructure and National Competitiveness

A well-planned array of functioning and efficient civil infrastructure is also key to national competitiveness. The World Economic Forum (WEF 2013) notes that a nation's infrastructure is part of what the forum calls the *enabling environment,* pillar number 2 of its 12 "pillars" of national competitiveness.

The forum's factors in ranking a nation's infrastructure include

- Quality and extensiveness of infrastructure networks that impact economic growth and reduce income inequalities;
- Location and kinds of economic activities or sectors that can develop within a country;
- Efficient, economical connections of products to markets;
- Movement of workers to suitable jobs; and
- Telecommunications networks that provide a rapid and free flow of information for business communication.

In 2019, the forum ranked the United States second in national competitiveness (WEF 2019) with a score of 83.7 out of 100. Singapore scored slightly higher at 84.8. The US infrastructure component ranked number 13 in the world, falling from number 9 in the previous year.

The Work of Civil Engineers

Ancient builders, precursors to civil engineers, enabled the emergence and development of human settlements, beginning when human societies transitioned from roaming hunter-gatherers to community-centric groups at fixed locations. They established permanent forms of shelter and multiple modes of transport. Even back then, these early engineers were solving society's problems and improving human quality of life through better understanding and applying scientific principles. Civil engineering, as distinguished from military engineering, focuses on the needs of the civilian population, creating the supporting structures and systems upon which economies and societies are built.

Civil engineers plan, design, build, operate, and maintain civil infrastructure, the foundation of modern society (Figure 2-3). Civil infrastructure includes almost everything in the built environment. In this role, civil engineers have a special responsibility to deliver structures, processes, and facilities that are not only cost-efficient but also durable, reliable, and protective of the public using them. Indeed, the Preamble of the Code of Ethics (ASCE 2020) reads as follows.

> Members of The American Society of Civil Engineers conduct themselves with integrity and professionalism, and above all else protect and advance the health, safety, and welfare of the public through the practice of Civil Engineering.

Figure 2-3. Civil engineers at work.
Source: Panumas Yanuthai/Shutterstock.com

Engineers govern their professional careers on the following principles:

- Create safe, resilient, and sustainable infrastructure;
- Treat all persons with respect, dignity, and fairness in a manner that fosters equitable participation without regard to personal identity;
- Consider the current and anticipated needs of society; and
- Utilize their knowledge and skills to enhance the quality of life for humanity.

Infrastructure Projects: Components, Requirements, Relationships, and Dependencies

A general description of an infrastructure project, along with its components, relationships, and dependencies, is presented in Figure 2-4. The finished infrastructure asset has multiple elements, components, and stakeholders beyond its physical and operational aspects.

Each infrastructure project has a project owner, who is responsible for its planning, design, construction, and operation. The owner, in consultation with stakeholders, determines the project need, the project performance requirements and specifications, and the required level of quality. The owner secures financing and oversees the project's planning, design, construction, and operation.

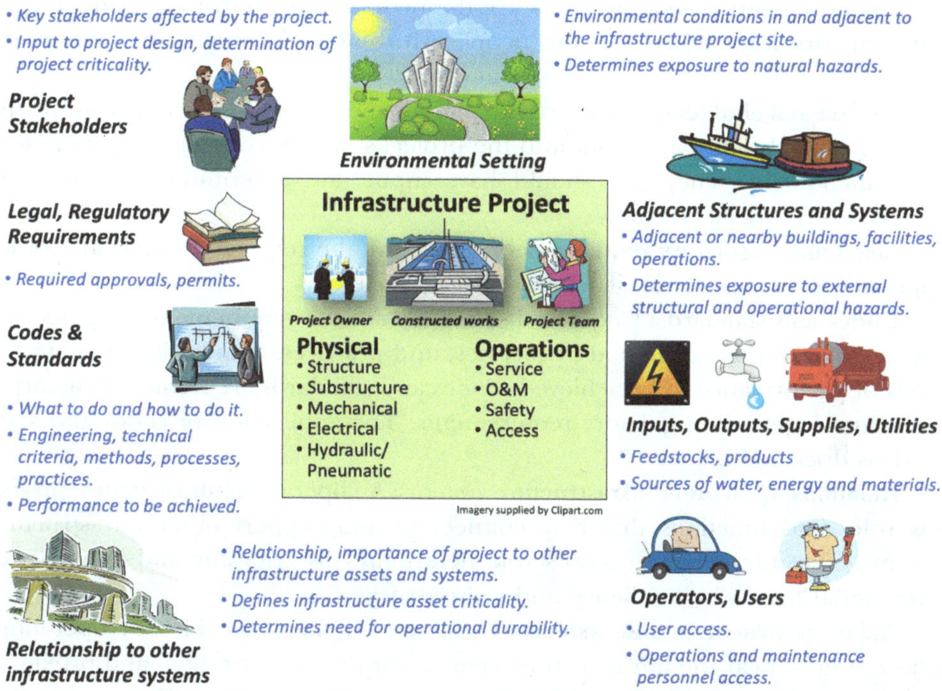

Figure 2-4. General model of an infrastructure project and its components, relationships, and dependencies.

The project team consists of one or more groups that plan, design, build, and operate the constructed works in accordance with the project owner's requirements and specifications. A project team can reside inside or outside the project owner's organization.

The infrastructure project must comply with the owner's requirements and specifications, and the project team is expected to deliver the project in accordance with applicable codes and standards and a defined level of quality. A description of those components follows along with the project's relationship to other infrastructure systems and assets, stakeholders, legal and regulatory requirements, and codes and standards.

Physical: The physical aspects include a structure, substructure, and assorted components, such as electrical, mechanical, and hydraulic/pneumatic systems.

Operations: The finished infrastructure project (constructed works) provides one or more services. It also allows access to users plus operations and maintenance personnel. Operator and user access, and safety are elements of asset operations.

Inputs, outputs, supplies, and utilities: Inputs are the feedstocks needed for the process. An infrastructure asset such as a water treatment plant will have inputs (raw water, water treatment chemicals) and outputs (treated water, spent materials, wastes). Its operations and maintenance crews will need supplies and utilities such as electricity, water, and sanitation.

Operators and users: Operators are the people who control and maintain the infrastructure asset. Users are the people who use and benefit from the asset's functions.

Project stakeholders: These are the people and organizations impacted by the project. They should have input into the project's design, construction, operation, and functionality. They also should have input into determining the finished infrastructure asset's criticality.

Legal and regulatory requirements: These are the required permits and other approvals given by federal, state, and local agencies.

Codes and standards: Codes and standards are the requirements, engineering and technical criteria, methods, processes, and practices. Standards may state the levels of performance to be achieved. Codes and standards are frequently incorporated into legal and regulatory requirements. The application of codes and standards is discussed below.

Relationship to other infrastructure systems: An operating infrastructure project has roles and functions that may connect to and support other infrastructure systems. The infrastructure asset's role determines its criticality and its need for operational durability, resilience and redundancy.

Adjacent structures and systems: These are buildings, facilities, or operations adjacent to or near the infrastructure asset. Their design, operation, and proximity can expose the asset to external structural and operational hazards.

Application of Codes and Standards

To fulfill their roles and responsibilities in planning, designing, and delivering infrastructure projects, civil engineers are obligated to follow the applicable codes and standards developed jointly and adopted by industry, government, and stakeholders. In general, codes are rules and specifications on what is needed to achieve minimum quality, safety, reliability, productivity, and efficiency. They can be approved and made enforceable by local, state, or federal governments. Standards are the "how-to" requirements that establish uniform engineering or technical criteria, methods, processes, and practices to achieve the required level of performance (ASCE 2023). In simple terms, codes tell the engineers what to do. Standards tell the engineers how to do it.

Standards establish uniform engineering or technical criteria, methods, processes, or practices to ensure acceptable levels of quality, reliability, and safety (MSOE University, Walter Schroeder Library 2022). Standards may also be performance-based (i.e., stating what performance levels an infrastructure project must meet). Such standards are in contrast to prescriptive standards that describe how the project is to be designed and constructed. The rationale behind prescriptive standards is this: if a particular design approach has been shown to deliver the desired result reliably, safely, and economically, then that approach can and should be followed by engineers working on similar projects.

Codes and standards are based upon a blend of historical knowledge and engineering judgment regarding how best to design and deliver projects that will perform as required safely, reliably, and cost-effectively under the expected range of environmental conditions. By closely following the applicable climate-related codes and standards, engineers, project owners, and the public had reasonable assurance that the completed projects would work effectively and safely across an expected range of climate variable values. This assurance was based on the assumption that climate conditions are stable and predictable, as evidenced by a long history of empirical performance data. The key proviso is that the climate conditions are *stable and predictable.*

To deliver safe, sound, and reliable infrastructure projects, the engineering profession has produced a cadre of experienced engineers educated in the codes, standards, practices, and tools, and tested and licensed for their competency in application. Licensed professional engineers are responsible for their work and the people affected by it (NSPE 2021). As such, they are held to high ethical standards. They are also liable for any errors that cause damage or losses. In such cases, engineers are judged against a standard of care that compares their actions against what other engineers in the same locale would have done in a similar design. However, as noted in Chapter 1, the standard of care may no longer be sufficient as a defense against claims of negligence if those claims are for damages caused by changing climate conditions and extreme weather events.

Climate Variables, Climate Stressors, and Climate-Derived Variables

Understanding the role of climate in infrastructure design requires understanding the applicable climate variables and their derivatives. *Climate variables*—temperature, precipitation, and wind—are environmental variables that can affect the function and performance of infrastructure assets through their components. Sea level is also included. Although not directly climate-related, sea level rise, especially when combined with high tides and storm surges, can cause flooding.

As noted in Chapter 1, climate variables, at some combination of intensity, duration, and frequency, can become *climate stressors.* Climate stressors at extreme values can degrade, i.e., disrupt, deteriorate, or damage, an infrastructure asset if it is not designed to handle those extreme values. For example, the climate variable precipitation is a variable that is known to affect the function and performance of infrastructure components. Precipitation affects infrastructure performance when rain falls at a level of intensity that creates a stormwater flow that overwhelms the design capacity of stormwater control structures. A table of climate variables, their associated climate stressors, and their effects are listed in Table 2-1. An extensive list of climate stressors and their effects on infrastructure assets organized by infrastructure category is provided in Appendix A.

Table 2-1. Examples of Climate Variables, Associated Climate Stressors, and Possible Effects on Some Infrastructure Assets.

Climate variable	Climate stressor	Example effects on infrastructure assets			
		Bridges	Roads	Rail	Ports/waterways
Temperature	High temperatures. Heat waves	Overheating of equipment. Worker heat stress	Softening of asphalt pavement. Buckling of concrete pavement	Railroad track buckling. Overheating of electrical equipment	Overheating of electrical equipment. Worker heat stress
Precipitation	Flooding. High-velocity water flow	Scouring of bridge support structures	Frequent flooding weakens road base. Service interruption because of high water	Flooding of underground transit systems. Flooding, washout of rail beds	Channel closings. Restrictions because of debris and sedimentation
Sea level	Extreme high tides. Sea level rise	Scouring of bridge support structures. Decrease clearance under bridges	Frequent flooding weakens road base. Service interruption because of high water	Flooding of underground transit systems. Flooding, washout of rail beds	Inundation of buildings and other port facilities
Wind	Wind, storm surge	Potential for vessel collision with bridge structures	Flooding. Scour and erosion of roadways	Flooding of underground transit systems. Flooding, washout of rail beds	Damage of facilities because of wave action

Source: US Climate Resilience Tool Kit (2014).

Climate-derived variables are variables that can be used as surrogates for climate stressors and are used in infrastructure design and operation. For example, the number of days greater than 100 °F at the location of an infrastructure project can be used by designers to evaluate extreme heat impacts on pavement deformation, thermal expansion, worker safety, and other heat impacts on infrastructure (Table 2-2).

Determining and accounting for harsh climate conditions have always been part of civil infrastructure design. Civil engineers have always identified, assessed, and made allowances for climate stressors. The range and probability distribution of these climate stressor variable values have been calculated based on historical climate data and used in infrastructure designs (Figure 2-5). In this figure, the probability distribution is depicted as a normal distribution, but other distributions are possible.

The range of values of another example climate stressor variable distribution is presented in Figure 2-6. For climate variables, there is a range of values that engineers and the public would consider plausible but highly improbable. For example, it is plausible for a 20 in. rainstorm event to occur in the southwestern United States, but it is highly improbable. Likewise, there is a range of probable climate stressor variable values that engineers factor into their designs, recognizing that the values at the extreme ends are possible but unlikely to occur. The importance and criticality of the finished and operating infrastructure project to the health, safety, and economy of the host and user communities determine the degree to which the possibilities of extreme climate stressor variable values are incorporated in the project design and operation.

Taking Climate Stressors into Account in Infrastructure Design

Physical and operational infrastructure components are designed to operate efficiently within the most probable range of values, considering the chances of extreme low-probability climate stressor variable values and their consequences. For example, since the late nineteenth century, asphalt has been a preferred road surfacing material. However, engineers have known that multiple days of high temperatures can soften asphalt-paved roads. Standards for asphalt composition have been formulated using climate-derived variable values, in this case, the expected magnitude and duration of extreme temperature events in different parts of the country. Knowing the probability of these events along with other design variables (e.g., traffic loads and road noise) enables engineers to specify the optimum asphalt mix, making road surfaces functional, maintainable, and cost effective.

Similarly, intense storms combined with high tides can create sea surges and flood coastal neighborhoods. But sea walls built to a sufficient height can prevent this from happening. Tide tables, plus an established history of local storm events, provide the information needed to design flood protection for all but the rarest of

Table 2-2. Example of Climate Variables, Climate-Derived Variables, and Their Use.

Climate variable	Climate stressor	Climate-derived variable	Purpose
Temperature	High temperature. Extended heat wave	Annual maximum temperature	To evaluate extreme heat impacts on the electrical and mechanical components of a bridge, on rail infrastructure, and on construction windows
		Average daily maximum temperature	To evaluate extreme heat impacts on pavement deformation, thermal expansion, and worker safety; increasing temperature impacts on roads; impacts to permafrost formation and thawing
		Number of days min daily temperature < 32 °F and max daily temperature > 32 °F	To evaluate impacts of freeze–thaw events on transportation and other infrastructure assets. Also to evaluate impacts on foundations and slope stability
		Number of days greater than 90 °F, 95 °F, or 100 °F	To evaluate extreme heat impacts on pavement deformation, thermal expansion, worker safety, and other heat impacts on infrastructure
Precipitation	Intense rainfall event. Drought	Annual average precipitation	To evaluate changes in precipitation on erosion, bridge scour, localized flooding, drought, overtopping, and corrosion impacts on infrastructure
		Max 1-, 2-, 5-day precipitation	To evaluate extreme precipitation impacts on highway, other infrastructure assets
		24-h precipitation depths by percentile (95th, 98th, 99th), or return period (2-, 5-, 10-, 25-, 50-, 100-, 500-year storms)	To evaluate extreme precipitation impacts on a drainage system for infrastructure assets, or extreme rainfall intensity on transportation infrastructure, or extreme precipitation impacts on bridge scour, roadway, and bridge overtopping. To evaluate corrosion impacts on infrastructure and flash flooding risks

(Continued)

Table 2-2. Example of Climate Variables, Climate-Derived Variables, and Their Use. *(Continued)*

Climate variable	Climate stressor	Climate-derived variable	Purpose
		Annual average number of drought events	To evaluate the frequency of drought impacts on infrastructure
		Keetch–Byram Drought Index (KBDI)	To help characterize future wildfire scenarios
Sea level	Sea level rise. Storm surge	Storm surge elevation	To evaluate storm surge elevation and flooding impacts on coastal infrastructure and electrical components; drainage system of a roadway; structural, electrical, and mechanical components of a bridge
		Wave height	To evaluate wave height impacts on coastal infrastructure
		Storm surge inundation depth	To help characterize depth of inundation at particular locations
Wind	High wind velocity	Annual average number of wind events (strong wind, high wind, thunderstorm wind) by county	To evaluate the frequency of wind event impacts on infrastructure

Source: FHWA (2017b).

high tide-storm combinations. These and other standards and practices, framed by a vast historical record of climate conditions at known locations, have been developed, tested, and used by practicing engineers.

Climate Stressors in Infrastructure Design

Designers determine the probable range of climate stressor values for each known climate stressor. Then, they design in a level of robustness to maintain functionality across a range of values that the infrastructure asset is expected to experience. The limits of that design range (i.e., the threshold of values beyond which the stressor will overwhelm some infrastructure element or component) are determined by the criticality of the application and the tradeoff between the cost of the additional range of functionality, the chances of failure, and the costs of recovery from failure. Designs

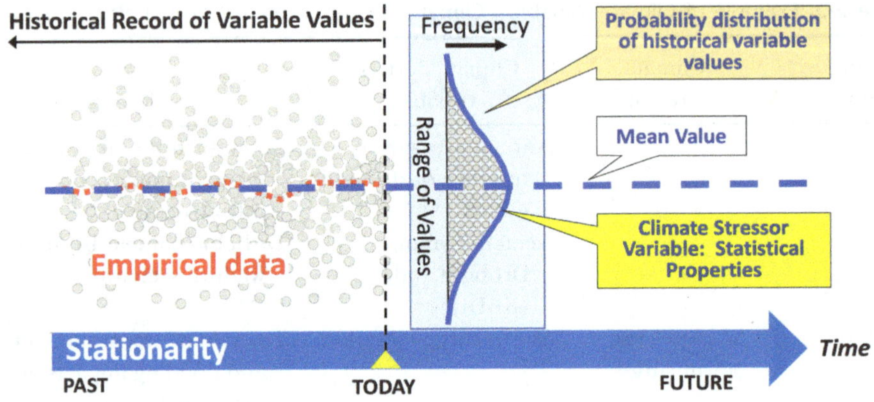

Figure 2-5. Development of a probability distribution for a generic climate stressor variable based on conditions of stationarity. Under stationarity, the statistical properties of these variables will be the same in the future as they were in the past.

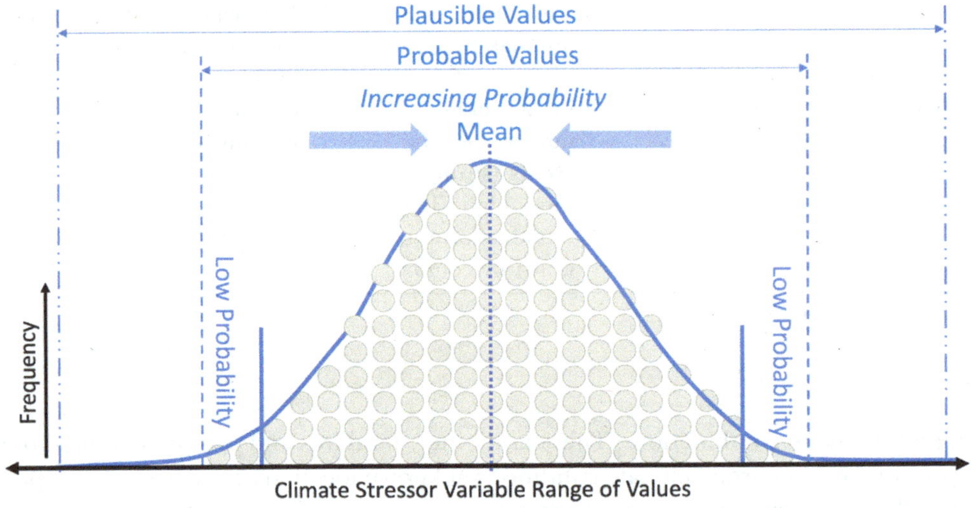

Figure 2-6. Range of historical values for a generic climate stressor variable. The number of circles in each column represents the frequency of variable values relative to others for a particular variable.

are fail-safe to the extent feasible (i.e., public health, safety, and welfare) and are protected in the event of failure. Infrastructure design should also be resilient to the extent practicable. Resilient design anticipates the possibility of failure, minimizes damage, and allows the infrastructure asset to recover quickly and cost effectively.

Climate Stressor Variable Range of Values

For the purposes of this book, a general model of a climate stressor variable probability distribution is presented in Figure 2-7. For this distribution of a climate stressor's range of values, what is of interest are the following value groups.

- **Plausible Values:** A range of plausible values that is physically possible. However, the extreme ends are so improbable that they are not considered in design. An example of a plausible but improbable temperature would be a summer temperature in San Francisco reaching 120 °F.
- **Probable Values:** Enclosed within the range of plausible values are probable values. Plausible values outside this range could occur, but as noted, the probability of their occurrence would be too small to warrant consideration in the design, comparing the cost of additional design robustness against the degree of risk reduction achieved.
- **Most Probable Values:** Within the range of probable values is a subset of values in which the infrastructure component will operate most frequently. Designers will seek to optimize performance in this range.
- **Occasional Outliers:** Just outside the band of most probable values are the infrequent outliers, stressor variable values that are less but still sufficiently probable to warrant attention. Infrastructure component performance in this region may be less than optimum. In establishing design standards, decisions

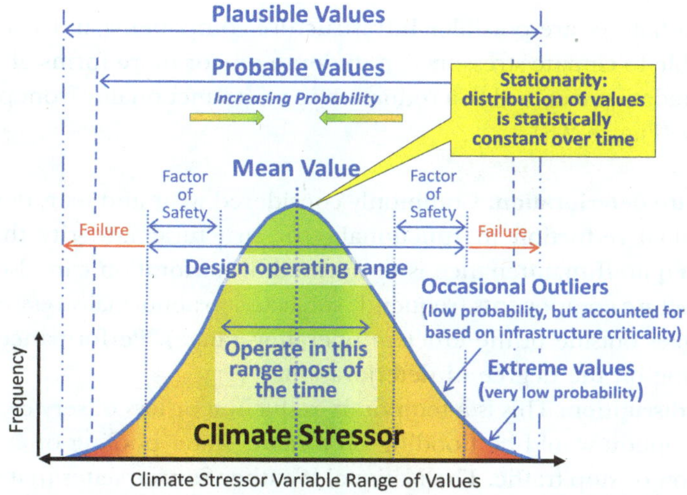

Figure 2-7. A general model of a climate stressor variable probability distribution applied to a component of an infrastructure project. The factor of safety is determined by balancing the cost of an additional range of functionality versus the cost of component failure. Occasional outliers and extreme values can appear on either side of the distribution.

must be made on the degree to which component functionality should be maintained across what spectrum of stressor variable values.

- **Extreme Values:** Moving toward the outer ends of the probability distribution, stressor values become less and less probable, continuing until they go beyond the range of plausible values. Mathematically, the extreme values have no limits in a continuous probability distribution of a variable. However, in a physical system, it is reasonable to assume, based on human experience, that the chances of reaching or exceeding specific values (i.e., beyond plausible) are essentially zero. Again, San Francisco's summer temperature reaching 120 °F is a good example, at least for now.
- **Factor of Safety:** In its simplest form, the factor of safety for a structure or component is the ratio of the expected ultimate stress (range of failure) on a component from a climate stressor to the working stress (design operating range). For example, the design operating range for the Apple iPhone is between 0 and 35 °C. Still, it will operate up to a high-temperature threshold, at which point it will present a temperature warning screen stating the device needs to cool down. Emergency calls may be allowed, but restoring from a backup will be paused (Apple Support 2023). The range of optimal performance and the required factor of safety will vary depending on the component and the stressor in question. The more critical the component (or infrastructure asset) is to the safe and reliable operation of the infrastructure asset (or system), the greater the factor of safety.

Forms of Infrastructure Degradation

What sorts of failures are possible? Infrastructure components and systems exposed and vulnerable to climate stressors can undergo one or more forms of degradation. In turn, degradation can lead to a reduction in asset functionality from partial to zero functionality (Figure 2-8).

- **Premature deterioration:** Commonly considered wear and tear, deterioration is the gradual reduction in functionality or structural integrity that continues unless required maintenance is performed. Deterioration can also result from infrastructure components frequently subjected to abnormal stressors (i.e., values of variables outside of the efficient operating range). Performance can decline depending on the degree of deterioration.
- **Service disruption:** This is a temporary reduction or loss of service. An example of a disruption would be flooding on a road. Water pooling on a road surface could slow or stop traffic. The National Weather Service states that 6 in. of water on a roadway is enough to touch the bottom of most cars; 1 ft of water will float many vehicles (NOAA, National Weather Service 2009).
- **Physical damage:** This is physical, electrical, or mechanical harm to an infrastructure component. An example would be a storm surge that washes away a

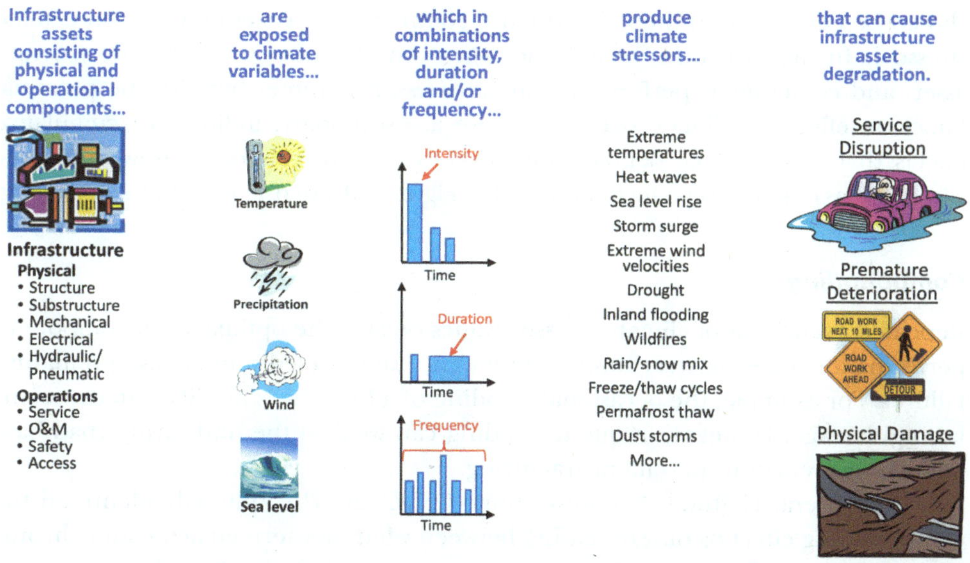

Figure 2-8. Climate stressors and their effects on infrastructure.

section of a road. Damage can cause significant loss of functionality until it is repaired (FHWA 2017a).

Whatever the form of potential infrastructure degradation, the designers should make them fail-safe (i.e., in the event of failure, the component or system will fail predictably to a "safe state"). A safe state is one that keeps harm to the system, other components, the environment, or people to a minimum (Herena 2011). For example, like the iPhone, digital video recorders (DVRs), the devices commonly provided with cable television subscriptions, have a fail-safe mechanism that prevents damage from overheating. A DVR operating in an unventilated cabinet may overheat and damage the internal electronics and circuitry. The device will shut down to avoid damage if temperatures exceed some predetermined level. In the case of infrastructure, designs are typically made fail-safe (i.e., designed to limit damage in case of failure). For example, reservoir dams are designed with spillways to allow floodwater to pass in a controlled way over, around, or through the dam. Another example is the design of semaphore signals in railway signaling. If the signal arm actuating cable fails, gravity pulls the signal arm into a stop position, halting the flow of rail traffic.

Knock-on Effects

Based on a history of infrastructure asset performance, engineers have also learned a great deal about what happens at extreme values for a wide assortment of stressors. One region of interest is the limit of functionality (i.e., what range of operability should be designed into the infrastructure component or system for a given climate stressor). In infrastructure design, engineers consider not only how infrastructure assets and components perform at climate stressor extremes but also the potential knock-on effects of climate stressors. These are secondary, indirect, or cumulative effects that can enhance degradation, produce new or reinforce known stressors. Three forms of knock-on effects, depicted in Figure 2-9, are discussed in the following.

Compounding

Repeated excursions of climate stressor values outside the optimal asset component performance zone can wear down components unexpectedly and cause premature failures. For example, the occasional flooding of a highway is not likely to cause any lasting damage. However, frequent flooding can weaken the underlying road base and cause premature pavement fracturing.

The Federal Highway Administration (FHWA 2017b) indirectly identified the compounding effect by differentiating between what they termed acute and chronic climate stressors. On the one hand, acute stressors cause immediate damage during short-duration events like intense storms or heat waves. An intense storm can wash away roadways. Heat waves can buckle pavements. Chronic stressors, on the other hand, involve gradual changes in climate stressors such as increasing frequency of

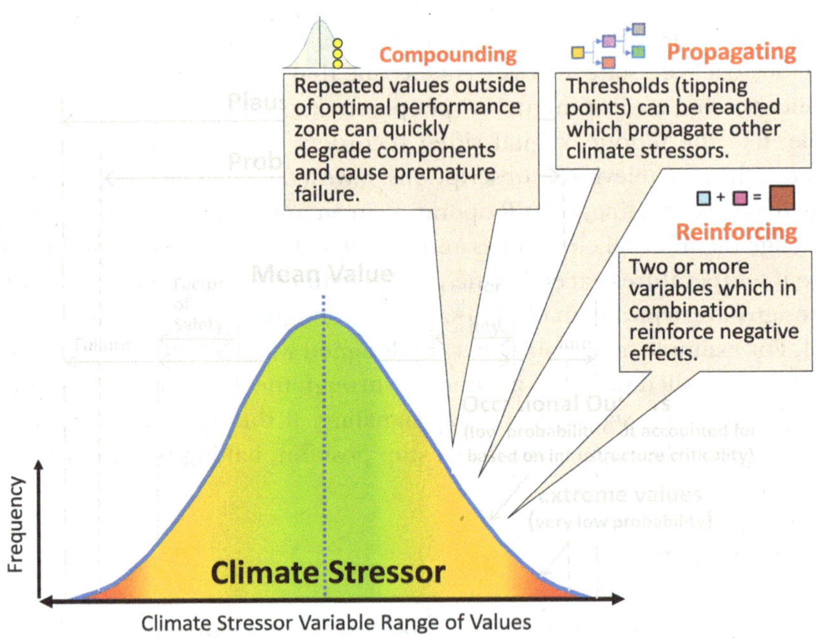

Figure 2-9. Climate stressor variable knock-on effects.
Note: These effects can be at either side of the probability distribution curve.

flooding or heat waves. Over time, these can accelerate asset deterioration, resulting in a shorter asset life. Frequent flooding or multiple extended heat waves can also stress the social health of communities. This repetition of the events is a compounding attribute of climate stressors.

Propagating

At extreme values, a climate-derived variable may reach a tipping point at which new stressors are triggered. For example, the United States is increasingly experiencing dry spells—extended periods of extraordinarily low precipitation. When these periods become lengthy, wildfires can become more frequent and devastating.

Wildfires can be viewed as the propagated effects of an extended drought. In 2019 and again in 2020, northern California experienced record-breaking wildfires. Canada experienced extensive wildfires in 2023 (Livingston 2023). If conditions exist where climate stressor propagation is significant, owners and engineers should avoid these locations or devise ways to reduce the propagation possibilities.

In the case of wildfires, California and other western states experiencing such dry conditions are limiting construction in wildfire-prone areas and shutting off electrical power during high wind events. The California Department of Forestry and Fire Protection (CAL FIRE) is also clearing brush and removing dead trees around communities that are in what is called the "wildland-urban interface." CAL FIRE is also establishing fire breaks, limiting vegetation around homes, and recommending fire-resistant home construction (Mulkern and E&E News 2019).

Reinforcing

Extreme values of climate-related variables in combination can reinforce adverse effects. A familiar example is ambient temperature and humidity. Days that are both hot and humid take a toll on human health. Also, when high heat is accompanied by drought conditions, asphalt concrete pavement can crack, making it vulnerable to water when it rains (US Climate Resilience Tool Kit 2014). Also, freeze–thaw events occurring shortly after a rainstorm can significantly affect slope stability (FHWA 2017b).

Another example is a forest fire quickly followed by an intense rainstorm. Debris from the fire can be picked up by flowing stormwater and transported to rivers and lakes downstream. In 2002, such a sequence of events happened in Colorado. Peter Binney, former director of Aurora Water for the City of Aurora, Colorado, describes the problem as "multiple black swans." A "black swan" event is an event that comes as a surprise, has a significant effect, and is often inappropriately rationalized after the fact with the benefit of hindsight. It is based on an old saying which presumed that black swans did not exist but was rewritten after they were discovered. In Aurora's case, a huge forest fire was followed by an exceptionally intense rainstorm in the city's watershed. The rainstorm washed fire debris along with sediment and fire retardants into one of the rivers supplying raw water to a water treatment plant for Aurora. The result was a massive increase in turbidity from the fire debris plus high phosphorus concentration from the fire retardants in Aurora's raw water supplies. This turbidity dropped the water treatment plant's production rate from 80 million gallons per day (MGD) to 10 MGD (Binney 2014, personal communication).

How Knock-On Effects Have Been Handled

Through many years of observing and cataloging performance in a wide range of climate conditions, engineers have recognized and documented instances of compounding, propagating, and reinforcing stressor expressions or situations. Using that record of experience, engineers have strengthened the components or avoided exposure to specific climate stressors, singly or in combination.

Safety factors are incorporated into the component's operating range to handle the occasional outliers: infrequent stressor variable values outside normal operating conditions. Safety factor boundaries are set at extreme values with a very low probability of occurrence, values deemed too remote to worry about and spend money to manage.

Types of Failure

If an infrastructure component degrades (i.e., deteriorates, disrupts service, or is damaged to the point where it stops doing what it is supposed to), then it has failed. Failure can be catastrophic or resilient.

Catastrophic Failure

Catastrophic failure of a component or system is a major or total loss of function that results in lengthy downtime and costly repairs. An example is the flooding of the New York City subway system from Hurricane Sandy (FHWA 2017a). In this case, designers had not considered a storm of such magnitude that would cause seawater to overtop barriers, flood the subway tunnels, and damage the electrical equipment. As of this writing, the city's subway system is still recovering from that 2012 storm.

City officials are developing plans to improve the overall resiliency of its transit and other critical infrastructure by, for example, improving flood barriers. However, work still needs to be done. In July 2021, heavy rains, this time from Tropical Storm Elsa, again flooded New York City roads and subway stations (Shanahan and Wong 2021).

Resilient Failure

A resilient failure of a component or system is a loss of function caused by extreme values of one or more climate stressors in which downtime is short and repairs are relatively inexpensive. The advantage of designing for resilient failure is that if the component or system fails, it can be brought back into operation quickly and cost effectively.

For this discussion, resilient failure is limited to components or systems within an infrastructure asset and the finished infrastructure asset. In comparison, community resiliency is the ability of a community to recover from a disaster, climate-related or otherwise. Community resiliency involves additional properties such as resourcefulness and rapidity. Resourcefulness is the capacity and capability of the leadership and residents of the affected community to identify problems resulting from the disaster and prioritize and mobilize the resources (human and monetary) to solve them. Rapidity is the community's ability to do so promptly (Bruneau et al. 2003). In the case of communities, the total cost of catastrophic failure is more than just the cost of rebuilding. This cost also includes the social costs of failure: loss of service, jobs, economic disruptions, and more.

Infrastructure Resiliency

Focusing on climate and individual infrastructure assets, the *overall resiliency of an infrastructure asset* is one that has the following properties:

- **Robustness:** The degree of robustness is the extent to which an infrastructure asset can withstand extreme climate stressor values without loss of function. Component or system robustness is provided by increasing safety factors in anticipation of extreme climate stressor values during operation.
- **Resiliency:** The extent to which design for resilient failure is incorporated into crucial infrastructure asset components and systems. Resiliency is the ability of an asset to recover quickly and cost effectively.
- **Redundancy:** The extent to which the infrastructure project design incorporates, when practical, other components or systems capable of taking over the functions of a failed component or system while that component or system is being repaired or replaced.

An example of resilient design is a flood protection system that, if overtopped by flood waters, channels those waters into holding ponds with sufficient capacity to contain and absorb those waters without causing further damage or loss of service. In addition, the system can be designed to be erosion-resistant on both sides so that overtopping waters do not erode the landside of the system and cause damage to overall structural integrity. The capacity of the holding ponds (i.e., the limit of resilient-failure design) is determined by the point at which the cost of the resilient design and operation reaches the cumulative costs of a catastrophic failure multiplied by the probability of that failure.

Designing Infrastructure Assuming Conditions of Stationarity

Building from the general probability distribution model of a climate stressor variable shown in Figure 2-7, a general model of a design standard for that particular variable can be constructed (Figure 2-10). The various elements are described as follows.

- **Efficient Range of Performance:** The range of climate stressor variable values for which performance is most cost-efficient. The range is based on the infrastructure component or system affected by climate stressor variable values and currently available technologies. As technologies improve, the range of efficient operation may also improve.
- **Effective Operating Range:** The range of climate stressor variable values for which performance will be maintained. This range includes the efficient range of performance plus the additional robustness provided by the safety factor.
- **Factor of Safety:** Economies depend on smooth-running infrastructure systems to drive economic growth and development. For that reason, the cost of lost functionality in infrastructure is high. To ensure that these critical infrastructure

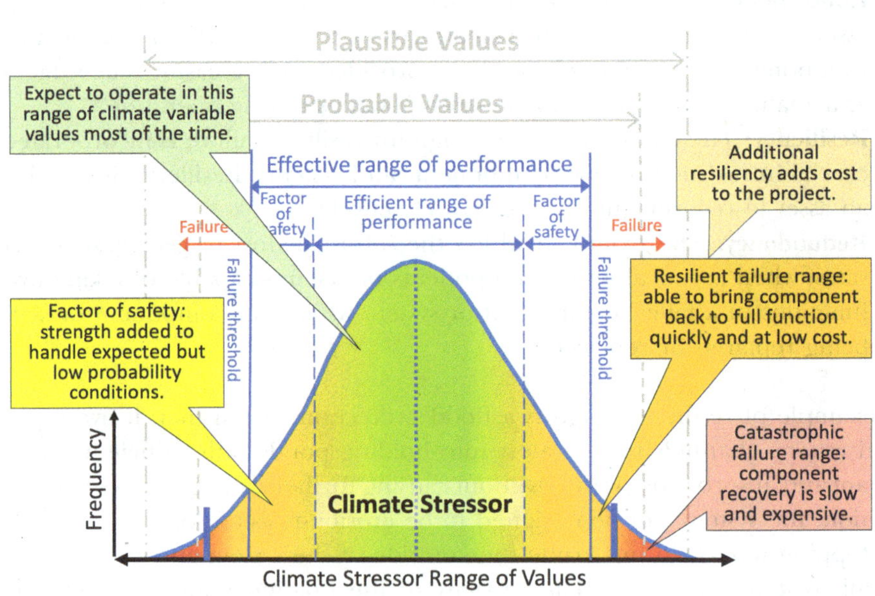

Figure 2-10. Conventional infrastructure design under the assumption of stationarity.

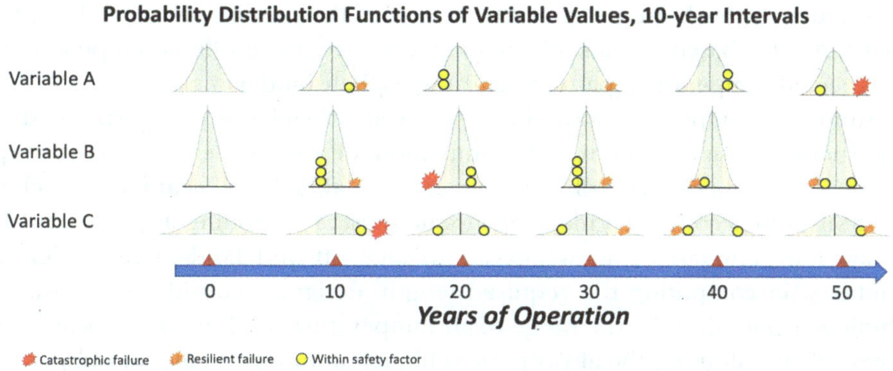

Figure 2-11. Probability distribution functions of three generalized climate stressor variables that follow the assumption of stationarity, assuming 50 years of operation shown with 10-year accumulated incidents.

systems keep running, designers incorporate additional safety factors to reduce the probability of failure of infrastructure components vital to overall system functions. Technically, the safety factor is the maximum load a component can handle without failure divided by the design load.

Figure 2-11 shows probability distribution functions for three generalized climate stressor variables accumulated at 10-year intervals over 50 years of operation. Variable values inside the efficient performance range are excluded. Under conditions of stationarity, the probability distributions of the variables are unchanged. Therefore, it is expected that the frequency of excursions outside the efficient performance ranges is virtually unchanged.

Using Climate-Derived Variables

Infrastructure designs depend on local physical and environmental conditions, including climate and weather. For example, to design new roadways, engineers refer to accepted standards for highway design. Given design constraints such as location, capacity, vehicle loads, and community budget, engineers can apply established standards tuned to the desired road specifications. Standards for roads in a given locale could specify, for example, the components of the asphalt mix that will support a range of vehicle loadings across the expected range of temperatures.

Based on historical performance experience, engineers have determined that an essential climate-derived variable for asphaltic pavement design is the likelihood of high ambient temperatures sustained over an extended period. Knowing the average ambient temperature at this location is not sufficient information. What is needed is the likelihood of high ambient temperatures extending over multiple days, the temperature of the hottest day of the year, and the number of days above 95 °F (FHWA 2017b). Based on this information, correlations can be developed between the expected temperature patterns and the asphalt binder mix.

Another example of how climate-derived variables are incorporated into infrastructure design is in the determination of runway length for an airport (Figure 2-12). The design of airport runways takes local weather and climate conditions into account. To make economic sense, functioning airports must have runways long enough to allow aircraft to take off and land safely under most conditions. In computing the required length, designers consider several factors, including airport altitude, the range of air temperatures at the airport location, and the type of aircraft using the airport. Air at higher altitudes and higher temperatures is less dense. As a result, aircraft lift is reduced.

In a stationary climate regime (i.e., statistically constant over time), it becomes possible to plot the frequency distribution (also called *probability distribution function* or *probability density function*) of some climate-derived variable of interest using historical values and use that plot to predict future values. That curve provides a useful picture of the range of values distributed according to the frequency at which they are likely to occur. From that information, a correlation can be developed that

Figure 2-12. Small aircraft landing on an airport runway. The runway length is based on local conditions, such as expected high temperatures and the altitude of the airport runway. The type of aircraft using the airport is also a factor.
Source: Nadezda Murmakova/Shutterstock.com.

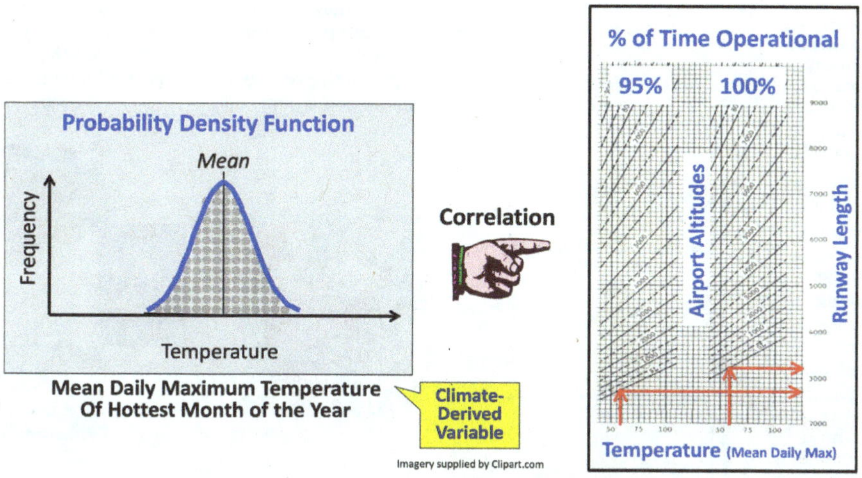

Figure 2-13. Relationship of the probability distribution of the climate-derived variable "mean daily maximum temperature for the hottest month of the year" to the FAA methodology for calculating airport runway length.
Source: FAA (2005).

relates the climate-derived variable to the design information needed. In the case of airport runway length, the Federal Aviation Administration has developed a nomograph (FAA 2005) that connects a climate-derived variable (the mean daily maximum temperature of the hottest month of the year) to the required runway length for 95% and 100% airport operation for various airport altitudes (Figure 2-13).

Stationarity Worked . . . At Least Until Now

As illustrated in Figure 2-14, designing and delivering civil infrastructure projects based on the assumption of stationarity was working fine, at least up until now. So, what happened? As it turns out, climate conditions are stable as long as the Earth's ecological systems remain in relative equilibrium. Unfortunately, that is no longer the case. In the 200-plus years since the Industrial Revolution, the Earth's human population has grown exponentially, now exceeding 8 billion. Moreover, the population is increasing in affluence and demanding more resources and services to fulfill its needs. Meeting this ever-increasing demand severely taxes the Earth's ecological systems, pushing them beyond their ability to function effectively. Increasing atmospheric concentrations of greenhouse gases have disrupted the Earth's climate-regulating ecosystem services.

Climate-regulating ecological services, those natural services that maintain climate and weather conditions within known and manageable boundaries, are

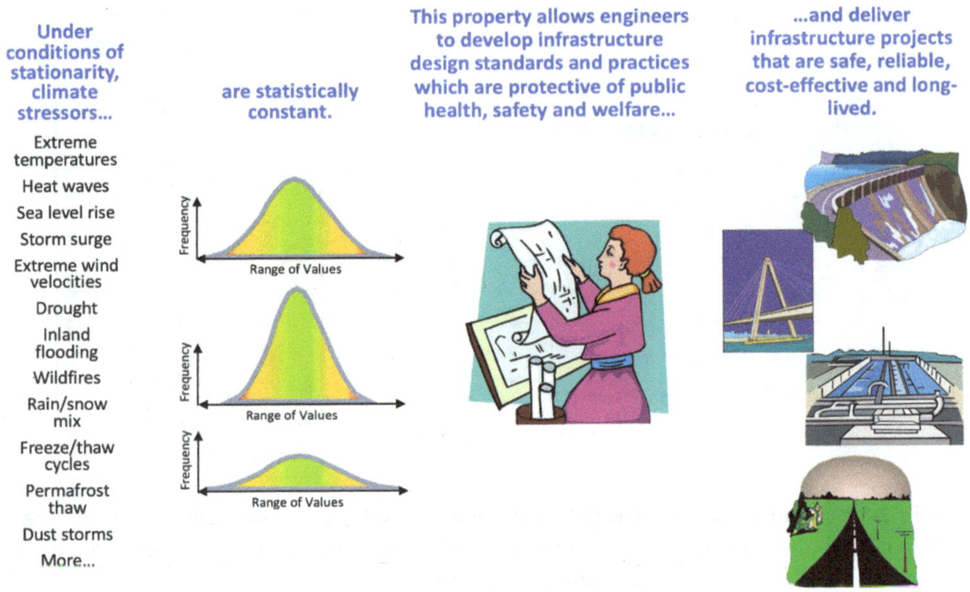

Figure 2-14. Relationship of climate stressors and civil infrastructure design under conditions of stationarity.

critical to civil infrastructure design. Therefore, it is essential to understand how these services work and what happens when they are pushed past their limits. Furthermore, when the assumption of stationarity is no longer reliable, civil engineers must modify designs to protect from, adapt to, or accommodate the changing climate conditions. As noted in Chapter 1, the climate is no longer stationary, the consequence of a disruption in the Earth's natural climate-regulating systems. These and other natural systems and their relation to sustainability are covered in Chapter 3.

References

Apple Support. 2023. "If your iPhone or iPad gets too hot or too cold." Accessed January 13, 2023. https://support.apple.com/en-us/HT201678.

ASCE. 2020. "Code of ethics." Accessed December 21, 2021. https://www.asce.org/career-growth/ethics/code-of-ethics.

ASCE. 2021. "Report card for America's infrastructure." Accessed December 29, 2021. https://infrastructurereportcard.org/.

ASCE. 2023. "Codes and standards." Accessed January 13, 2023. https://www.asce.org/publications-and-news/codes-and-standards.

Bruneau, M., S. E. Chang, R. T. Eguchi, G. C. Lee, et al. 2003. "A framework to quantitatively assess and enhance the seismic resilience of communities." *Earthquake Spectra* **19** (4): 733–752. https://doi.org/10.1193/1.1623497.

FAA (Federal Aviation Administration). 2005. "Runway length requirements for airport design." Advisory Circular 150/5325-4B. Accessed February 29, 2024. https://www.faa.gov/documentLibrary/media/Advisory_Circular/AC_150_5325-4B.pdf.

FHWA (Federal Highway Administration). 2017a. *Post Hurricane Sandy transportation resilience study in NY, NJ, and CT.* FHWA-HEP-17-097. Washington, DC: FHWA.

FHWA. 2017b. *Synthesis of approaches for addressing resilience in project development.* FHWA-HEP-17-082. Washington, DC: FHWA.

Herena, P. 2011. "The principle of fail-safe." Accessed December 23, 2021. https://www.aiche.org/chenected/2011/02/principle-fail-safe.

Livingston, I. 2023. "Canada's unprecedented fire season burns into fall." Accessed October 15, 2023. https://www.washingtonpost.com/weather/2023/09/25/canada-smoke-fire-british-columbia-alberta/.

MSOE University, Walter Schroeder Library. 2022. "Mechanical engineering codes and standards." Accessed January 2, 2022. https://libguides.msoe.edu/me/codes.

Mulkern, A. C., and E&E News. 2019. "California tests new strategies to prevent deadly wildfires." Accessed January 22, 2024. https://www.scientificamerican.com/article/california-tests-new-strategies-to-prevent-deadly-wildfires/.

NOAA, National Weather Service. 2009. "Turn around don't drown." Accessed January 13, 2023. https://www.weather.gov/tsa/hydro_tadd.

NSPE (National Society of Professional Engineers). 2021. "What is a PE?" Accessed August 29, 2021. https://www.nspe.org/resources/licensure/what-pe.

NYC DOT (New York City Department of Transportation). 2023. "Brooklyn bridge." Accessed March 23, 2023. https://www.nyc.gov/html/dot/html/infrastructure/brooklyn-bridge.shtml.

Shanahan, E., and A. Wong. 2021. "Heavy rains pound New York city, flooding subway stations and roads." Accessed January 22, 2024. https://www.nytimes.com/2021/07/08/nyregion/flooding-subways-nyc.html.

US Climate Resilience Tool Kit. 2014. "Transportation climate change sensitivity matrix." Accessed January 19, 2024. https://toolkit.climate.gov/tool/transportation-climate-change-sensitivity-matrix.

WEF (World Economic Forum). 2013. *The global competitiveness report: 2013–2014.* Edited by K. Schwab, 5–6. Cologny, Switzerland: WEF.

WEF. 2019. *The global competitiveness report.* Edited by K. Schwab, 582–585. Cologny, Switzerland: WEF.

Weingardt, R. 1998. *Forks in the road: Impacting the world around us.* Denver: Palamar.

CHAPTER 3

Climate Change, Stationarity, and Sustainability: How Are They Connected?

"I recognize the right and duty of this generation to develop and use the natural resources of our land; but I do not recognize the right to waste them, or to rob, by wasteful use, the generations that come after us."

Theodore Roosevelt, "The New Nationalism," speech at Osawatomie, Kansas, August 31, 1910

The Foundational Assumption of Stationarity

Chapter 2 discussed the work of civil engineers and their dependence on the assumption of stationarity: although day-to-day weather conditions change, the overall climate, statistically speaking, remains constant. This assumption is one of the key foundations of civil engineering. It enables engineers to design and deliver robust, resilient, reliable, and long-lived civil works projects using historical climate conditions to estimate future climate conditions. These estimates are embedded in assorted engineering codes, standards, manuals, and guidance documents.

This condition of climate stationarity has enabled society to flourish. In his book, *Eaarth: Making a Life on a Tough New Planet*, Bill McKibben (2010) notes, "For the last 10,000 years that constitute human civilization, we've existed in the sweetest of sweet spots." Average temperatures had leveled off, staying between 58 and 60 °F (14 and 16 °C). The world's oceans are tame and level. Heat and rainfall are predictable. Resources are plentiful. McKibben renamed our planet of today "Eaarth" because it is not the same planet we had 50 years ago.

This extended period of climate stability is unique. Ice core records over the last 350,000 years (Figure 3-1) show natural warming and cooling cycling periods. Called Milankovitch cycles, they vary in intensity and time. Periods range between approximately 10,000 and 100,000 years.

Over those periods, the Earth's average temperature had varied, something on the order of 10 °C across cycles (Jouzel et al. 2007). While these ambient temperature changes can be large, the rate of temperature change is slow. A shift of 1 °C (1.8 °F) generally takes several thousand years.

Figure 3-1 shows that over the last 350,000 years, natural atmospheric concentrations of CO_2 have also varied, staying within the range of 170 and 300 parts per million (ppm). As a greenhouse gas (GHG), CO_2 molecules in the atmosphere

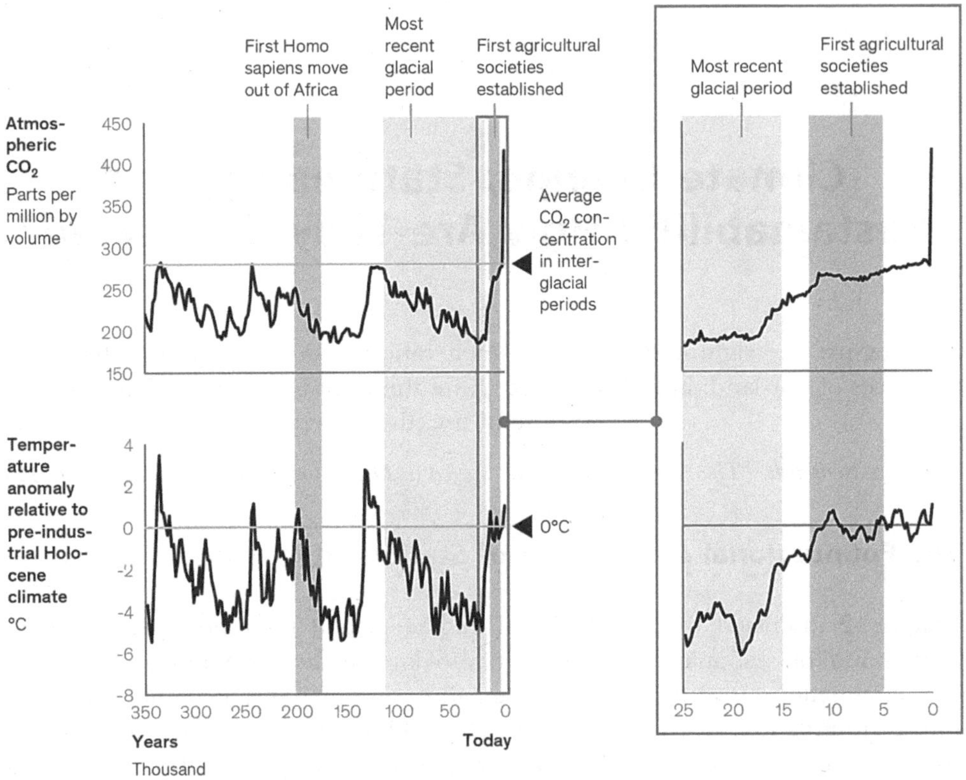

Figure 3-1. Comparing atmospheric concentrations of carbon dioxide (CO_2) and temperature anomaly relative to pre-industrial levels over the last 350,000 years.

Source: © McKinsey & Company (2020). All rights reserved. Reprinted by permission.

absorb heat energy, some of which is re-radiated back toward Earth. Had it not been for naturally occurring atmospheric CO_2 in the atmosphere, average temperatures would be about 0 °F (−18 °C) (University of Calgary 2012).

Today, climate conditions are no longer stable. Since the beginning of the Industrial Age (approximately 1760 to 1840), atmospheric concentrations of CO_2 have increased by 50%. CO_2 concentrations have risen to over 420 ppm (NASA 2022). Concurrently, the Earth's temperature has increased by 1.3 °C (2.3 °F) over pre-industrial times. This additional heat energy not only warms the Earth but also increases the likelihood of severe climate and weather events (IPCC 2021), with more droughts and heat waves, stronger and more intense storms, longer wildfire seasons, and more.

Is Our Form of Economic Development Sustainable?

Climate regulation is just one of the Earth's natural ecosystem resources and services from which society benefits. Its current instability raises some important questions. How is society using this and other natural ecosystem resources and services? What will be available for future generations? Will they be sufficient? This brings us to the overarching question: Is our form of economic and social development sustainable over the long term?

The Brundtland Commission Report

As the opening quotation in this chapter suggests, the mandate for using the Earth's resources in a way that is sustainable has been around for some time. However, it took the Brundtland Commission Report (Figure 3-2), *Our Common Future* (UN 1987), to raise the issue globally. The report stated,

> Humanity has the ability to make development sustainable – to ensure that it meets the needs of the present without compromising the ability of future generations to meet their own needs.

In 1983, the United Nations established the Commission and appointed its chair, Gro Harlem Brundtland, a former prime minister of Norway. The Commission's charge (UN 1983), in part, was to look into the processes and progress of human development and its relation to the environment and to "propose long-term environmental strategies for achieving sustainable development by the year 2000."

Figure 3-2. Gro Harlem Brundtland, chair of the World Commission on Environment and Development, and the Commission's report, "Our Common Future," published in 1987. *Source*: Photo, Harry Wad, Creative Commons BY 2.5, via Wikimedia Commons. Book cover, Sigurður Kaiser, Creative Commons Attribution-ShareAlike License 3.0.

The key findings of the report (relevant to infrastructure) were

- Economics, environment, and society are inextricably linked,
- Damage to the environment also damages the economy and society,
- Resources are finite, and the environment is not infinitely repairable, and
- Problems from human activities are no longer compartmentalized.

The report's finding that problems are no longer compartmentalized is particularly important. Because of the strength and scale of our technologies and the extent of our interconnectedness, problems are no longer isolated to a particular geography, environment, or industry. For example, acid rain produced by emissions from coal-burning power plants in the midwestern United States devastated forests downwind in the eastern states. Also, as depicted in Figure 3-3, runoffs, heavily contaminated with pesticides and fertilizers in the Midwestern United States, have reached the Mississippi River watershed. This contamination has created a dead zone in the Gulf of Mexico (NOAA 2023), killing fish populations and harming local economies. Today, the most critical and pervasive example is global warming: nations around the world are powering their development by burning fossil fuels and, by doing so, disrupting the Earth's climate.

Are We Sustainable?

The notion that we as a society are not sustainable and that future economic development needs to be sustainable has confused many people. There were many interpretations:

- What do you mean we are not sustainable? We have always respected the environment.
- This is socialism in sheep's clothing!
- This all sounds very expensive. We cannot afford it!
- What problems? Whatever they are, technology will fix them. We always find a way.

The assertion that the environment, economy, and society are linked challenged the existing views on these relationships.

Before Brundtland, many people, particularly those in business, asserted that the economy drives everything. Society and the environment are subordinate (Figure 3-4), or worse independent of one another (Figure 3-5). Moreover, the Earth's resources were assumed to be, in effect, infinite, with ecological systems restorable or self-healing.

This relationship was noted in 2009 by then ASCE president Wayne Klotz when, at a national meeting in Kansas City, he criticized the profession for treating natural resources carelessly (Klotz, personal communication, 2009).

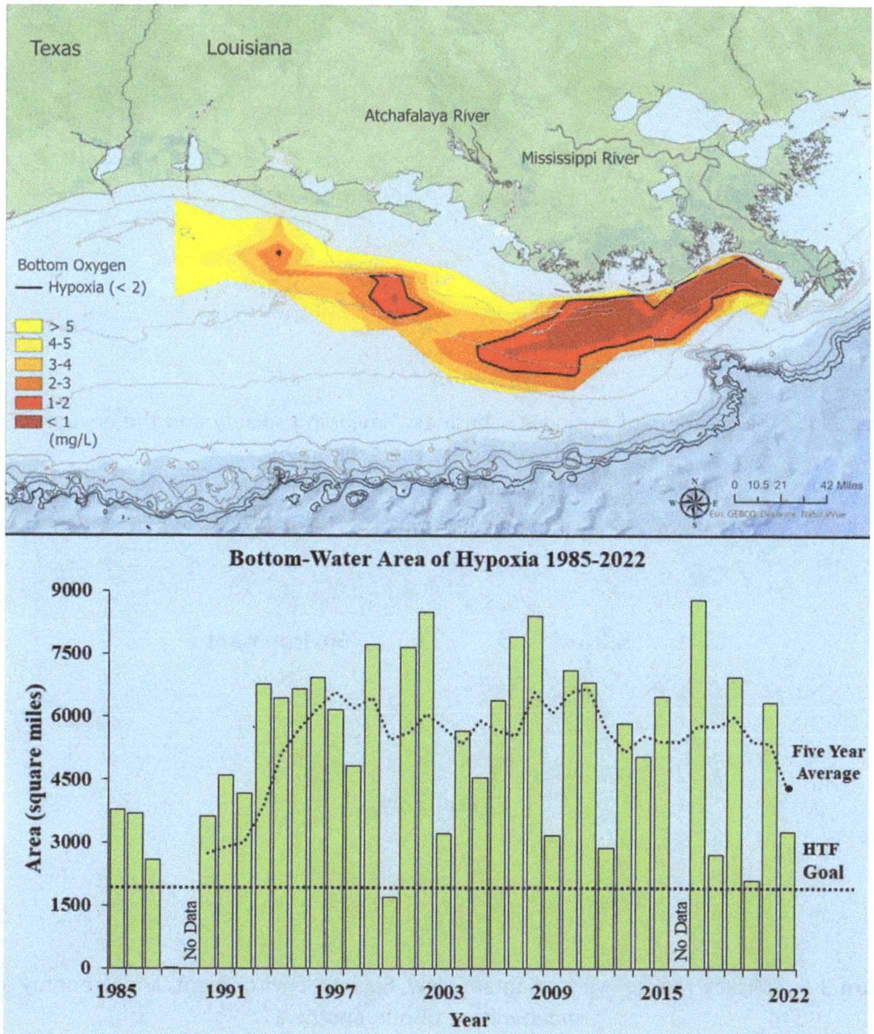

Figure 3-3. Contours of dissolved oxygen near the bottom on the Louisiana and Florida shelves, June 6 to July 18, 2019. Note the predominance of orange colors nearer shore in coastal Louisiana indicating widespread hypoxia, levels of dissolved oxygen not sufficient to support marine life. *Source*: Top: Louisiana Universities Marine Consortium (2022), see https://oceanservice.noaa.gov/facts/deadzone.html, bottom: NOAA (2023).

> "You engineers go about your business working as if our natural resources were infinite. They are not, and you have to stop treating them as if they were."
>
> *Wayne Klotz, President of the American Society of Civil Engineers, speech before the National Convention in Kansas City, Missouri, 2009*

The linkage of the environment, economy, and society took another twist with the publication of John Elkington's book, *Cannibals with Forks: The Triple Bottom Line*

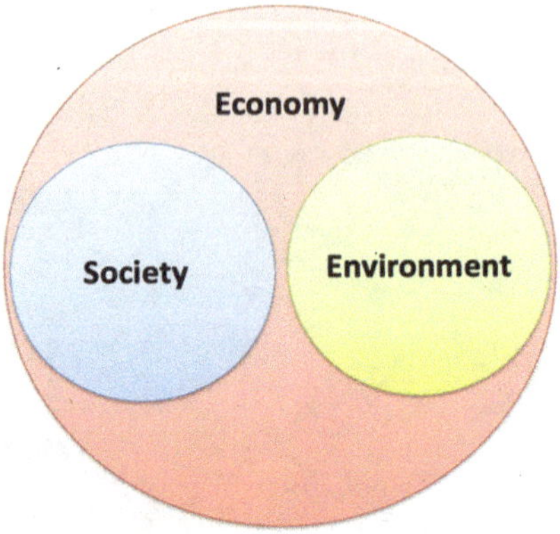

Figure 3-4. "The business of business is business," meaning society and the environment are subordinate to the economy.
Source: Statement attributed to economist Milton Friedman.

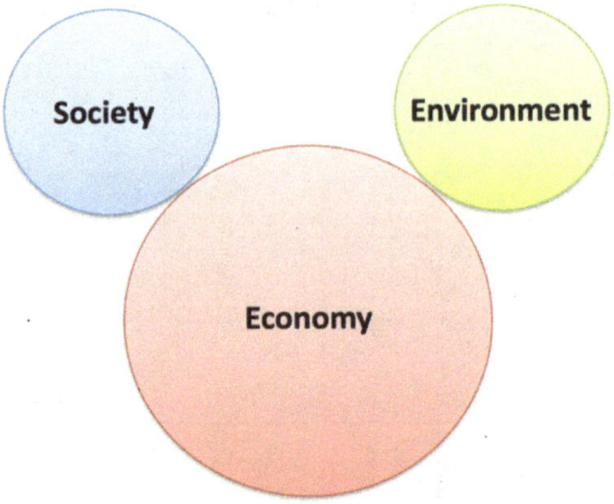

Figure 3-5. "Mickey Mouse" Venn diagram view. Society, environment, and economy are all independent of one another.

of 21st Century Business (Elkington 1998). The unusual title was derived from a quote by Polish poet Stanislaw Lec: "Is it progress if a cannibal uses a fork?" Elkington was chiding business for measuring their performance using financial metrics only (i.e., economic performance) when, in fact, business ought to be looking at a "triple bottom line," bringing in the equally essential measures of environmental and social performance. This metaphor took hold, giving rise to a graphic interpretation shown in Figure 3-6: The environment, economy, and society are represented as three intersecting circles of a Venn diagram, with the common intersection representing sustainability.

If a business were guided by a triple, not a single, bottom line, society might see it as a sustainable enterprise. However, the figure implies that being sustainable involves balancing all three goals, making tradeoffs among the three until all are

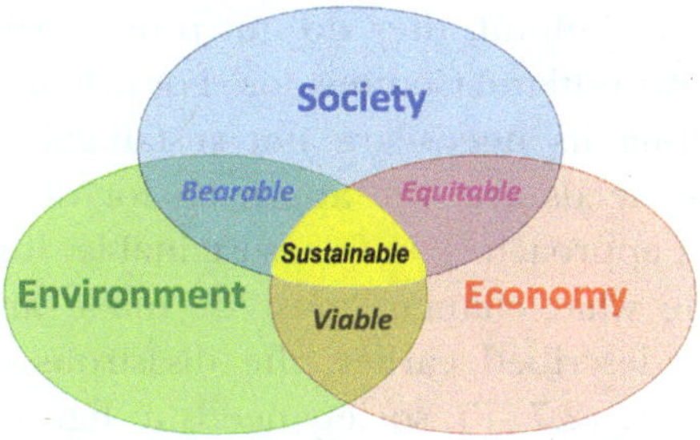

Figure 3-6. Representation of John Elkington's triple bottom line.

equally satisfied. Unfortunately, this concept has gained considerable traction, particularly among engineers who work on civil infrastructure projects where tradeoffs among stakeholder concerns are commonplace.

As it turns out, Elkington never intended the triple bottom line to be used this way. Writing for the *Harvard Business Review* 20 years later, Elkington (2018) stated,

> [T]he [Triple Bottom Line] wasn't designed to be just an accounting tool. It was supposed to provoke deeper thinking about capitalism and its future, but many early adopters understood the concept as a balancing act, adopting a trade-off mentality.

A better way of visualizing these relationships is as three nested interdependencies depicted in Figure 3-7. The environment provides the natural resources and ecosystem services society needs to survive and thrive. Society comprises human capital that provides the knowledge and institutions essential to the economy.

As the late environmentalist and businessman Ray Anderson stated, quoting former US Senator Gaylord Nelson (Senge et al. 2008),

> ... [B]usinesses need to wake up to the simple fact that "the economy is the wholly owned subsidiary of nature, not the other way around." Similarly, there can be no healthy economy without a stable and vibrant social order—just ask businesspeople trying to do business in corrupt, lawless, or extremely poor societies.

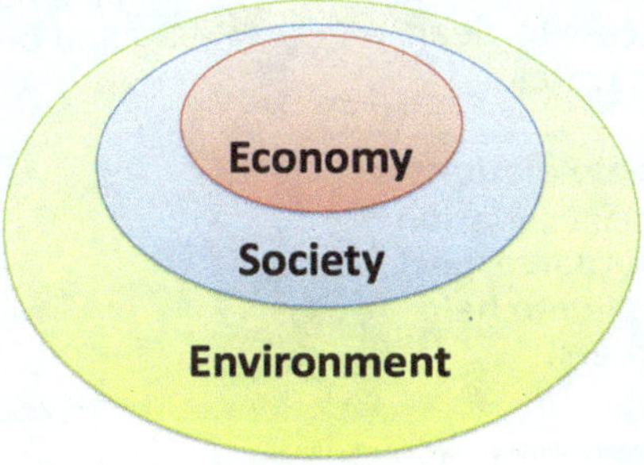

Figure 3-7. Environment, economy, and society as three nested interdependencies.

While these concepts are helpful, they do not provide sufficient insight into the concerns raised by the Brundtland Commission. For instance, if society's methods of development for meeting its needs are not sustainable, then it is essential to understand what those needs are, the approach for development used to meet them, and whether that approach is, in fact, sustainable. If they are not sustainable, society must determine what changes must be made and when. Recalling the warnings of the IPCC described earlier, the disastrous consequences of global warming exceeding 1.5 °C (2.7 °F), society needs to figure this out rather quickly.

What Are the "Needs" of Present and Future Generations?

The Millennium Ecosystem Assessment (MA) defines the needs of global society as a reasonably high state of human well-being: security, a fulfilling life, health, and social relations (MA 2003). Societies aspire to achieve sufficiency in all four, a state of human well-being. These societal needs and results for human well-being are summarized in Figure 3-8.

Figure 3-8. Meeting the needs of society and achieving human well-being.

Security involves not only personal safety but secure access to resources. A good life means having enough income, food, shelter, clothing, furniture, goods and services, and so on; in other words, all the material aspects for life fulfillment. Health means feeling wellness and strength and being in a clean environment. Good social relations involve respectful and productive interactions with family and others. Meeting these needs satisfactorily results in a state of well-being: freedom of choice and action, i.e., all individuals in societies having the opportunity to achieve what they value doing and being.

How Does Society Meet Those Needs?

The Earth's natural capital consists of our planet's vast supplies of natural capital and energy flows. Energy is supplied by the sun. Ocean tidal changes produced by the moon's gravitational pull create additional energy flows and water circulation (Figure 3-9). Society seeks to meet its needs by making efficient and intelligent use of the Earth's natural capital, extracting or harvesting natural resources (e.g., food, fiber, fresh water, minerals, metals, chemicals), and converting them into useful products. Society also leverages the Earth's ecological systems, using the services (e.g., erosion control, water purification, and storm protection) for its benefit. Society also enjoys the many cultural ecosystem services, the nonmaterial benefits such as the beauty of a wide variety of environments and habitats found in the natural world. Nature provides these resources and services for free.

For its long-term survival, society must achieve conditions of sustainability. Ecological economist Herman E. Daly (Daly and Thompson 1993) argues that ecological sustainability is achieved when (1) the use of renewable resources (ecological resources and systems) is less than their rate of regeneration, (2) the use of nonrenewable resources (minerals, metals, fuels) is less than the rate of

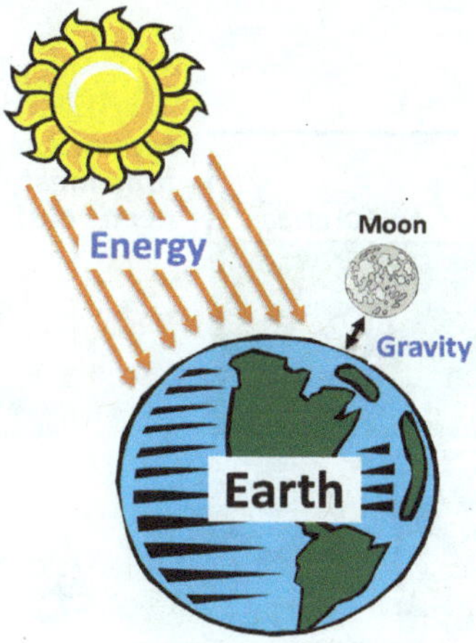

Imagery supplied by Clipart.com

Figure 3-9. The Earth's natural capital and energy flows: An overview.

deployment of renewable substitutes, and (3) pollution emission rates are less than the capacity of ecosystems to process them.

Sustainability: A Simple Example

Using a simple financial example, here is a way to visualize sustainability and sustainable development. Imagine two families, the Frugals (Figure 3-10) and the Spendies (Figure 3-11), each having a rich uncle who deposited for them a large sum of money, enough to live on generation after generation if they handled it wisely. Those funds are placed in investment accounts that earn various amounts of interest. Both families manage and draw money from their accounts to fund their chosen lifestyle, meeting their current needs and wants and preparing for emergencies and future expenditures, known or unknown.

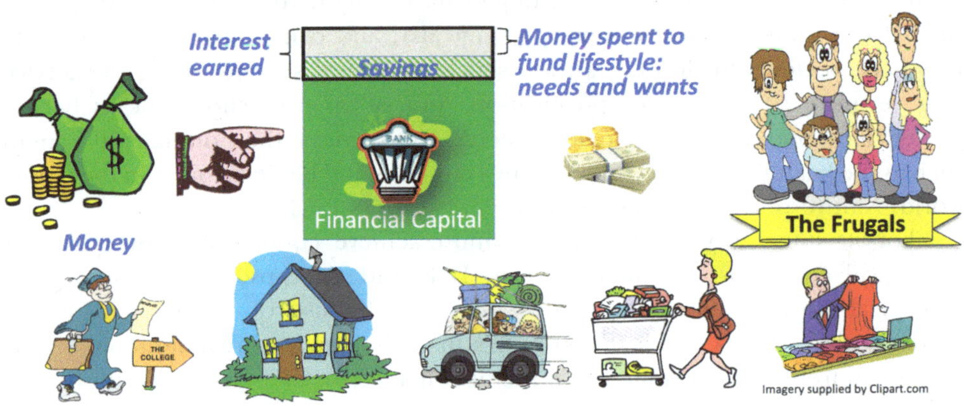

Figure 3-10. The Frugals: their lifestyle and investments.

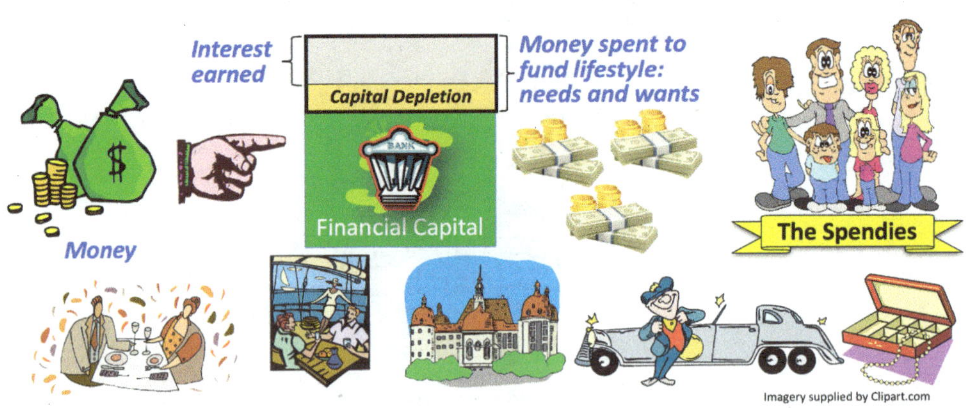

Figure 3-11. The Spendies: their lifestyle and investments.

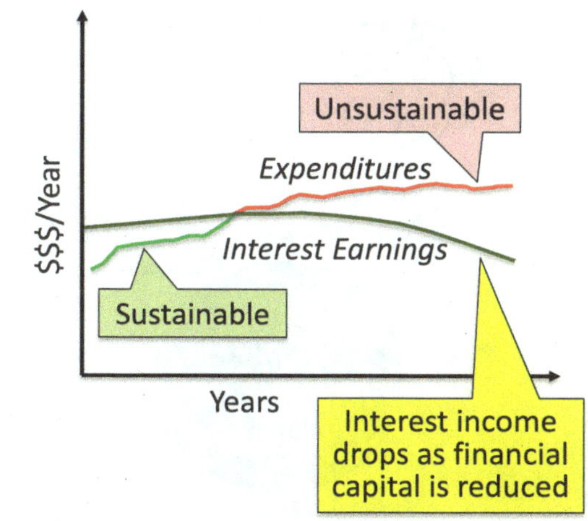

Figure 3-12. Graph of interest earnings and expenditures over time.

The Frugals take a sustainable approach, buying what they need but at a rate of expenditure that is less than the interest earned on their investments. They can maintain and even increase their financial capital by making smart investments, using some of the interest earned to support their lifestyle, saving the rest for emergencies, and increasing their capital. They can meet their needs and maintain a high state of well-being now and in the future.

The Spendies take a different approach. Their expenditures support a more extravagant lifestyle, more costly than the interest they earn on the money they received. And they do not give much thought to managing their investments. They can meet their current needs and maintain a high state of well-being for a while. However, it will not last. Their financial capital diminishes as they spend more than they earn on interest, lowering their capital-earning power (Figure 3-12).

Financial and Natural Capital: What's the Connection?

We can think of the Earth's natural capital as a series of bank accounts, some earning interest, some not (Figure 3-13). If we use this natural capital in ways that are not sustainable (i.e., using up renewable resources and damaging the services they provide), we are creating enormous problems for ourselves and for future generations. As in the case of the Spendies, reducing natural capital results in a corresponding reduction in the resources and services natural capital provides.

Civil engineering involves the use of the Earth's natural capital for the benefit of society. The Institution of Civil Engineers' founding charter (ICE 2023), released on June 3, 1828, defined civil engineering as

> being the art of directing the great sources of power in nature for the use and convenience of man, as the means of production and of traffic in states, both for external and internal trade, as applied in the construction of roads,

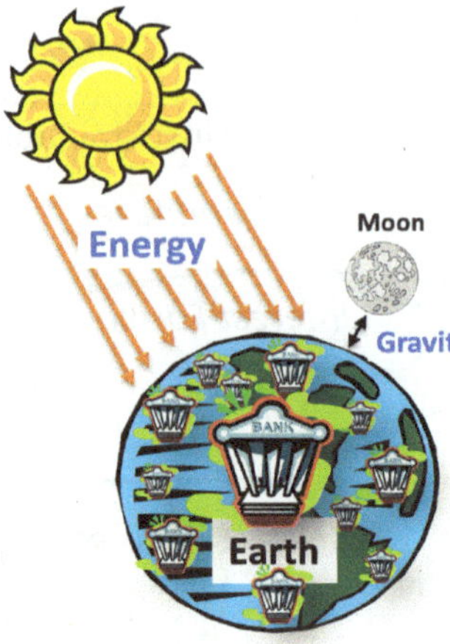

Figure 3-13. Natural capital as a series of investment accounts, some interest-bearing, some not.

> bridges, aqueducts, canals, river navigation and docks for internal intercourse and exchange, and in the construction of ports, harbors, moles, breakwaters and lighthouses, and in the art of navigation by artificial power for the purposes of commerce, and in the construction and application of machinery, and in the drainage of cities and towns.

Directing the sources of power in nature requires that engineers understand those sources of power and how they function. It also requires that with the use of nature's power comes the responsibility of protecting, maintaining, and even enhancing nature's resources, the Earth's natural capital.

Natural Capital Resources and Flows

Natural capital is a network of resources and flows, living (biotic) and nonliving (abiotic), and renewable and nonrenewable. As in the financial example, many are subject to overspending. The value they provide human society in products and services essential for maintaining and enhancing human well-being has been roughly estimated (Costanza et al. 1998) at over $61 trillion per year in 2021 constant dollars, about two-thirds of the 2021 global gross domestic product (GDP) of $97 trillion (World Bank 2023). There are also distortions. Depleting natural capital by overharvesting natural resources is counted as income, an addition to a nation's GDP. Nations also value as GDP the design and construction of systems built to replace depleted natural system services. Examples of natural capital resources and flows are presented in Figures 3-14 and 3-15.

Abiotic Resources and Flows

Abiotic resources and flows provide resources and energy while supporting the structure and condition of biotic (ecological) systems. Abiotic resources include air

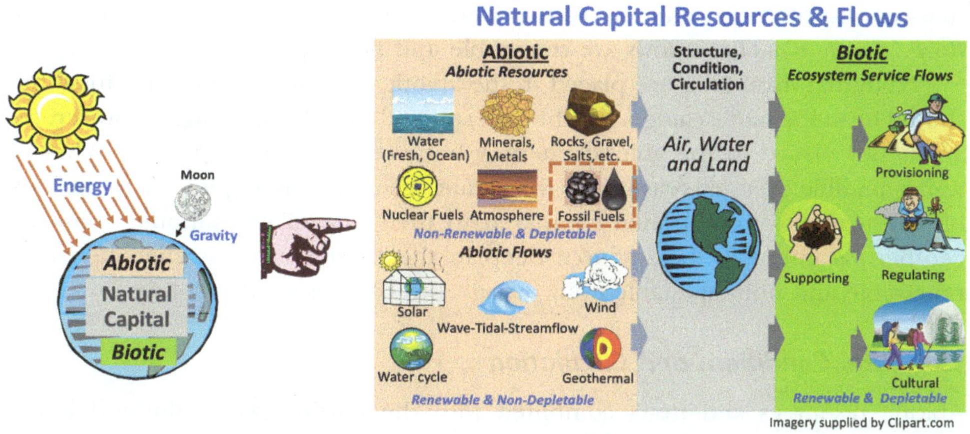

Figure 3-14. Natural capital resources and flows.

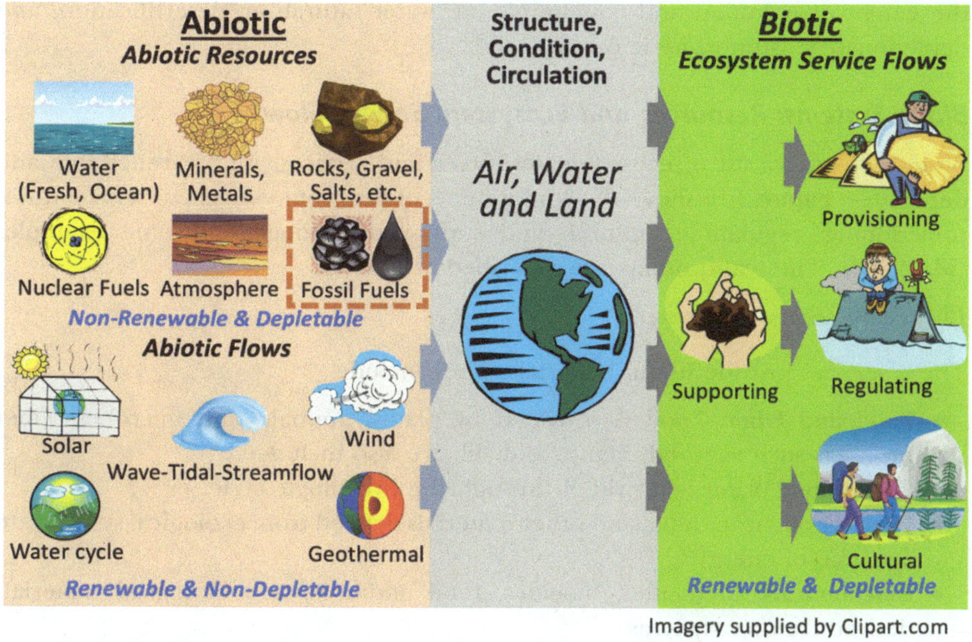

Figure 3-15. Natural capital resources and flows: abiotic and biotic.

(atmosphere), fresh water, ocean water, minerals, metals, rocks, gravel, and salts. They also include fossil fuels such as coal, oil and gas, and nuclear fuels. In addition, abiotic resources provide buffering, shielding, and storage. The Earth's oceans provide a buffer for heating and cooling along with storage of excess carbon dioxide. The atmosphere is the foundation of the Earth's climate and provides radiation protection. These resources are nonrenewable and depletable. Abiotic flows consist of energy and circulation (e.g., solar, geothermal, wave, tidal, streamflow, and wind). These flows are renewable and nondepletable.

Note that fossil fuels are placed in the abiotic category. Technically, fossil fuels should be in the biotic category as they are sourced from living things. However, they are better placed in the abiotic category because they behave as such. They are nonrenewable over any reasonable time scale. Also, they are depletable as their rate of production is infinitesimal compared to their rate of consumption. Similarly, geothermal energy, a heat source deep in the Earth's crust, is characterized as renewable and nondepletable even though the Earth is slowly cooling.

Structure, Condition, and Circulation

Abiotic resources and flows combined with the Earth's geography and human activity create the structure and condition for biotic ecosystem service flows. The structure includes land and water use, weather patterns, and urbanization. The condition of those resources is marked by their physical and chemical properties, the extent of pollution and contamination, water saturation, desertification, and depletion (e.g., stratospheric ozone).

Biotic Systems: Resources and Ecosystem Service Flows

Biotic systems consist of ecosystem resources and services. All are renewable and depletable. Figure 3-16 shows the components of the biotic ecosystem service flows: provisioning, regulating, cultural, and supporting, along with some examples (MA 2003). These and others are described as follows.

Provisioning Ecosystem Services

These services provide useful materials. They include

- **Food and Fiber:** Food is provided by plants, animals, and microbes. Fiber products such as wood, hemp, and silk are also included.
- **Freshwater:** Water is provided through the hydrologic cycle.
- **Fuel:** Wood, peat, dung, and other materials derived from ecological systems are a source of energy.
- **Biochemicals:** Medicines, biocides, food additives, and biological materials derived from ecosystems.

Regulating Ecosystem Services

These services control and buffer other ecosystem processes and flows. They include

- **Air Quality Maintenance:** Contribution of chemicals to and the extraction of chemicals from the atmosphere.
- **Climate Regulation:** This service controls the sequestration and flux of GHGs (including carbon dioxide, methane, and nitrous oxide), cloud formation, surface albedo, aerosol formation, and heat and moisture transfer.

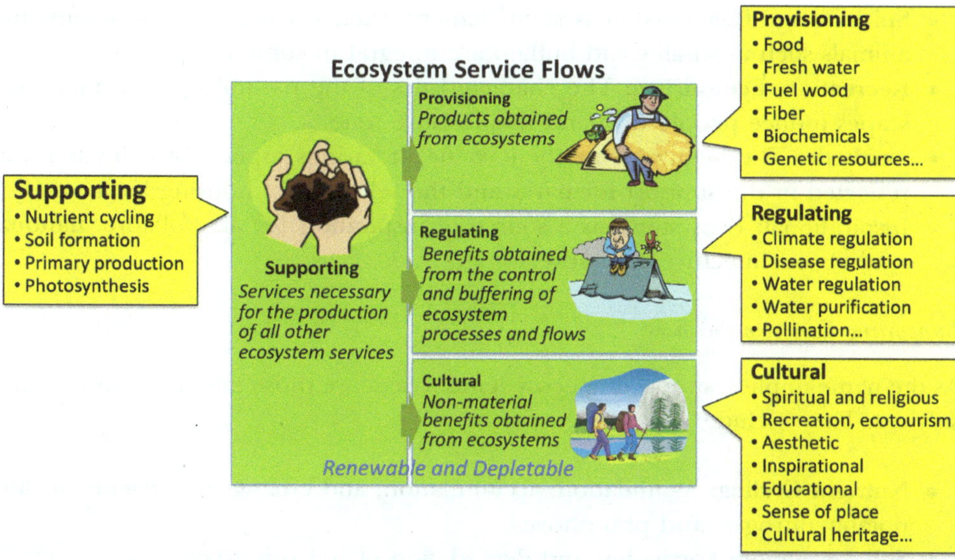

Figure 3-16. Ecosystem service flows: products and services nature provides for free.

- **Water Regulation:** Controlling the timing and magnitude of runoff, flooding, and aquifer recharge, influenced by changes in land cover.
- **Erosion Control:** Vegetative cover for soil retention and the prevention of landslides.
- **Disease Regulation:** Changes in the abundance of human pathogens (e.g., cholera) and altering the abundance of disease vectors (e.g., mosquitoes).
- **Genetic Resources:** Sources of genes and genetic information are used to improve the productivity, quality, and resiliency of crops, livestock, forestry, and fisheries.
- **Water Purification and Waste Treatment:** Filtration and decomposition of organic wastes introduced into inland waters and coastal and marine ecosystems.
- **Biological Control:** The prevalence of crop and livestock pests and diseases.
- **Storm Protection:** Reduction in damage caused by hurricanes or large waves by coastal ecosystems such as mangroves and coral reefs.
- **Pollination:** The distribution, abundance, and effectiveness of pollinators.

Cultural Ecosystem Services

These are the nonmaterial benefits obtained from ecosystems. They include

- **Cultural Heritage:** Societal value systems, cultural identity, and language interactions with the environment.

- **Spiritual and Religious:** Ecosystem elements such as rivers and mountains, and animals such as whales and buffalo are integral to some religions.
- **Recreation, Ecotourism:** The characteristics of the natural or cultivated landscapes induce people to spend their leisure time.
- **Aesthetic:** The beauty or aesthetic value in various aspects of ecosystems, as reflected in the support for parks and the selection of housing locations.
- **Inspirational:** Ecosystems as a source of inspiration for art, folklore, national symbols, architecture, and advertising.

Supporting Ecosystem Services

As the name implies, supporting ecosystem services are those necessary for all other services. They include

- **Nutrient Cycling:** Assimilation, accumulation, and cycling of nutrients for life, notably nitrogen and phosphorus.
- **Soil Formation:** Formation and degradation of soil resources.
- **Primary Production:** Assimilation and accumulation of energy and nutrients by organisms.
- **Photosynthesis:** Plants and other organisms use sunlight to produce food and oxygen from carbon dioxide and water.

Resources and Ecosystem Service Flows: Example

An example of how ecosystem services are delivered is depicted in Figures 3-17 and 3-18. In this example, the earthworm biomass contributes to many ecosystem services. Through waste decomposition, it recycles soil nutrients and contributes to soil formation, soil structure maintenance, and water flow regulation. In doing so, it increases agricultural production and reduces the need for chemical fertilizers. If this biomass operates in an agrarian setting, it could be compromised by poor farming practices, land use, and pollution. The earthworm biomass is one element of a vast collection of ecosystems, networked to and supporting each other, all delivering various valuable services to society.

Question: How Did We Get to this Point in Human Development?

The history of human development, as depicted in Figure 3-19, is a history of how human societies made use of natural capital resources and flows to improve their standard of living. Throughout most of history, advances in human development have been slow. The world's population was small, dispersed, and disconnected. Communities were constantly threatened by disease, hunger, and other neighboring tribes. Technologies were weak, and most people were living at a subsistence level.

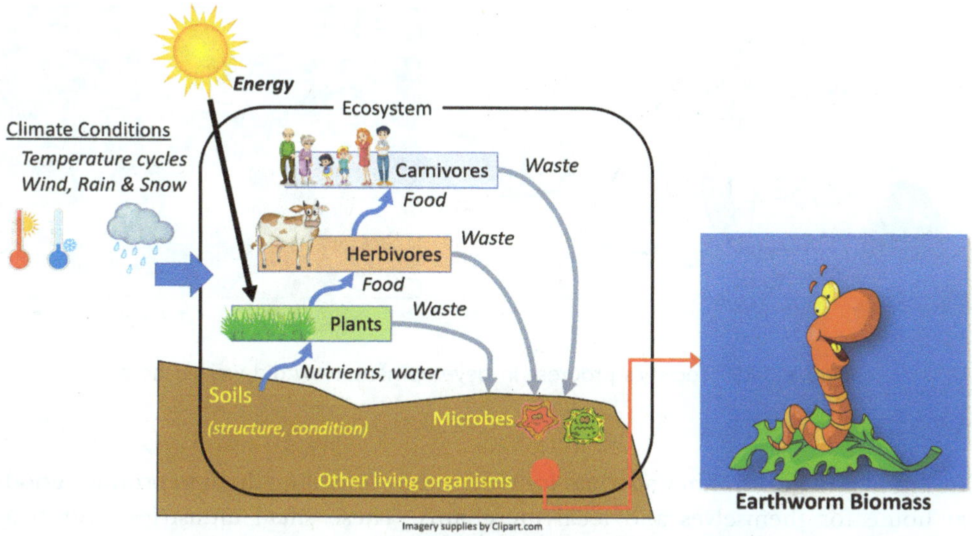

Figure 3-17. Relationships of living organisms and ecosystems.

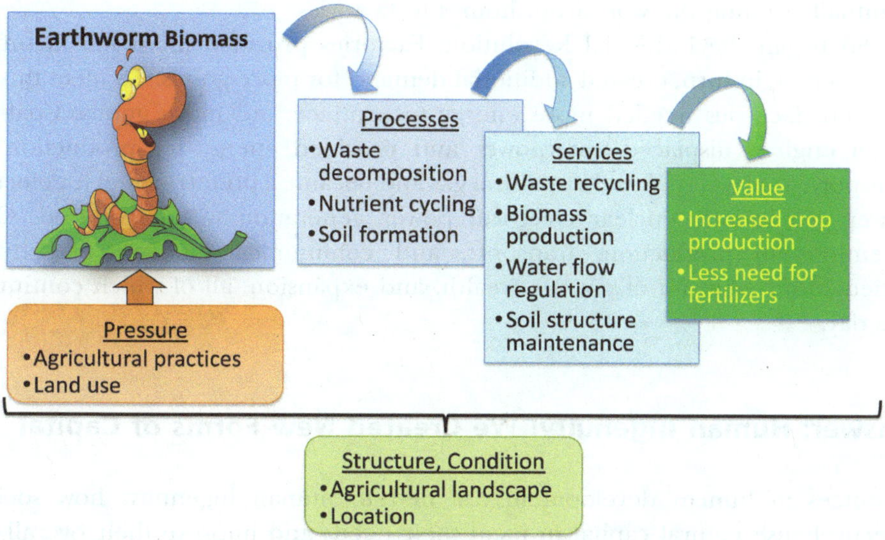

Figure 3-18. An example of an earthworm biomass performing processes and delivering multiple ecosystem services.

Starting in the seventeenth century, things began to change. The Agricultural Revolution introduced crop rotation, breaking away from traditional farming methods and significantly improving agricultural productivity. With improved productivity, food became less expensive, and families were able to improve their

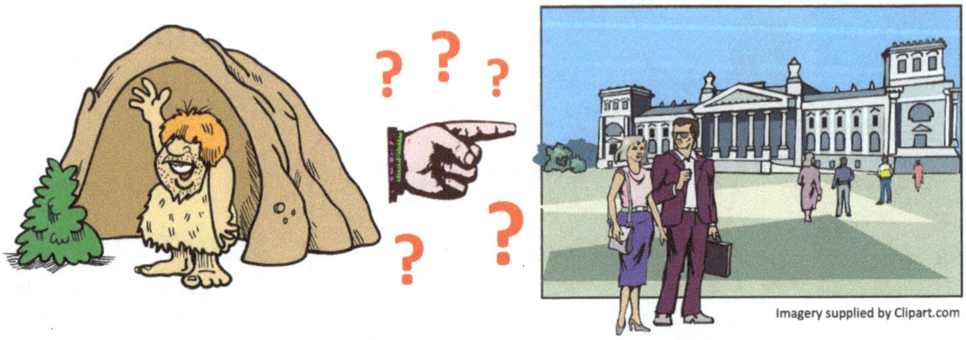

Figure 3-19. Society's progress in development: How did we get here?

nutrition and health. They also had more free time, enabling them to produce goods at home for themselves and local merchants. These small industries, known as cottage industries, broke the grip of the craft guilds by producing finished products for merchants at much lower costs. At the same time, these industries provided families with additional income to purchase goods and services, thereby increasing demand. That new demand spurred the development of more efficient machinery, eventually shifting the work from homes to factories.

So began the Industrial Revolution. Factories provided increased income to families that, in turn, created additional demand for more goods. To meet this new demand, factories needed more energy to produce and move goods. Coal-fired steam engines displaced waterpower and provided energy for production and transport. Coal was replaced by natural gas and became a preferred fuel for electrical power generation. Nuclear electrical power generation was introduced. Other inventions in production, transport, and communications launched human society into a new era of growth, wealth, and expansion, all of which continue to this day.

Answer: Human Ingenuity! We Created New Forms of Capital

Advances in human development are tied to human ingenuity: how societies effectively use natural capital to meet their needs and improve their overall well-being. As author and business advisor Jonathon Porritt describes it, humans created new forms of capital: produced, social, financial, and human. "These five forms of capital, judiciously combined by entrepreneurs, are the essential ingredients of modern industrial productivity" (Porritt 2007).

Figure 3-20 shows and summarizes all five capitals. As shown in Figure 3-21, the five capitals align well with the nested interdependencies concept of sustainability (FFF 2020).

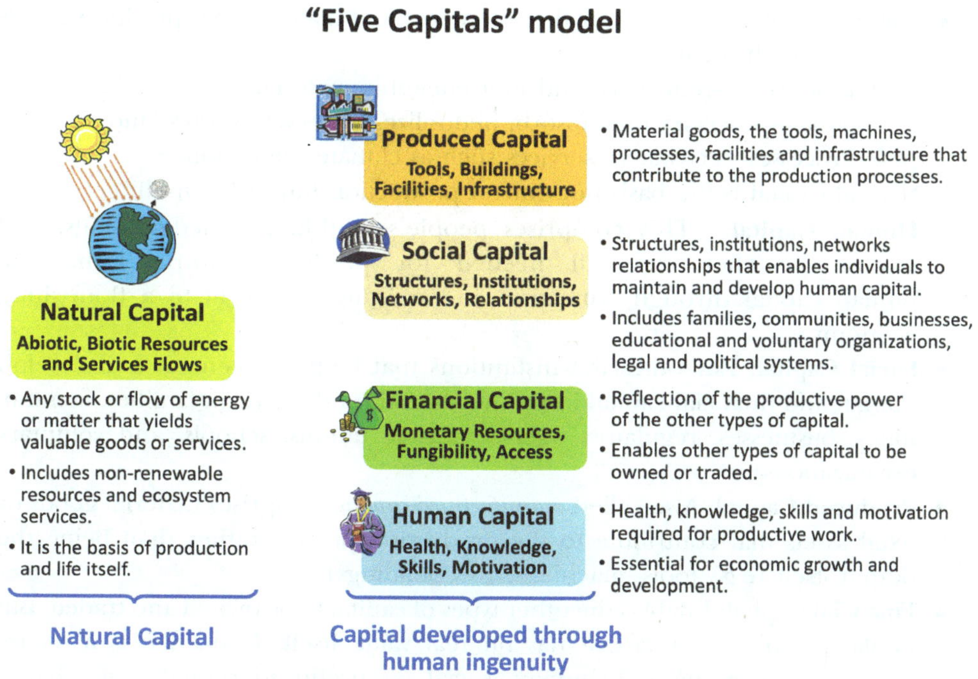

Figure 3-20. The five capitals model.

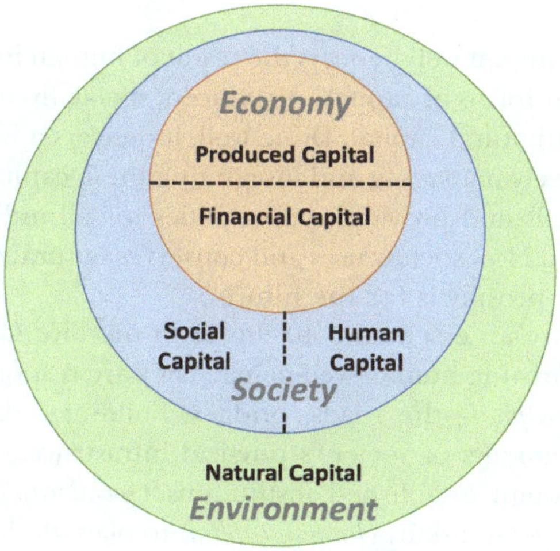

Figure 3-21. The five capitals model.
Source: Adapted from FFF (2020).

- **Natural Capital:** Any stock or flow of energy and material that produces goods and services. It includes
  - Resources—renewable and nonrenewable materials;
  - Sinks—processes that absorb, neutralize, or recycle wastes; and
  - Processes—ecosystem services such as climate regulation.

  Natural capital is the basis not only of production but of life itself.
- **Human Capital:** This comprises people's health, knowledge, skills, and motivation. These are all needed for productive work. Enhancing human capital through education and training is central to a flourishing economy.
- **Social Capital:** Embodies the institutions that help us maintain and develop human and financial capital in partnership with others (e.g., families, communities, businesses, regulatory agencies, trade unions, schools, and voluntary organizations).
- **Produced Capital:** Also called *manufactured capital*, comprises material goods or fixed assets that contribute to the production process rather than being the output itself (e.g., tools, machines, and buildings).
- **Financial Capital:** Enables the other types of capital to be owned and traded. But unlike the other capitals, it has no real value itself. Instead, it is a secure representative of natural, human, social, or produced capital (e.g., shares, bonds, or banknotes).

Putting It All Together

The current state of human well-being is the result of human ingenuity: creating and managing these other forms of capital—produced, social, financial, and human—to make efficient use of natural capital. Done well, it delivers a high degree of societal well-being. By effectively managing and integrating these capitals, societies can meet essential human needs and provide opportunities for all individuals to thrive and prosper (Figure 3-22). How society uses and conserves natural capital is a measure of its development and prospects for the future.

Civil engineers are an essential contributor in the Five Capitals framework for maintaining and improving human well-being. They are responsible for a significant portion of *produced capital*—the roads, bridges, buildings, dams, water treatment plants, and other elements of society's physical infrastructure. Using *natural capital* and working through established institutions (*social capital*), they provide the engineering knowledge and skills (*human capital*) to plan, design, build, and operate the physical infrastructure: roads, bridges, buildings, dams, water treatment plants, and more. A sound *financial capital* system provides the monetary resources to procure the required materials and engineering services.

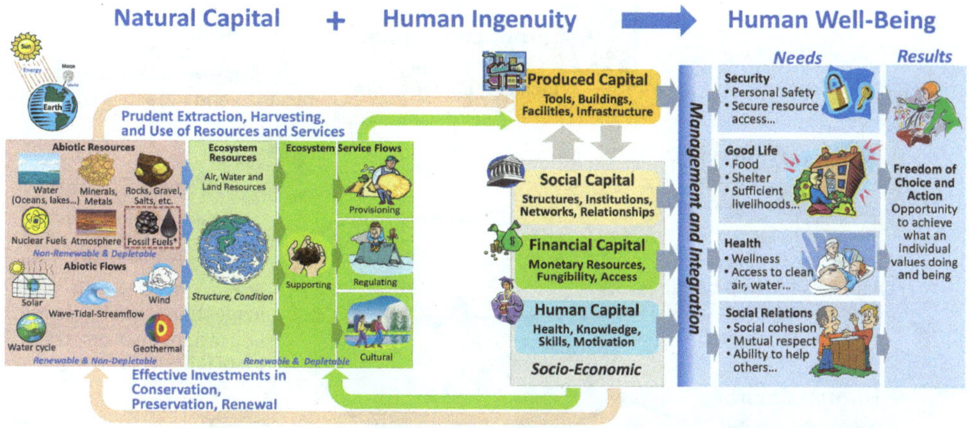

Figure 3-22. A sustainable society: An ideal functioning relationship between the five capitals and human well-being.

Multiple structures and institutions (*social capital*) educate, regulate, communicate, and enhance human and financial capital. If well-managed and integrated, the five capitals support society's needs, resulting in a satisfactory level of human well-being. Examples of its application to fishing are offered in Figures 3-23 and 3-24.

Is Society's Approach to Development Sustainable?

If society's approach to development is sustainable, then

- Natural capital is sufficient for current and future generations to meet their needs;
- Human ingenuity is applied broadly and effectively to benefit all people in all nations; and
- Human well-being is sufficiently high, fairly achieved, and equitably distributed.

To make these determinations, it is necessary to know the current state of human well-being (i.e., is it sufficiently high, fairly achieved, and equitably distributed?). It is also necessary to understand the current state of and prognosis for the Earth's natural capital (i.e., how much has been used compared to what is available, and how has it been maintained for current and future generations?).

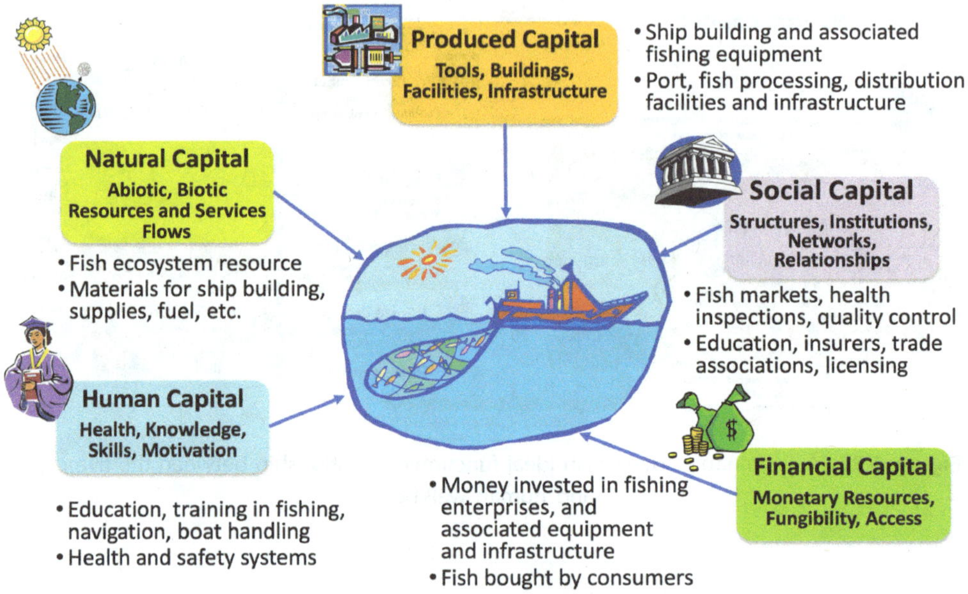

Figure 3-23. Application of the five capitals model to modern fishing.

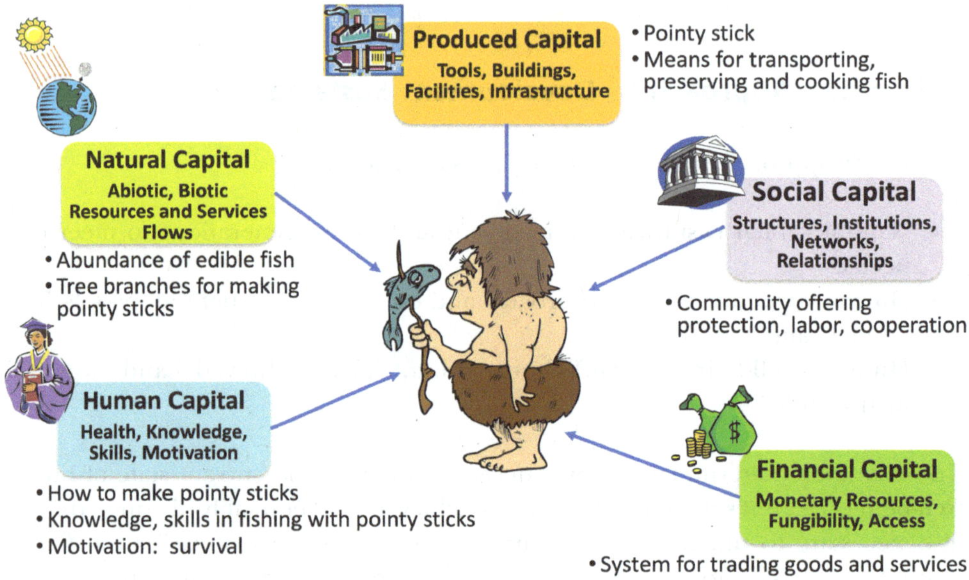

Figure 3-24. Application of the five capitals model to early fishing.

Human Well-Being: What Has Been Achieved?

Since 1990, the United Nations Development Program (UNDP 2021) has published the Human Development Index (HDI), a rough measure of the economic development and well-being of the world's nations. The HDI is a combination of three factors: life expectancy, knowledge and education, and income scaled from 0 to 1. Fundamentally, it is an indicator of

- Health sufficiency to expect a long and healthy life,
- Education sufficiency to pursue a higher quality of life, and
- Income sufficiency to pursue that quality of life with dignity.

The latest information shows a wide range of national scores, from a low of 0.385 (South Sudan) to a high of 0.962 (Switzerland). Examples are presented in Figure 3-25. Note that the United States did not come out on top. In 2021, the United States placed 21st, with an HDI of 0.921, below countries such as Germany, Ireland, Sweden, Singapore, Denmark, and others. The UNDP has established country classifications of low, medium, high, and very high based on country values of HDI. Figure 3-25 shows that HDI disparities among countries are vast.

In terms of overall accomplishment in human development, the national differences are significant. More than 1 billion people live in countries classified as high or very high development, while 5.5 billion are in conditions of medium development. About 1 billion are in low development conditions.

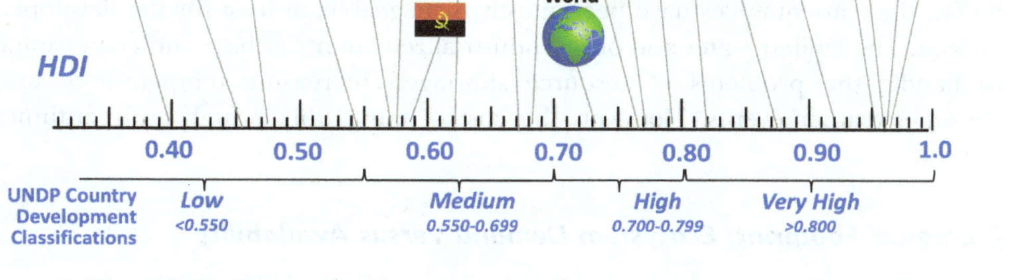

Figure 3-25. Human Development Index of nations, some country examples.
Source: UNDP (2021).

What Is the State of the Earth's Resources and Ecosystems?

If natural capital's resources and ecosystem functions are essential to society, then it is important to know their present state and future prospects for natural capital. Enter the Millennium Ecosystem Assessment (MA 2003). Initiated by the United Nations in 2001 and published in 2005, the Assessment offered a comprehensive scientific appraisal of the condition and trends of the Earth's ecosystems. It engaged more than a thousand experts from across the world, the largest group of social and natural scientists ever assembled for such a study. Their findings, presented in eleven technical and synthesis reports, documented the scientific community's current knowledge of the condition of the world's biotic systems. The Assessment also offered options for the restoration, conservation, and sustainable use of the remaining resources.

The Assessment considered in detail the global and regional condition of the biotic systems in terms of ecosystem flows: provisioning, regulating, cultural, and supporting. What the experts found was remarkable. On a positive note, they found that over the last 50 years, society's use of these systems has resulted in substantial net gains in economic development and human well-being. However, the cost of those gains has been the unprecedented degradation of ecosystem resources and service flows, the foundations for human existence. Furthermore, the extent of that degradation to the benefit of some is jeopardizing the economic development and well-being of others across geographies, cultures, and generations. Unless changes are made, this degradation will worsen and may become irreversible.

Some of the findings of the Assessment (MA 2005) were

- Five percent to 25% of freshwater use exceeds available supplies,
- Significant loss of ecosystem capacity (wetlands, forests, mangroves) as buffers from extreme events,
- Twenty-five percent of marine fish stocks are overharvested,
- Increasing number of dead zones (currently more than 100) in coastal areas,
- Global decline in the abundance of pollinators, and
- Declining air quality, especially in major cities.

Repairing this situation is possible but few substantive actions are being taken. So far, the consequences have been mostly manageable, at least for the developed nations. The United States and other industrialized countries have sufficient capital to handle the problems of resource shortages, increasing temperatures, and extreme weather events. However, the cost of mitigation is high and continues to grow.

Ecological Footprint: Ecosystem Demand versus Availability

While the Assessment documented the degradation of the Earth's ecosystem resources, it did not establish how many resources society uses versus how much

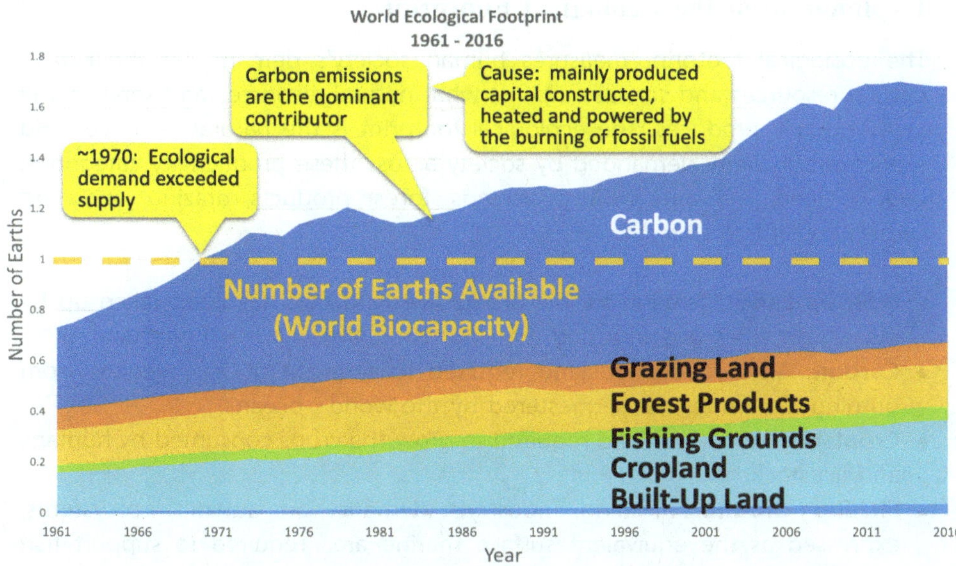

Figure 3-26. Historical world ecological footprint.
Source: Global Footprint Network (2023), Creative Commons BY-SA 4.0.

is available. Only recently has society begun to pay attention to Earth's balance sheet of resources and services: ecosystem demand versus availability.

The Global Footprint Network (GFN 2023) is a nonprofit organization that provides data resources and tools to measure society's natural resource consumption (i.e., society's ecological footprint). This organization created and regularly updates such a balance sheet by determining how much biologically productive area it takes to meet human society's demands. These demands are measured as the area required for agriculture, fiber, timber, buildings, and infrastructure, and for the absorption of carbon emissions. This measure is compared to biocapacity, the amount of biologically productive land and sea area available to provide the resources a society consumes and uses to absorb its wastes.

Historical values of the world's ecological footprint versus the available biocapacity are presented in the graph of Figure 3-26, showing biocapacity in terms of the number of Earths available (one) versus how many Earths are in demand. To put this another way, to meet society's demand for natural resources and ecosystem services, the graph shows how many Earths worth of natural systems productivity would be required. As it turns out, we have been living in an ecological deficit since the early 1970s. Also, as Figure 3-26 shows, carbon (GHG emissions) is by far the largest contributor, approximately 60% of the world's ecological footprint. How our infrastructure has been designed, built, and operated is the primary source of that footprint contribution.

Explanation of the Ecological Footprint

The ecological footprint measures human society's demand for the Earth's natural resources and services against what natural resources and services the Earth actually produces. The ecological footprint is the natural resources and service productivity demanded by society across these productivity categories: built-up land, cropland, fishing grounds, forest products, grazing land, and carbon absorption (forests).

- **Built-up Land:** Demand for bio-productive land that has been taken up by human activities (e.g., housing, transportation, and industrial production).
- **Carbon:** Area of forested land required to sequester CO_2 emissions from burning fossil fuels not sequestered by the world's oceans.
- **Cropland:** Amount of land required to grow the crops consumed by humans and livestock.
- **Fishing Grounds:** Fishing industry's demand on aquatic ecosystems, expressed as the equivalent surface marine area required to support fish harvesting sustainably. Inland fishing and aquaculture are included.
- **Forest Products:** Demand for products of wood forests: wood for fuel and for timber and pulp used for wood products.
- **Grazing Land:** Demand for grazing land to feed livestock, including the embodied energy requirements for concentrate feeds, forage crops, and crop residues.

Biocapacity is the Earth's productive capacity to supply these resources (biological materials) and services (absorb wastes) across these categories.

- **Built-up Land:** Land taken out of production because it is occupied by buildings and infrastructure.
- **Cropland:** Amount of land available to grow crops consumed by humans and livestock.
- **Fishing Grounds:** Productivity of fisheries: marine, inland, and aquaculture.
- **Forest Products:** Capacity of forests to produce products and sequester carbon.
- **Grazing Land:** Productivity of grazing land to feed livestock.

Source: Global Footprint Network (2023).

The common measure is an area called a *global hectare*, a biologically productive hectare with the world average productivity in a given year in terms of production and waste assimilation. It is a unit of measure that averages the productivity of the various categories of productivity and demand. For example, an area of desert land has a different level of biological productivity than an equal area of land in, say, the wheat-growing region of the midwestern United States. For comparability, ecological footprint and biocapacity are both measured in global hectares.

The ecological footprint accounts of individual nations can also be calculated by determining the national footprints of consumption and production and the net ecological footprint of trade (i.e., imports less exports). Today, according to the Global Footprint Network (GFN 2023), society as a whole is operating as if it had about 1.7 planets to work with. The United States is operating at a 4.7 planet level; Europe operates at 2.3. Note that the largest contributor by far to the ecological footprint is carbon, primarily the carbon dioxide emissions from burning fossil fuels. If society is to become sustainable, reducing carbon emissions is the obvious target.

The Sustainability Quadrant: Where Nations Need to Be

Knowing these values for individual nations, its HDI (what level of development a nation achieved), and its ecological footprint (how much of the Earth's natural resources and services that nation consumes to achieve that level of development), the relative position of each nation can be plotted (Figure 3-27).

Here, we can define the sustainability quadrant, the area bounded by high human development (the upper quartile of nations, the approximate level of

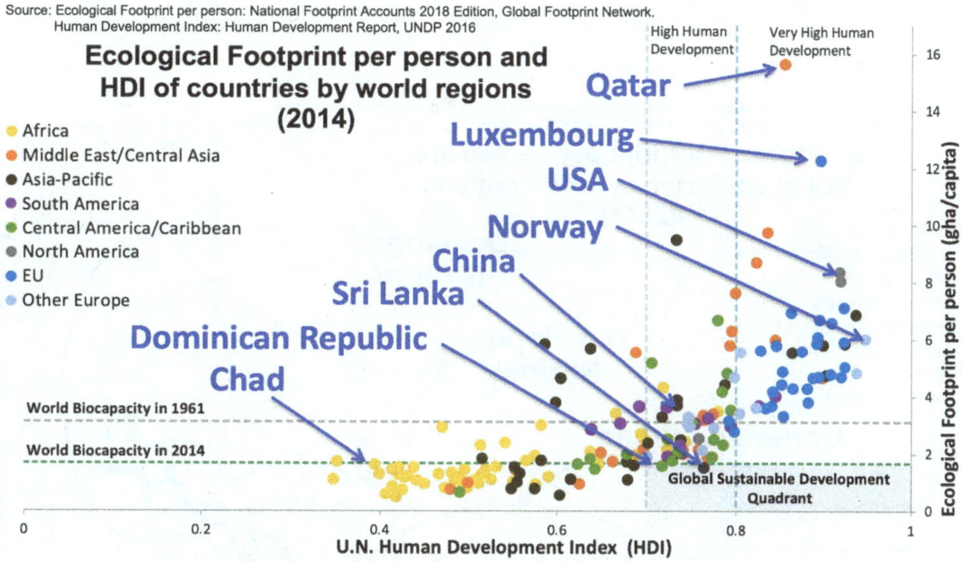

Figure 3-27. Ecological demand (footprint) of nations in biologically productive hectares or global hectares (gha) per person compared to their level of human development, 2014. The Global Sustainable Development Quadrant in the lower right is the target, where human development is high or very high and ecological demand is within the world's available biocapacity. *Source*: Global Footprint Network (2023), Creative Commons BY-SA 4.0.

development societies strive to achieve) and available biocapacity of the Earth (how much the Earth provides). Only a few nations even come close to residing in the sustainability quadrant. Moreover, as shown in Figure 3-27, world biocapacity per person has diminished substantially, about 40%, in the last 40 years.

Figure 3-27 also shows the position of the United States relative to other nations, a very high level of human development, made possible by a high level of natural resource consumption. Residing in the same cluster are the nations of North America, Europe the Middle East, and Central Asia, all achieving very high development through heavy use of ecological resources. One can infer that human development improvement in nations is directly related to an increase in their ecological footprint. Figure 3-28 presents an approximate breakout of nations by level of development.

Figure 3-29 presents the historic US ecological footprint. Since 1961, when the first ecological footprint measurements were published, the United States has been in a state of ecological overshoot (i.e., about 2.6 times its available biocapacity). Today, the US carbon footprint exceeds all its other footprint components combined.

How did the United States and the other developed nations get to their positions of high or very high human development? Human development progress can be traced by following the expansion and advancements of civil infrastructure—infrastructure that is heated and powered primarily by fossil-fueled energy. As we have seen, carbon (i.e., GHG emissions) is by far the largest contributor to the world's ecological footprint.

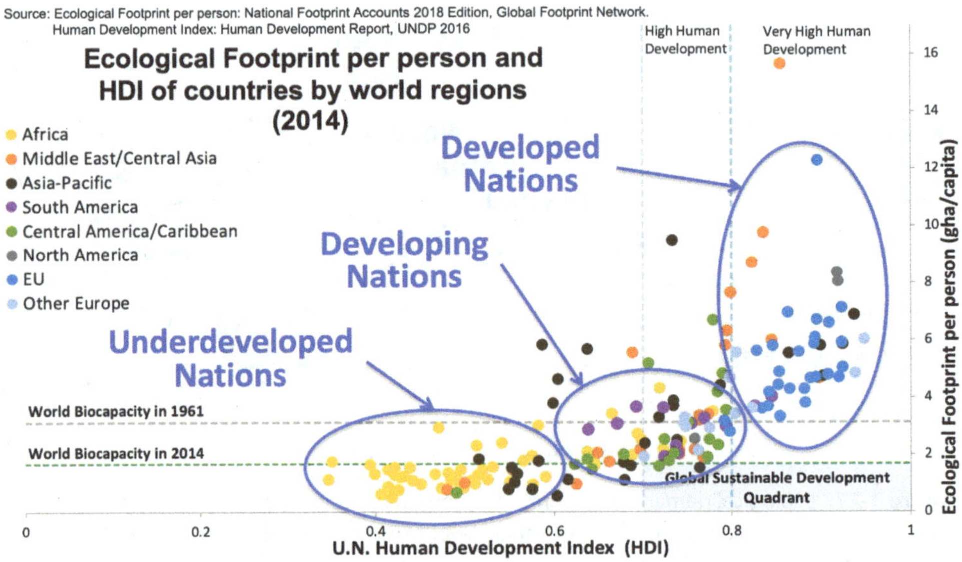

Figure 3-28. Approximate breakout of nations by level of development.
Source: Global Footprint Network (2023), Creative Commons BY-SA 4.0.

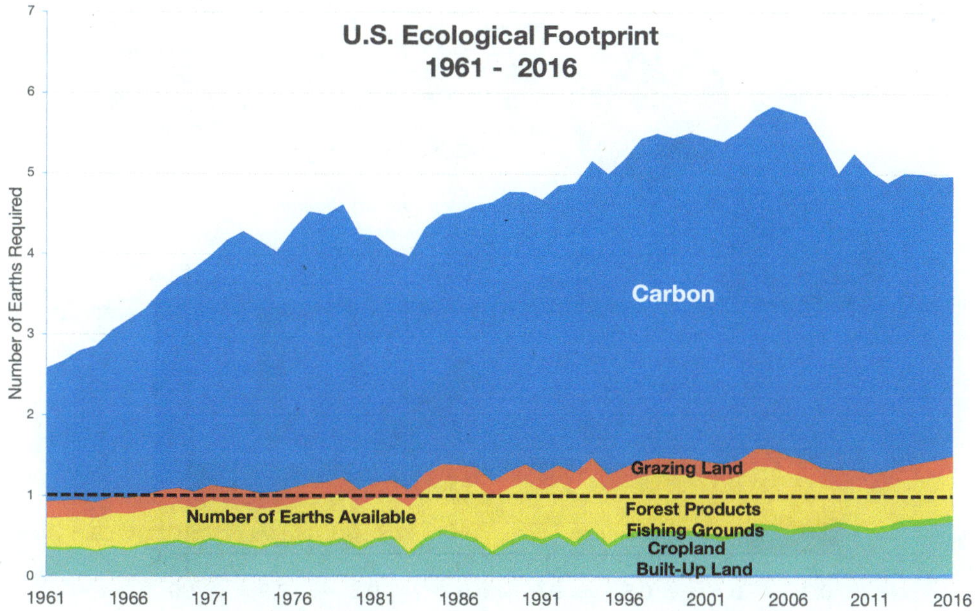

Figure 3-29. Historical ecological footprint of the United States.
Source: Global Footprint Network (2023), Creative Commons BY-SA 4.0.

Moving Toward the Sustainability Quadrant

For nations in that high to very high development cluster, moving toward the sustainability quadrant is a matter of reducing their national ecological footprint while maintaining their level of human development. As carbon is the largest footprint component, carbon emissions reduction is the logical place to start.

Developed Nations: High and Very High Income

Figures 3-30 and 3-31 show the world's nations color-coded by development level. The high- and very high-income nations (Figure 3-32) are the developed, post-industrial nations with a population of 1.1 billion. The United States, Canada, France, Norway, the United Kingdom, and Australia fall into this category. The overall ecological footprint for these nations is high, requiring 3.6 planet Earth equivalents to satisfy its demand for ecological resources.

Relatively speaking, the citizens in the developed nations enjoy a high standard of living, a lifestyle that is the envy of the rest of the world. Their infrastructure supports a strong economy and has a high degree of resiliency.

Because of their economic success, most people in these nations like things just the way they are. However, the increasing number of weather and climate disasters and the associated costs are starting to change opinions. A majority of the US public

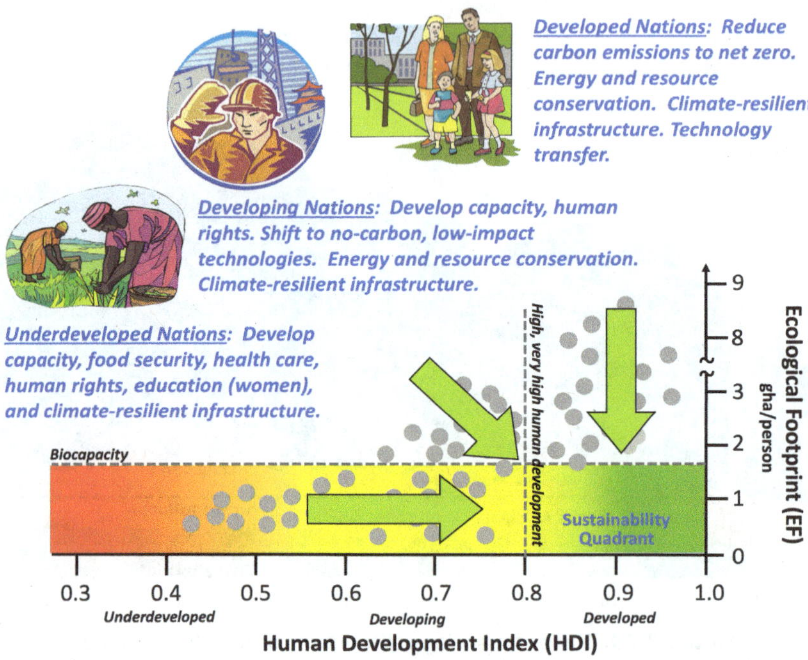

Figure 3-30. Moving toward the sustainability quadrant. Paths toward conditions of sustainability.

now sees climate change as a real existential threat and wants governments to take meaningful action (Krosnick and MacInnis 2020).

Developing Nations: Lower and Upper Middle Income

The developing nations (Figure 3-33) have the largest net population, currently at 5.3 billion. Examples include China, India, Russia, and Brazil. In the decades since the Cold War ended, these countries have experienced tremendous growth in both population and income. The combination of an educated and industrious population willing and anxious to work for relatively low wages (compared with the developed nations) has spurred growth as businesses from the developed nations relocate their factories to these countries.

This industrialization has resulted in a dramatic increase in per capita disposable income, enabling these countries to improve their standard of living. To them, continued economic growth is what matters. The overall ecological footprint for these nations, 1.5 Earths, is lower than for the developed nations. However, it is increasing as these nations pursue growth through carbon-intensive energy technologies.

One of the products of this dramatic rise in income growth is the creation of megacities, defined as cities with populations of 10 million or more. New factories are drawing people away from rural areas into urban areas, promising jobs and

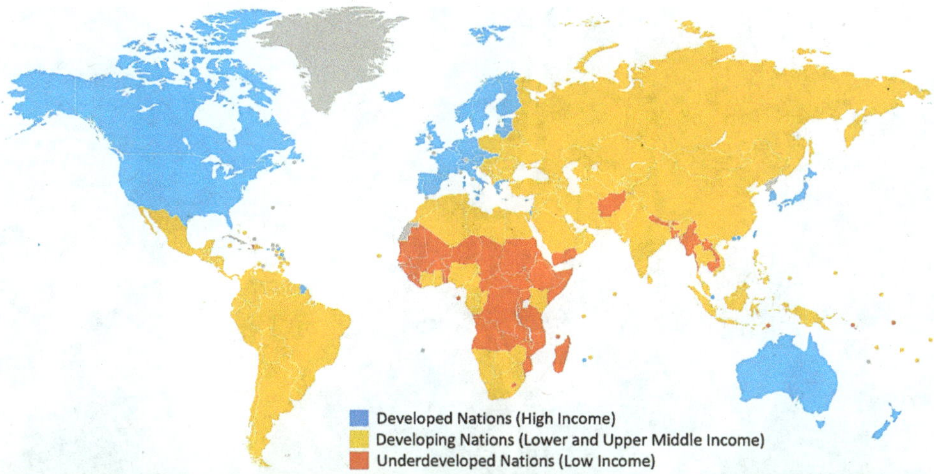

Level of Development	Population (billions)	Annual Population Growth Rate	Annual Gross National Income (GNI)	Annual Gross Domestic Product (GDP) Growth Rate	Annual GNI per Person
Developed	1.1	0.6%	US$59.5 trillion	5.1%	US$47,904
Developing	5.8	1.2%	US$15.8 trillion	6.4%	US$3,927
Underdeveloped	0.7	2.7%	US$0.51 trillion	0.3%	US$722

Figure 3-31. Range of economic development.
Source: Data from World Bank (2023), map from Allice Hunter, Creative Commons Attribution-Share Alike 4.0 International.

Figure 3-32. Developed nations snapshot.
Source: FamVeld, F11photo, NDAB Creativity/Shutterstock.com.

Figure 3-33. Developing nations snapshot.
Source: Frame China, AsiaTravel, Lewis Tse/Shutterstock.com.

higher wages. This migration has produced an enormous increase in urban populations, outstripping the pace of infrastructure development. As a result, clean water and sanitation in megacities are in short supply. Severe traffic congestion and air pollution are commonplace. Built rapidly and generally below established standards, infrastructure in developing nations is highly vulnerable to disasters, both natural and human-constructed.

In addition to providing financial aid, developed nations have encouraged developing nations to abandon fossil energy sources, replacing them with renewable energy sources. The response from the developing nations was not unexpected. They ask, "How can you ask us to spend money on relatively untried technologies when you have done nothing of the sort? Your growth and development are based on the same polluting and resource-wasting technologies you are asking us to abandon."

Underdeveloped Nations: Low Income

For the low-income group of nations, the underdeveloped nations (Figure 3-34), what matters is survival. Examples include Senegal, Rwanda, Tanzania, and Ethiopia. These nations have little infrastructure, and many people struggle to survive day to day. As a result, their income growth is very low, about 0.2% per year. However, this group has the largest population growth rate, 2.4% per year, creating enormous challenges for development.

For underdeveloped nations, the impacts of climate change have been more severe. Prolonged and severe droughts are especially devastating to farmers trying to

Figure 3-34. Underdeveloped nation snapshot.
Source: Valeriya Anufriyeva, Harmattan Toujours, Tatsiana Hendzel/Shutterstock.com.

make a living through rain-fed agriculture. Intense storms damage crops and housing, forcing people to move to cities in search of work and safe shelter. In turn, megacities are being created. Most of these cities suffer from overcrowding, unsafe structures, and inadequate supporting infrastructure.

Global Efforts to Reduce Poverty

In its call to make society's economic development sustainable, the Brundtland Commission recognized that meeting the needs of present and future generations required particular emphasis on the basic needs of the world's poor. In response to this and other critical concerns, 30,000 people, including leaders from more than 100 nations, came to Rio de Janeiro in 1992 for the United Nations Conference on Environment and Development, also known as the Earth Summit.

The purpose of the summit was to take a fresh look at economic development in light of the findings of the Brundtland Commission. One of the summit outcomes was *Agenda 21*, an 800-page report calling for a broad set of programs for achieving sustainable development on a global scale. *Agenda 21* laid out a comprehensive set of goals and priorities accompanied by 120 action programs. Although nonbinding, there were implied commitments by the signatory nations to take meaningful action (Wallace 2005).

Despite the commitments, it soon became clear that the programs and actions proposed in *Agenda 21* were going nowhere. So, in 2000, the world's nations came

Figure 3-35. United Nations Millennium Development Goals.
Source: MintArt/Shutterstock.

together again, this time to add focus and make real commitments to eliminate extreme poverty. Nation representatives met in New York at the United Nations (UN) headquarters for what was called the Millennium Summit. The outcome of this meeting was the adoption of the United Nations Millennium Declaration, a new commitment to reduce extreme poverty. This summit also produced the Millennium Development Goals: eight goals, 18 measurable targets, and 48 indicators. The goals and targets were to be achieved by 2015 (Figure 3-35).

Although all the goals were only partially achieved by 2015, substantial progress was made (UN 2015). Since 1990, the number of people living in extreme poverty has dropped by more than half, falling from 1.9 billion to 836 million. In that same period, childhood mortality was reduced by one-half, and access to safe drinking water nearly doubled. Releases of ozone layer-depleting substances were virtually eliminated. Although these accomplishments are remarkable, there were significant shortfalls. More than 800 million people still live in extreme poverty, with women being most at risk. Although primary education enrollment has improved, there are still wide disparities between rich and poor children completing primary education. And there are still problems with gender equity. Finally, the sources of many problems were found to be political, cultural, or religious, not technical.

Based on the achievements of the Millennium Development Goals and recognizing their shortcomings, the UN expanded the goals and their coverage. In its new 2030 *Agenda for Sustainable Development,* the UN issued Sustainable Development Goals (SDGs) for 2030, 17 goals along with 169 targets covering the multiple dimensions of sustainable development. These goals cover more territory, addressing inequity, economic growth, good jobs, cities, industrialization, energy, climate change, sustainable consumption and production, peace, and justice (Figure 3-36).

Figure 3-36. United Nations Sustainable Development Goals: 2030 agenda.
Source: Andy Duarte/Shutterstock.

While the approach for the MDGs was about rich nations helping poor nations out of poverty, the SDGs emphasized all nations working together to achieve conditions of sustainability. In addition, the SDG program recognized the importance of high-quality, reliable, and detailed data for making decisions.

Sadly, progress toward these goals has had multiple setbacks. The United Nations (UN 2022) reports that because of the COVID-19 pandemic, 4 years of progress in poverty eradication were lost. The war in Ukraine has caused food, energy, fuel, and fertilizer prices to soar. Amid these setbacks, climate change acts as a "crisis multiplier," adding to the miseries of poverty, hunger, and instability.

References

Costanza, R., R. D'Arge, R. De Groot, S. Farber, et al. 1998. "The value of ecosystem services: Putting the issues in perspective." *Ecol. Econ.* **25** (1): 67–72. https://doi.org/10.1016/S0921-8009(98)00019-6.

Daly, H. E., and K. N. Townsend. 1993. *Valuing the earth: Economics, ecology, ethics*, 267. Cambridge, MA: MIT Press.

Elkington, J. 1998. *Cannibals with forks: The triple bottom line of 21st century business.* Gabriola Island, BC, Canada: New Society Publishers.

Elkington, J. 2018. "25 years ago I coined the phrase 'triple bottom line.' Here's why it's time to rethink it." Accessed January 22, 2024. https://hbr.org/2018/06/25-years-ago-i-coined-the-phrase-triple-bottom-line-heres-why-im-giving-up-on-it.

FFF (Forum for the Future). 2020. "The five capitals—A framework for sustainability." Accessed November 6, 2022. https://www.forumforthefuture.org/the-five-capitals.

GFN (Global Footprint Network). 2023. "Calculate your footprint." Accessed January 15, 2023. https://www.footprintnetwork.org/.

ICE (The Institution of Civil Engineers). 2023. "Royal charter and by-laws." Accessed January 15, 2023. https://www.ice.org.uk/media/ytpnalrr/ice-royal-charter-and-by-laws-2022.pdf.

IPCC (Intergovernmental Panel on Climate Change). 2021. "Summary for policymakers." In *Climate change 2021: The physical science basis. Contribution of Working Group I to the Sixth Assessment Report of the Intergovernmental Panel on Climate*, 19. Geneva: IPCC. https://www.ipcc.ch/report/ar6/wg1/chapter/summary-for-policymakers/.

Jouzel, J., Y. Masson-Delmotte, O. Catanni, G. Dreyfus, et al. 2007. "Orbital and millennial Antarctic climate variability over the past 800,000 years." Accessed January 22, 2024. https://www.science.org/doi/10.1126/science.1141038.

Krosnick, J. A., and B. MacInnis. 2020. "Climate insights 2020: Overall trends." Accessed January 15, 2023. https://www.rff.org/publications/reports/climateinsights2020/.

MA (Millennium Ecosystem Assessment). 2003. "Ecosystems and human well-being: A framework for assessment." Accessed January 18, 2023. https://www.millenniumassessment.org/en/Framework.html.

MA (Millennium Assessment). 2005. "Ecosystems and human well-being: Synthesis." Accessed January 18, 2023. https://www.millenniumassessment.org/documents/document.356.aspx.pdf.

McKibben, B. 2010. *Eaarth: Making a life on a tough new planet*, 1. New York: Henry Holt.

McKinsey & Company. 2020. "Climate risk and response: Physical hazards and socioeconomic impacts." Accessed January 22, 2024. https://www.mckinsey.com/capabilities/sustainability/our-insights/climate-risk-and-response-physical-hazards-and-socioeconomic-impacts.

NASA (National Aeronautics and Space Administration). 2022. "Carbon dioxide. Latest measurement: November 2022." Accessed January 16, 2023. https://climate.nasa.gov/vital-signs/carbon-dioxide/.

NOAA (National Oceanic and Atmospheric Administration and National Ocean Service). 2023. "What is a dead zone?" Accessed March 23, 2023. https://oceanservice.noaa.gov/facts/deadzone.html.

Porritt, J. 2007. *Capitalism as if the world matters*, 140. London: Earthscan.

Senge, P. M., B. Smith, N. Kruschwitz, J. Laur, and S. Schley. 2008. *The necessary revolution: How individuals and organizations are working together to create a sustainable world*, 103. New York: Crown Publishing Group.

UN (United Nations). 1983. "Process of preparation of the environmental perspective to the year 2000 and beyond." https://digitallibrary.un.org/record/65594?ln=en

UN. 1987. *Our common future: Report of the world commission on environment and development.* New York: Oxford University Press. https://sustainabledevelopment.un.org/content/documents/5987our-common-future.pdf.

UN. 2015. "The millennium development goals report." Accessed April 12, 2022. https://www.undp.org/library/millennium-development-goals-report-2015.

UN. 2022. "The sustainable development goals report 2022." Accessed March 14, 2023. https://unstats.un.org/sdgs/report/2022/The-Sustainable-Development-Goals-Report-2022.pdf.

UNDP (United Nations Development Programme). 2021. "Human development reports, human development data 2021." Accessed January 15, 2023. https://hdr.undp.org/data-center/human-development-index#/indicies/HDI.

Wallace, B. 2005. *Becoming part of the solution: The engineer's guide to sustainable development*, 34–35. Washington, DC: American Council of Engineering Companies.

Wolfson, R. 2012. *Environment and climate*, 335–337. 2nd ed. New York: Norton.

World Bank. 2023. "GDP (current US$)." Accessed January 15, 2023. https://data.worldbank.org/indicator/NY.GDP.MKTP.CD.

Additional References (Not Cited)

IEA (International Energy Agency). 2022. "Global CO_2 emissions rebounded to their highest level in history in 2021." Accessed May 22, 2022. https://www.iea.org/news/global-co2-emissions-rebounded-to-their-highest-level-in-history-in-2021.

Le Quéré, C., R. B. Jackson, M. W. Jones, A. J. P. Smith, et al. 2020. "Temporary reduction in daily global CO_2 emissions during the COVID-19 forced confinement." *Nat. Clim. Change* **10** (7): 647–653. https://www.nature.com/articles/s41558-020-0797-x.

Lindsey, R. 2022. "Climate change: Atmospheric carbon dioxide." Accessed January 16, 2023. https://www.climate.gov/news-features/understanding-climate/climate-change-atmospheric-carbon-dioxide.

OSS Foundation. 2020. "Global warming natural cycle." Accessed September 30, 2020. http://ossfoundation.us/projects/environment/global-warming/natural-cycle.

UN (United Nations). 2020. "United Nations department of economic and social affairs." Accessed January 15, 2023. https://www.un.org/development/desa/dpad/wp-content/uploads/sites/45/publication/SDO2020_Book.pdf.

CHAPTER 4

The Consequences of Unsustainable Development

"Nature provides a free lunch, but only if we control our appetites."

William Ruckelshaus, former US Attorney General and first administrator of the US Environmental Protection Agency

Human Development Progress and Its Consequences

Since the Industrial Revolution, human development has increased generally across all nations, some more rapidly than others. Figure 4-1 presents a plot (Roser 2014) of selected countries across a range of development using the Historical Index of Human Development, the historical values of the United Nations Development Programme's Human Development Index from 1870 to 2015.

Unfortunately, as discussed in Chapter 3, those achievements are closely correlated with the size of national ecological footprints. Nations that are ecologically sustainable (i.e., produce ecological footprints that are less than the world's biocapacity) generally score low on the human development scale. Conversely, nations with high human development ratings mostly have very high ecological footprints, well beyond the Earth's biocapacity. Those nations in between, the developing nations that make up the bulk of the world's population, aspire to achieve the same level of human well-being as the developed nations. Regrettably, most are following the same carbon-intensive pathways of the developed nations.

What About Natural Capital?

The general situation is summarized in Figure 4-2. Natural capital withdrawals to meet society's demands for goods and services have outstripped efforts to conserve, preserve, or restore these resources and systems. Pollutant emissions have degraded ecosystem functions. Disparities among nations in social, financial, and human capital have impeded progress in meeting societal needs. The result is a global potpourri: 195 nations with wide variations in progress toward human well-being and equally wide variations in their ecological footprint.

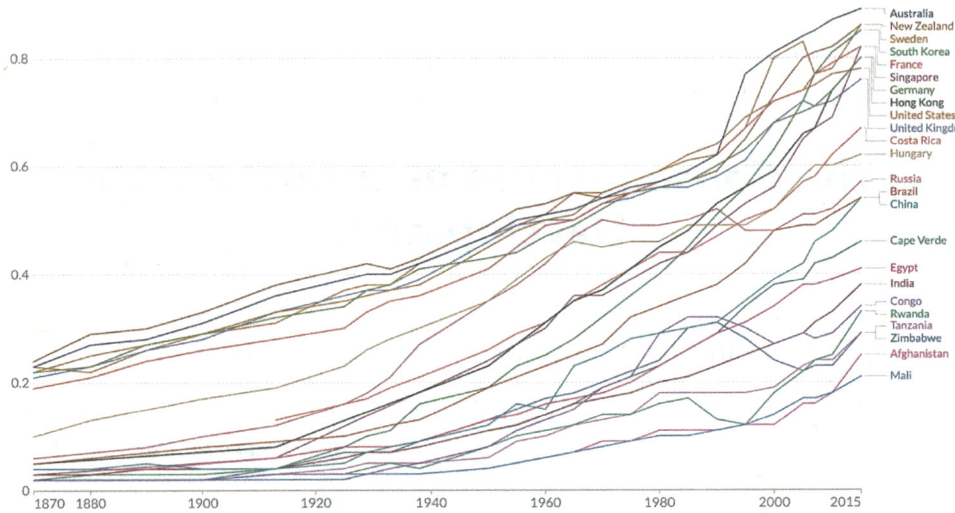

Figure 4-1. Example of Historical Index of Human Development (HIHD) for a range of nations. HIHD is a summary measure of average achievement in three key dimensions of human development: a long and healthy life, being knowledgeable and having a decent standard of living. *Source*: Roser (2014).

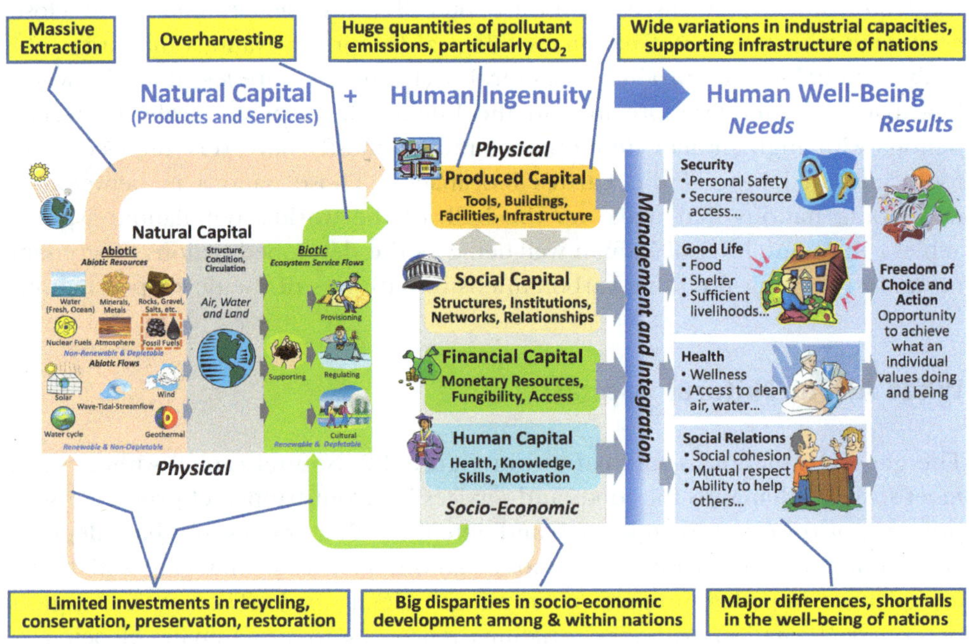

Figure 4-2. Issues in the quest to achieve a high level of human well-being.

The IPAT Equation

As nations work to meet the needs and wants of their citizens, the impact of their efforts can be expressed as

$$I = P \times A \times T \tag{4-1}$$

This equation was developed in the early 1970s by Paul Ehrlich and John Holdren (Holdren 1993) to address deteriorating environmental conditions. These terms are defined as

- **Impact (I):** Effects on the Earth's natural capital.
- **Population (P):** Number of people seeking to achieve a satisfactory level of well-being.
- **Affluence (A):** Level of prosperity or the degree to which society's perceived needs are currently being met.
- **Technology (T):** Means being employed by society to achieve that level of affluence.

The impact on the Earth's natural capital (I) is a function of the size of the population (P), amplified by both the level of affluence (A) being sought and the technology (T) being employed. If the net impact is greater than the Earth's ability to absorb, it will diminish the ability of future generations to meet their needs.

The IPAT equation has been criticized for its simplicity. Still, it is a useful concept for describing the environmental impact of an increasing population in pursuit of a better quality of life. The influence of technology is a function of the kind of technology used. To the extent that technologies diminish natural capital, for example, through excessive harvesting and extraction of renewable and nonrenewable resources, the value of T increases, as does the impact.

Furthermore, the impact is not confined to natural capital. Heavy-handed extraction, harvesting, processing, and use of natural resources can significantly harm other forms of capital. Strip-mining lands and clear-cutting forests not only reduce their regulating services functions but also diminish the cultural value of these environments. Civil infrastructure (manufactured or produced capital) heated and powered by fossil fuels not only wastes a nonrenewable resource but also disrupts the Earth's climate-regulating systems. Production processes that use and abuse workers reduce the value of social and human capital.

In his book *Mid-Course Correction*, the late business leader and environmentalist Ray Anderson (1998) argued that the equation needed to be changed. High-impact technologies should be termed T_1, not T. Following the recommendation of architect and environmentalist Bill McDonough, Anderson then introduced the technology term T_2, which refers to technologies that use renewable energy, produce no waste, are benign, and perhaps enhance natural capital. He proposes that the equation be rewritten, moving T_2 to the denominator.

$$I = (P \times A)/T_2 \tag{4-2}$$

Climbing the Heat Ladder

Of all the technologies used by society, those that use fossil fuels have had the most negative impact. Yet ironically, the rise of human civilization can be traced mainly by its use of these fuels in their various forms (Figure 4-3). Hydropower is a possible exception, although hydropower dams are known to have impacted fish migration and habitats (USGS 2018).

Educator and anthropologist Loren Eiseley (1954) noted in his essay "Man the Fire-Maker" that human progress can be tracked as a climb up the heat ladder. "Each move to the next [rung] is based on finding a source [of heat and energy] that is more convenient, potent, or economical than the last. No one lugs firewood to warm a high-rise apartment building in Chicago" (Revkin 2007).

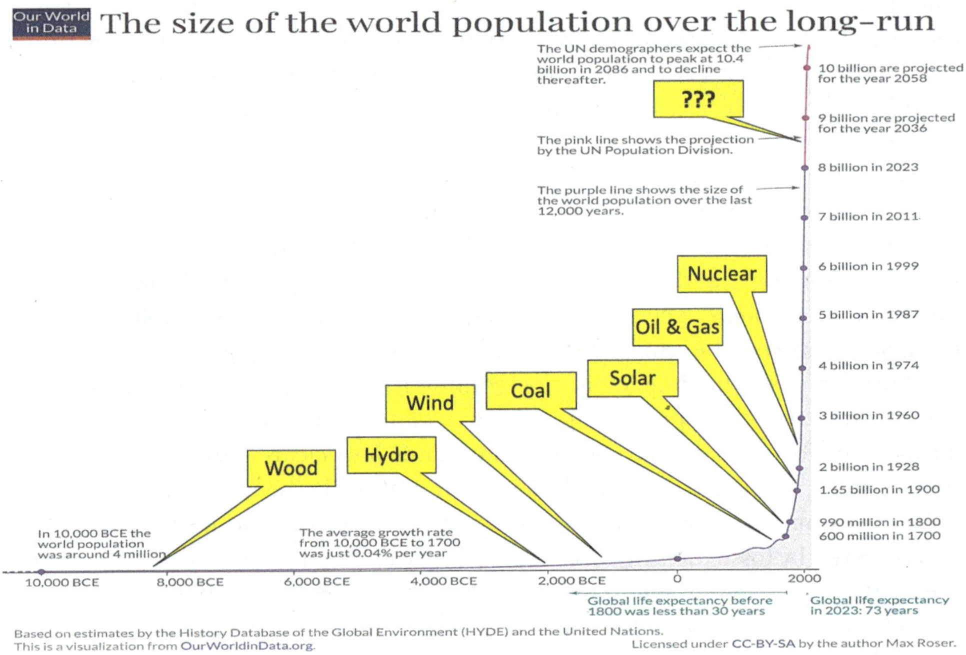

Figure 4-3. Climbing the heat ladder. Population growth and the advancements in society's sources of energy. The plan was to switch over to nuclear energy, but nuclear accidents (Three Mile Island, Chernobyl, and Fukushima Daiichi) have dimmed nuclear energy prospects. However, renewable energy sources are becoming cost-competitive with fossil energy sources. *Adapted from*: Roser and Ritchie (2023).

As noted in Chapter 3, the Earth's population remained low until the Industrial Revolution, a significant turning point in human history. Improvements in health and agricultural productivity freed up people to do other things. Moving away from subsistence living, men and women were able to provide goods and services to others and be paid for their efforts. With this extra income, they could improve their livelihoods, obtain better food, clothing, and living conditions, and have a better life overall. As health and productivity improved, populations grew, increasing the demand for more goods and services.

As populations and demand increased, so did the demand for fuels. Wood became scarce as a fuel source. Oak was required to construct sailing ships for commercial and military use, a higher priority than home heating or production. But homes and factories still needed to be heated and powered, so people turned to coal. Petroleum fuels came into use when vast quantities of these organic liquids and gases were discovered. Petroleum had multiple advantages over coal—a potent energy source much easier to transport and store.

Developed in the mid-twentieth century, nuclear power was the logical replacement for coal and petroleum. It is not polluting, at least in terms of emissions from nuclear power plants, and can provide enormous energy to power national electrical grids. However, safety concerns have stalled expansion. Nuclear accidents—Three Mile Island, Chernobyl, and Fukushima Daiichi—escalated public opposition to nuclear power. Also, securing locations for the safe disposal of nuclear waste remains problematic.

Consequently, fossil fuels continue to dominate as the world's energy source, accounting for about 80% of US energy consumption. Nuclear power supplies 8.0%. The remaining 13% is supplied by renewable sources (Figure 4-4).

Fossil Fuel Use and Its Consequences

A useful way of illustrating the consequences of using fossil fuels is to show them in relation to natural capital and human well-being. As presented in Figure 4-5, the extraction and burning of fossil fuels for heat, mobility, and electric power generation release enormous quantities of CO_2 and other greenhouse gases into the Earth's atmosphere. These gases are degrading the Earth's climate-regulating ecosystem services. We are now experiencing the results: a warming and increasingly harsh planet with more severe weather and climate events, all increasing in intensity, duration, and frequency. In terms of the Five Capitals, these events threaten and frequently overwhelm structures, facilities, and systems in the built environment (produced capital). Frequent and extended disruptions in daily life (e.g., losses in electrical power, water supplies, mobility, and shelter) place additional stresses on our socioeconomic systems and degrade human health and well-being (Figure 4-6). The developed nations are better equipped to handle severe weather and climate events through their financial and institutional strengths. Less so for the other

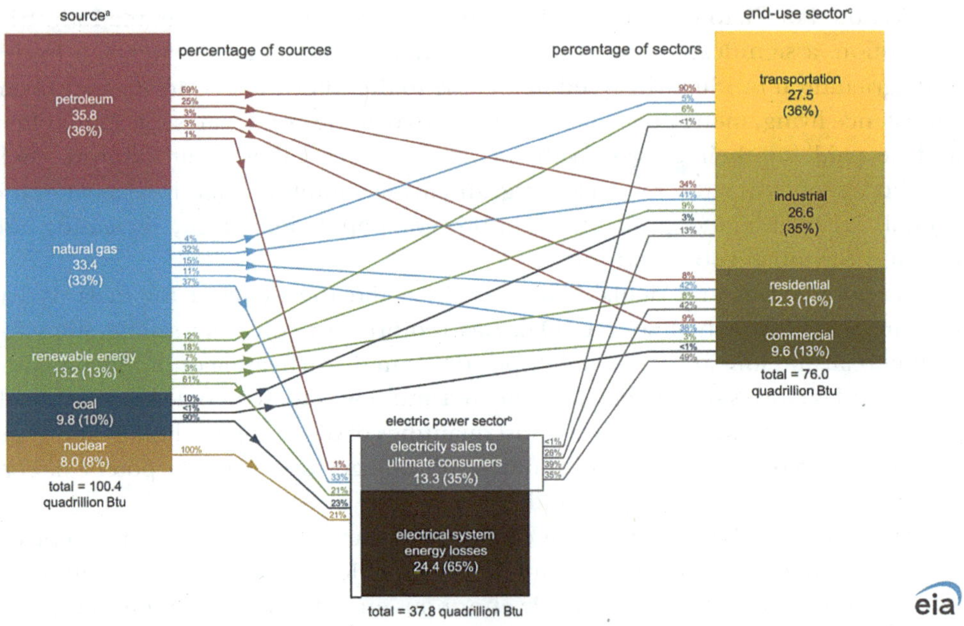

Figure 4-4. US energy consumption by source and sector, 2021.
Source: US Energy Information Administration (2023).

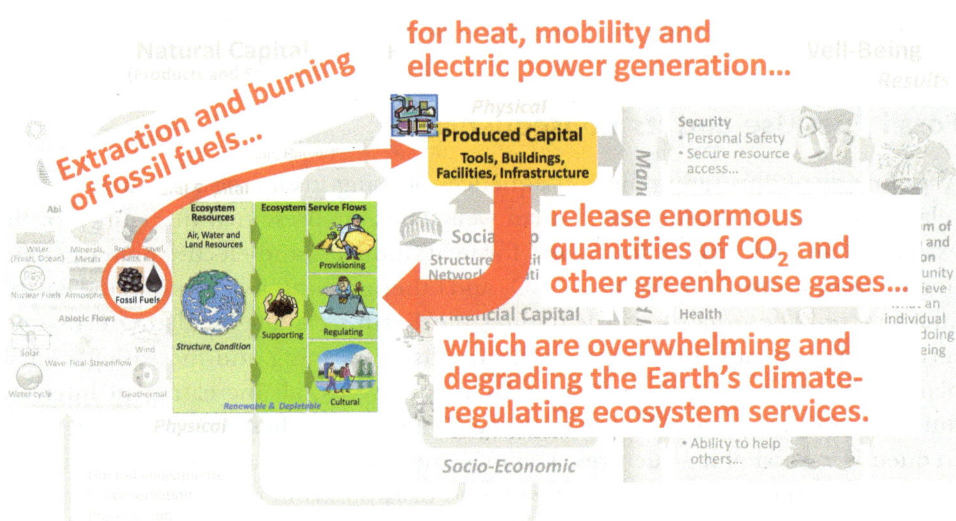

Figure 4-5. Relationship of the extraction and burning of fossil fuels for heat, mobility, and electric power generation to the Earth's climate-regulating ecosystem services.

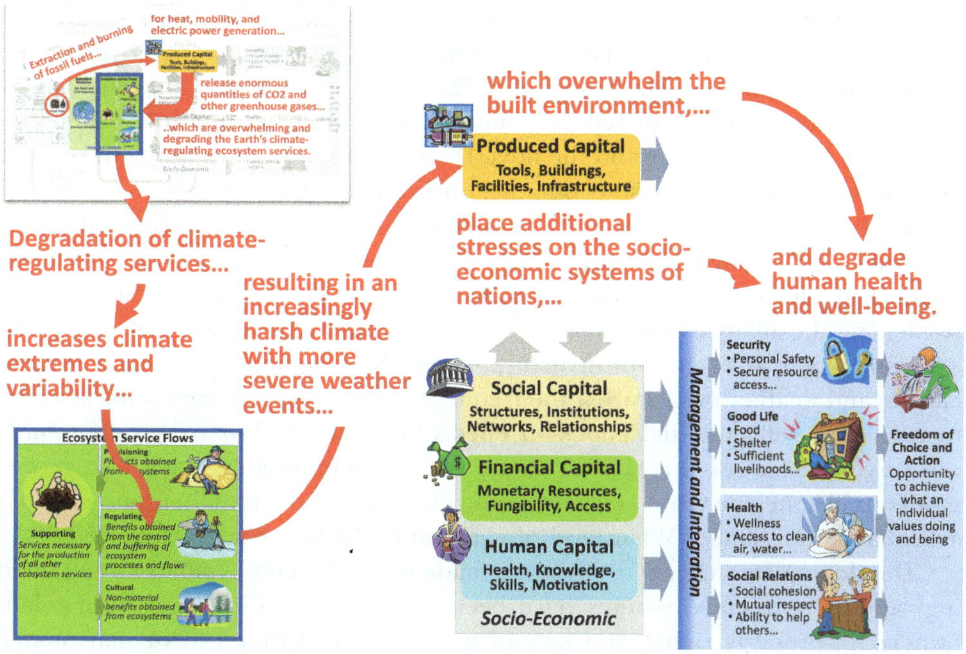

Figure 4-6. Fossil fuels use and its consequences: degradation of climate-regulating services on produced capital, socioeconomic capital, and human health and well-being.

nations. Ironically, the underdeveloped nations, having the least to do with the causes, are the most affected by the consequences.

A "New Normal" of Disasters

Susan Cutter (NASEM 2022) describes a "new normal" of disasters: disasters of increased intensity and frequency, making it more difficult for the affected communities to recover. Under climate stationarity, destructive storm events were possible. However, the chances of occurrence were very low. Building and infrastructure codes and standards, based on historical information on climate conditions, were devised to handle a broad range of climate and weather events except for extremes deemed too rare to worry about. Accordingly, disaster management systems were designed to handle rare, single destructive events. First responders mobilize and address immediate community needs. Second responders follow to start the recovery process, helping the affected communities return to normal. Because they were rare events, the time interval between disasters was expected to be long, allowing communities to recover sufficiently and return to their previous state of normalcy, likely better informed about their vulnerabilities and better prepared for future extreme events.

However, in a changing climate, extreme events are becoming more intense and more frequent. Now, communities must contend with recurrent acute disasters

(RADs), disasters that happen in rapid sequence at a specific locale, with each disaster creating conditions that affect subsequent disasters, potentially increasing the cumulative impact (Machlis et al. 2022) and reducing the time interval for recovery. Without sufficient recovery time, communities become increasingly vulnerable to the next disaster and the next. Disaster impacts and subsequent recovery are not only physical but also economic, environmental, and social.

In this new normal of disasters, developing effective community resilience is especially challenging. In a world where greenhouse gas emissions are still mostly unchecked, climate and weather hazards continue to rise. In this world, achieving resiliency—the ability to withstand or recover from these extreme events—becomes a moving target. How can infrastructure systems be designed to withstand extreme climate and weather events when the frequency and severity of the events continue to rise? How will potential knock-on effects be handled? What does community recovery look like in the face of RADs? When communities reach "No more!" moments like the one Cape Town experienced in 2018, how does their perception of acceptable risk change? What level of community resiliency is acceptable?

Developing community resilience is made more difficult by how existing buildings and infrastructure systems have been designed. Land-use regulations, land ownership, property rights, and lack of investment contribute to or may amplify disaster effects. The National Academies (NASEM 2022) note that "[R]ecovery requires more than getting back to normal, especially when what is considered normal may be a major contributor to a community's vulnerability to cascading disasters. How can communities recover from disasters when the normal they have lost was a major contributor to the disaster itself?"

Sustainability and Resilience

In the wake of multiple extreme storms, droughts, heat waves, and other unexpected difficulties, cities, states, and other public and private organizations have sought to make their infrastructure more resilient (i.e., able to withstand or recover quickly and at low cost from extreme weather and climate-induced damages). Somehow, the demand for additional civil infrastructure resiliency has been disconnected from the concept of sustainability. Here is the connection.

The reason cities and others are building in additional resilience is that society's approach to economic development has not been sustainable. For human society to be sustainable, it must maintain, preserve, and protect the Earth's ecological resources and services, leaving them in levels of quantity and productivity sufficient for future generations to meet their needs. Unfortunately, society has been using these resources and damaging services faster than they can be replaced, replenished, or repaired. One of those services happens to be the Earth's climate-regulating services, an array of ecological systems that, for the last 10,000 years, has maintained the climate in what Bill McKibben (2010) described as *the sweetest of sweet spots*: temperatures staying between 58 and 60 °F, seas tame and level, resources plentiful.

Unfortunately, Earth's climate is no longer in that sweet spot. Excess atmospheric concentrations of carbon dioxide and other greenhouse gases have degraded the Earth's climate-regulating system to such an extent that the climate is no longer stable. These heat-trapping gases continue to add thermal energy to the Earth's atmosphere.

Excess thermal energy has not only warmed the planet but has destabilized the climate. The result has been more frequent and intense storms and other extreme climate conditions and weather events. The intensity–duration–frequency combination of these events has overwhelmed many of society's existing infrastructure assets that were designed based on the assumption of a stable climate. Being stuck with numerous assets inappropriately designed for current conditions, cities, states, and other public and private organizations are trying to make these assets more resilient, protecting and adding more strength and durability to their existing assets, finding ways to recover quickly and at low cost, and rethinking how future assets need to be designed, built, and operated. Importantly, if carbon emissions are not drastically reduced, increasing resiliency will become a continuous activity.

The Impacts of Business-As-Usual

Figure 4-7 shows the relationship between ambient temperature increase and potential damage to essential resources, systems, and services. The figure was adapted from a chart in the 2007 UK publication (Stern 2007) that presented the costs of addressing climate change and its impacts on the world economy if no substantive actions are taken. This publication, known as the *Stern Review*, makes it clear that the higher the temperature increase from preindustrial times, the greater the possibility and severity of the impacts.

Dates for expected average ambient temperatures are consistent with the RCP 8.5 "Business-as-Usual" scenario and the limit and pursuit targets of 2.0 °C (3.6 °F) and 1.5 °C (2.7 °F), established by the 2015 Paris Climate Accords. Today, the increase in the Earth's average ambient temperature above preindustrial levels is 1.3 °C (2.3 °F). Since 1981, the rate of warming has been 0.18 °C (0.32 °F) per decade. Note that the early projected impacts—decreasing water supplies, melting of mountain glaciers, rising intensity of storms, forest fires, droughts, and flooding—are already appearing.

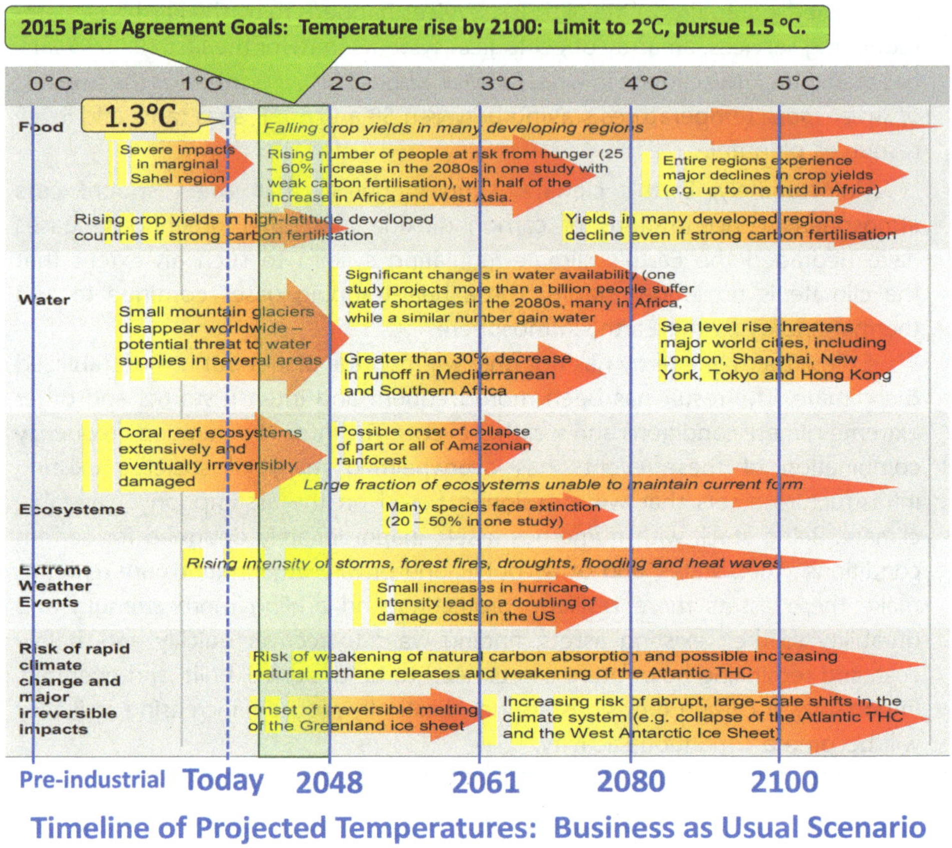

Figure 4-7. Projected impacts of climate change on essential services for the RCP 8.5 "Business-as-Usual" temperature scenario and expected range.
Source: Adapted from Godard (2008), Creative Commons Attribution 3.0 License.

A New Sense of Urgency

As noted previously, the Intergovernmental Panel on Climate Change (IPCC) 2018 Special Report issued a disturbing warning: unless society makes radical reductions in greenhouse gas emissions to keep global average temperatures from rising above 1.5 °C (2.7 °F), the damages caused by global warming will become far more devastating (Levin 2018) and more costly to fix (Table 4-1). Society must start now to make these reductions, reducing greenhouse gas emissions to about one-half by 2030 and achieving net-zero emissions by 2050. Recognizing that civil infrastructure is the primary source of greenhouse gas emissions, project owners and engineers are beginning to design or retrofit infrastructure assets to reduce embodied and operational carbon emissions. They are also starting to design or retrofit infrastructure assets or are instigating efforts for managed retreat, all geared to adapt to the conditions brought on by a changing climate.

Table 4-1. Climate Impacts: The Difference Between 1.5 °C and 2.0 °C.

Impacts	1.5 °C	2.0 °C	2.0 °C impacts
Extreme heat. Global population exposed to severe heat at least once every five years	14%	37%	2.6 × worse
Sea ice-free Arctic. Number of ice-free summers	1/100 years	1/10 years	10 × worse
Sea level rise. Amount of sea level rise by 2100	0.40 m	0.46 m	0.06 m more
Species loss: vertebrates. Vertebrates that lose at least half their range	4%	8%	2 × worse
Species loss: plants. Plants that lose at least half their range	8%	16%	2 × worse
Species loss: insects. Insects that lose at least half their range	6%	18%	3 × worse
Ecosystems. Amount of the Earth's land area where ecosystems will shift to a new biome	7%	13%	1.86 worse
Permafrost. Amount of Arctic permafrost that will thaw	4.8 million km^2	6.6 million km^2	38% worse
Crop yields. Reduction of maize harvests in the tropics	3%	7%	2.3 × worse
Coral reefs. Further decline of coral reefs	70%–90%	99%	Up to 29% worse
Fisheries. Decline in marine fisheries	1.5 million tonnes	3 million tonnes	2 × worse

Source: Levin (2018).

Opportunities in a Low-Carbon Economy

The resulting opportunities in moving to a low-carbon economy are not small. A recent study from the Global Commission on the Economy and Climate (WRI 2018) stated that shifting from business as usual to a low-carbon economy would produce US$26 trillion in cumulative economic benefits from 2018 to 2030. The benefits include avoidance of flooding in coastal cities, reductions in desertification, decreases in population migration, phase-out of fossil fuel subsidies, industrial energy and resource conservation gains, moving to renewable power, and monies from carbon pricing. The study also estimates the creation of 65 million additional jobs by 2030 and the avoidance of 700,000 deaths from air pollution (Table 4-2).

As noted in Chapter 1, 77 countries, 10 regions, and more than 100 cities have committed to net-zero carbon emissions by 2050. More than 4,400 companies across the world have also committed to work toward the goal of limiting warming to 1.5 °C (2.7 °F). This shift to a low-carbon economy will be accomplished through a major rework of the current civil infrastructure portfolio, replacing fossil energy with renewable energy systems, implementing resource and energy conservation measures, making new infrastructure assets more resilient, and relocating or refurbishing existing assets to increase their utility and survivability.

Table 4-2. Benefits from Shifting to a Low-Carbon Economy.

Category	Benefits
Air pollution	Avoid more than 700,000 premature deaths
Employment	Generate more than 65 million additional low-carbon jobs. Increase female employment
Revenue	Add US$2.8 trillion in carbon price revenues and fossil fuel subsidy savings to reinvest in public priorities
Gross domestic product	Achieve higher global GDP growth

Source: WRI (2018).

Impacts on Civil Infrastructure Design

Shifting to a low-carbon economy will drive massive changes in infrastructure systems in terms of the resources and energy they use, how they are operated and maintained, and how they will be adapted to changing climate conditions. For civil engineers working in this new economy, it is essential to understand how these changes in climate conditions will affect infrastructure assets and determine how to address these changes efficiently and effectively.

Changes in Probability Distributions of Climate Stressor Variables

Changing climate conditions can be depicted as changes to the probability distributions of the values of some climate stressor variables (Figure 4-8). Examples of the potential changes in the probability distribution of ambient temperatures are depicted in Figure 4-9 (IPCC 2012). As the climate changes, the probability

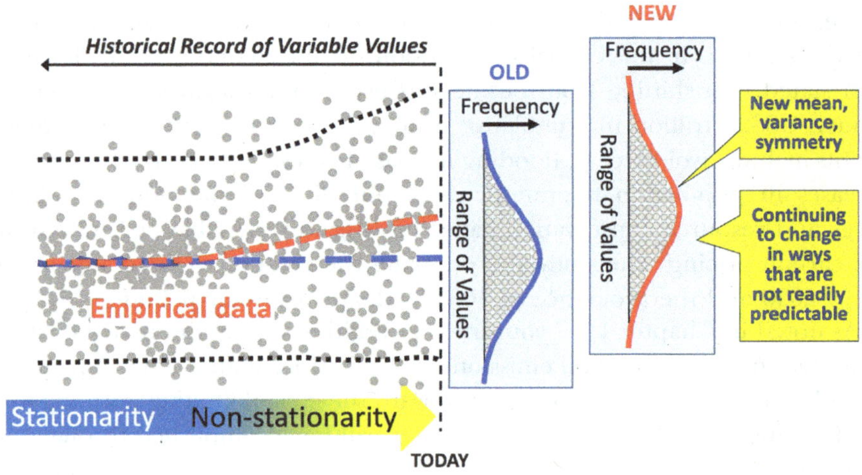

Figure 4-8. Potential changes in the probability distribution function for a climate stressor variable under conditions of nonstationarity.

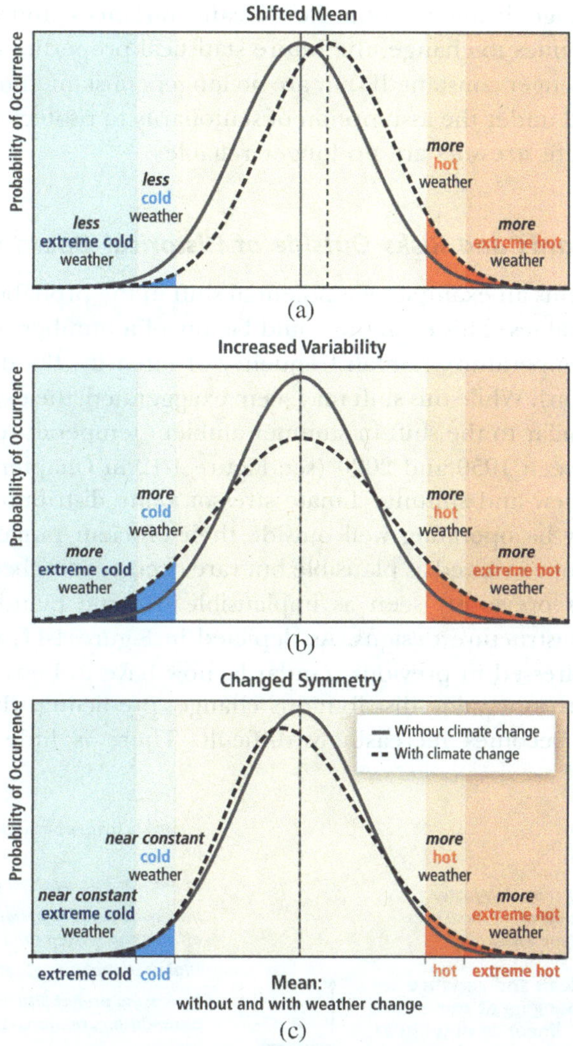

Figure 4-9. The effect of changes in temperature distribution on extremes. Different changes in temperature distributions between present and future climate and their effects on extreme values of the distributions: (a) effects of a simple shift of the entire distribution toward a warmer climate, (b) effects of an increase in temperature variability with no shift in the mean, (c) effects of an altered shape of the distribution, in this example a change in asymmetry toward the hotter part of the distribution. Furthermore, extreme values once thought to be highly improbable now become not only more likely to occur but also occur more often. Having little historical operating experience with these new extreme values, designers are faced with not only new conditions to contend with but also the possibility of secondary and tertiary effects and impacts, the possibility of reaching tipping points, and/or reaching values at which negative reinforcing functions come into play. As a result, making design decisions that will maintain resilient and reliable operation becomes highly problematic.

Source: IPCC (2012).

distributions change, bringing about new means, variances, and symmetries. Since the climate continues to change, the future statistical properties of climate-stressor variables are no longer constant. If they are no longer constant, then the engineering standards devised under the assumption of stationarity to protect civil infrastructure assets from climate stressors are no longer reliable.

Additional Hazards and Risks Outside of Historical Operating Experience

Figure 4-10 presents an example of a potential shift in the probability distribution of climate stressor values. This example could be any of a number of climate stressors (e.g., ambient temperature, rainfall frequency or intensity, flood level, drought or heat wave duration). While this shift may seem exaggerated, the extent of the change as depicted is similar to the shift in summer ambient temperatures in the northern hemisphere between 1950 and 2020 (see Figure 1-12 in Chapter 1).

Under this new and shifting climate stressor value distribution, infrastructure components may be operating well outside their efficient range of performance. Extreme values, once viewed as plausible but rare occurrences, become increasingly probable. Values previously seen as implausible become plausible and must be respected in infrastructure designs. As depicted in Figure 4-11, possible knock-on effects, once addressed in previous standards, now have to be reevaluated.

As climate stressor value distributions change, predicting what will happen at extreme values becomes increasingly difficult. There is little or no historical

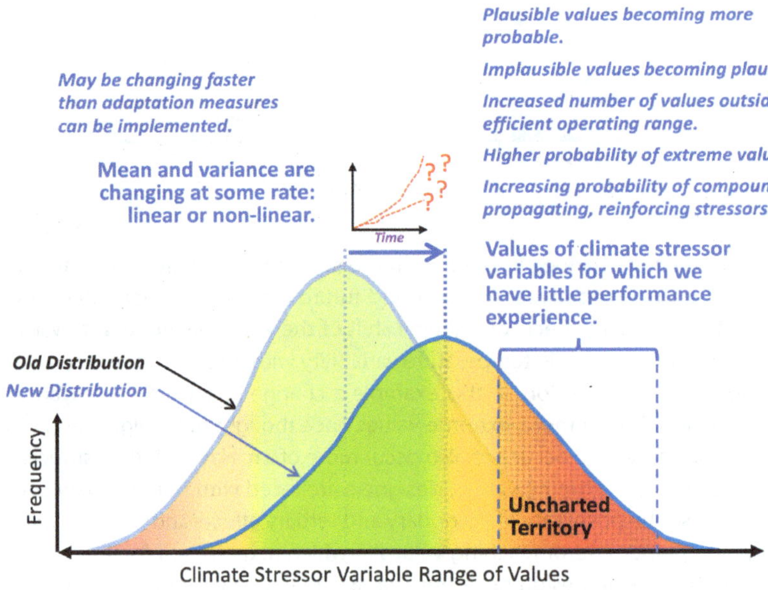

Figure 4-10. Implications of nonstationarity for climate stressor variables. Colors show the old distribution mapped onto the new distribution.

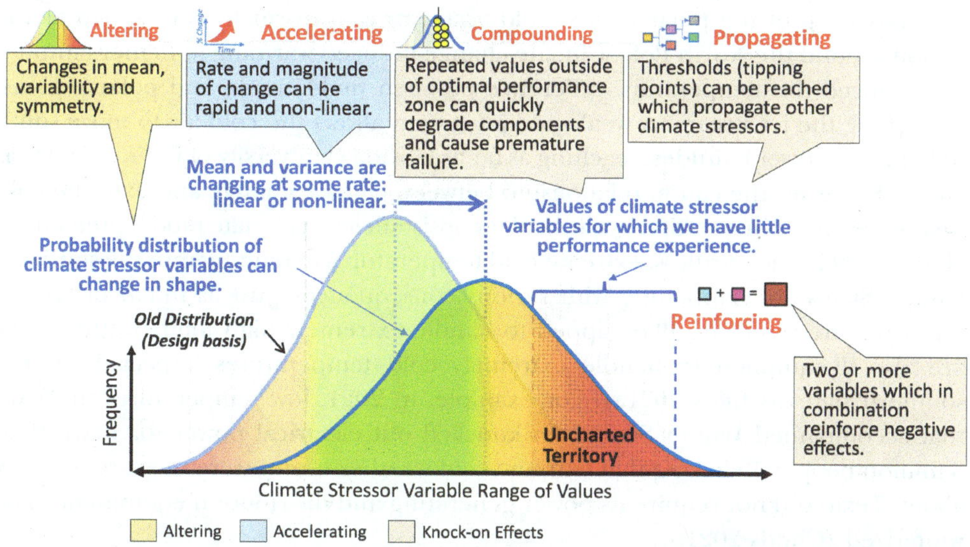

Figure 4-11. Changes in climate stressor distributions under conditions of nonstationarity.

performance information from which to determine possible effects. As Kim Cobb, a climate scientist and one of the lead authors of the IPCC AR6 report (2021), stated, "We're going into uncharted territory" (Dennis and Kaplan 2021).

Consequently, engineers face an additional set of challenges: how to design civil infrastructure that can operate safely and reliably in this uncharted territory with extreme values of climate stressors for which there is little or no operating experience. The operating experience that does exist mainly resides in after-action reports prepared in response to climate and weather disasters. Moreover, climate stressor value distributions will continue to shift to some degree throughout this century. The degree of shift is dependent on the extent to which carbon emissions are reduced, greenhouse gases are removed from the atmosphere and sequestered or destroyed, and solar irradiance is blocked or reflected back into space.

Effects on Infrastructure Assets

For project owners, planners, and designers, understanding the impacts of the various categories of infrastructure assets on changing climate stressor values is essential. Appendix A provides an extensive compilation of potential impacts organized by infrastructure categories aligned to the ASCE Infrastructure Report Card (ASCE 2021). Within each category, climate stressor impacts are separated into those affecting the physical aspects and those affecting asset operations. The grouping of climate stressors was modeled after the US Department of Transportation's Sensitivity Matrix (FHWA 2014).

Extreme cold was added to the list of climate stressors. It has been suggested that climate change has increased the chances of extreme cold weather events because of a destabilizing of the polar vortex, cold arctic air that normally is confined to the region around the North Pole. Climate change-induced warming in higher latitudes has reduced the temperature difference between mid-latitude and polar regions, weakening the jet stream. A weakened jet stream allows the cold air to move south well into the lower latitudes, reaching as far as Florida (University of California, Davis 2021). However, the causal relationship between climate change and extreme cold weather events is in question because of inconsistencies in climate model predictions (Lindsey 2021). Regardless, extreme cold temperatures extending into the southern United States are happening with serious consequences. Just as much of today's infrastructure assets are ill-equipped to handle extremely high temperatures, they are also ill-equipped to handle extremely cold temperatures, especially in the southern states (Childs 2021a). For example, in 2021, low temperatures in Texas that accompanied Winter Storm Uri knocked out electrical power for more than 3 million people. Believing that extreme cold weather events were too rare to worry about, Texas did not require its power generating and distribution equipment to be winterized (Childs 2021b).

Altering: Changes in the Probability Distributions of Climate Stressor Values

Figure 4-12 depicts what could happen over the design life of an infrastructure project as the climate stressor probability distributions change, deviating from the historic probability distributions. Infrastructure projects designed to existing (climate stationarity-based) standards will experience increasing degradation over

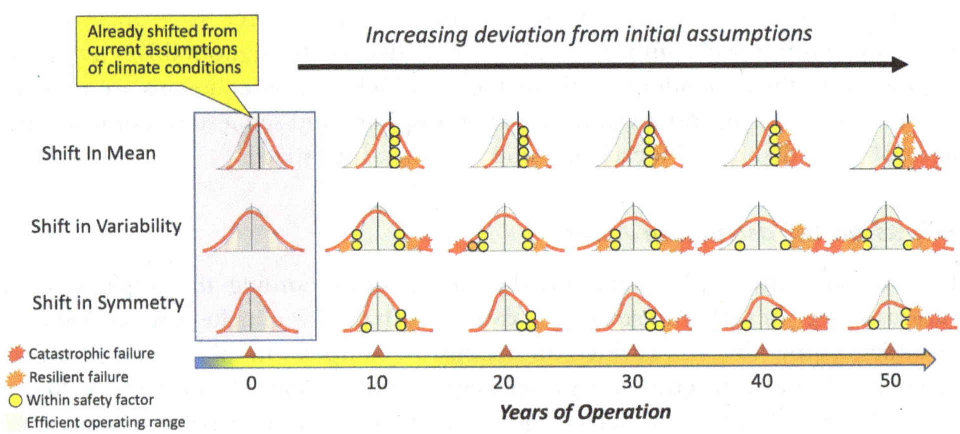

Figure 4-12. Example of expected operating experience of 50 years, shown in 10-year increments, for a civil infrastructure system starting operation today under conditions of nonstationarity but designed using the assumption of stationarity. The diagrams show different shifts in the probability distributions for three climate stressor variables, each experiencing a different type of change in value distribution.

time. As time passes and the climate changes, infrastructure elements designed to those standards will experience climate and weather conditions that fall increasingly outside the expected range. Values can become extreme enough to instigate infrastructure component failure, either resilient or catastrophic, depending on whether elements of resiliency were included in the design. Note that at project startup (year zero), actual conditions are shown as already different from design assumptions.

It should be noted that climate-driven changes are not always harmful. In all infrastructure categories, warming temperatures lower energy demands for heating, reduce the cost of snow and ice removal and improve outdoor work productivity. Ports can benefit as less Arctic ice opens up northern trade routes such as the fabled Northwest Passage. As temperatures warm, the atmosphere is able to hold more moisture, resulting in increased precipitation. Some areas in the United States are already experiencing more rainfall.

Nevertheless, the benefits of warming are slight compared to the harmful effects. Higher temperatures bring about a host of problems and are made worse as temperatures continue to rise. Concrete pavements and railroad tracks buckle. Asphalt pavements soften and develop ruts. Airport operations are curtailed as high temperatures reduce air density and, consequently, aircraft lift. Cooling demands increase for equipment, workers, and users. Worker productivity drops as the chances of worker and user heat stress increase. As precipitation increases, so do the chances of heavy rainfall events, raising the chances of flooding. High winds, extended droughts, wildfire dangers, freeze/thaw cycle changes, and thawing permafrost contribute to infrastructure damage and significant drops in operational efficiencies.

Here is a snapshot of climate-related challenges organized by ASCE infrastructure categories (ASCE 2021). Appendix A provides additional details.

- **Aviation:** This category covers aircraft, airfield buildings and structures, runways, taxiways, aprons, and navigational aids. As temperatures rise, air density decreases, resulting in less aircraft lift for takeoffs and landings. At high air temperatures, operations are curtailed. To maintain expected levels of operation, airports will need longer runways to offset rising temperatures.
- **Bridges:** Bridges include the bridge superstructure, substructure, deck, and operator houses. Thermal expansion can cause bridge deck pavements to buckle and may make bridge opening mechanisms inoperative. Heavy precipitation can increase stream or river flow, affecting local scour depth (i.e., the depth of erosion of soils surrounding bridge foundations). Also, more bridges are needed in Alaska. Rivers and lakes are freezing later and thawing earlier, making travel over ice unpredictable and risky.
- **Broadband:** Broadband consists of multiple connections of lines, cell towers, power supplies, data transfer stations, and other components, creating a network critical for education, communications, healthcare, and emergency

notification. These components are vulnerable to many of the same climate hazards as our power networks. There continues to be a drive to expand these networks quickly to provide high-speed internet coverage throughout the United States. Consequently, many networks were installed haphazardly and opportunistically, using approaches based on past climate conditions. Also, they are operated by private companies that tend to keep the locations of these components confidential.

- **Buildings:** Commercial, industrial, residential, and school buildings make up this category. Frequent flooding, high winds, and high temperatures lower usable building life. Melting permafrost may cause buildings to sink, resulting in structural damage.
- **Dams:** These structures contain water for storage, flood control, or electric power generation. Droughts, temperature fluctuations, storms, and flooding can cause additional stresses on dams.
- **Drinking Water:** This category includes the water utilities that collect, treat, and distribute clean water to homes and businesses. In drought conditions or in cases where source waters are heavily contaminated from stormwater debris, water treatment systems may not be able to meet drinking water standards. In addition, most water treatment systems are small and lack the funds to make the necessary modifications to handle a changing climate.
- **Energy:** The energy infrastructure in the United States generates, transmits, and distributes electrical power to industrial, commercial, and residential users. Demand for electric power is increasing, driven by demands for increased industrial, commercial, and residential cooling, expanded use of heat pumps, electric vehicle charging, and more. To access cooling water, many thermal and nuclear power plants are located near the coast, making them vulnerable to rising sea levels. Extended droughts threaten hydropower resources.
- **Hazardous Waste:** Hazardous waste sites are land areas that contain significant quantities of dangerous materials, which, if released, could endanger the public. If hazardous waste containment and treatment systems are flooded, contaminants can mobilize and move into the environment.
- **Inland Waterways:** These waterways, consisting of approximately 25,000 mi of navigable waterways, locks, and dams, are considered to be the water highways of the United States. In drought conditions, low water levels restrict vessel cargo carrying capacity. Heavy precipitation and flooding can also restrict waterway operations. In some places, storm surge and sea level rise combinations could destroy barrier islands, eliminating waterway systems entirely.
- **Levees:** Levees are earthen embankments or concrete floodwalls built next to rivers to prevent river overflow or divert water flow from a defined area. Levees protect critical infrastructure from flooding. Extreme rainfall can overtop or breach levees, resulting in a surge of flood waters. An example of such a breach is presented in Figure 4-13.

Figure 4-13. Levee breach on the Clear Creek Levee System along the Platte River. *Source*: USACE Omaha District, Clear Creek Levee Breach, March 22, 2019.

- **Public Parks:** Parks are open, natural spaces provided for recreational use. Urban parks are located in urban settings and serve as a recreational respite from city life. They are also an essential line of defense in reducing the impact of urban heat islands on local climates. Virtually all are affected by climate change.
- **Ports:** The United States has more than 300 coastal and inland ports, consisting of piers, wharves and berths, channels, and terminal buildings. The rise in sea level will affect dock levels and could require retrofitting. Hotter and drier conditions increase the risk of dust explosions in grain handling and storage facilities. Water-sensitive cargo can be damaged in flood events. Increased coastal erosion because of sea ice retreat and permafrost thaw can damage ports. Shorelines are increasingly exposed to direct wave action. Structure foundations are at risk. Climate change-related damage may also affect rail links to ports.
- **Rail:** The US rail infrastructure consists of freight and passenger rail. Infrastructure and equipment include rail cars, railroad ties, tracks, rail switches, ballast, and electrical equipment (e.g., gates/flashers and signal bungalows). High temperatures can cause the track to buckle (Figure 4-14). Railroad operators are required to reduce train speeds during high-heat events.

Figure 4-14. Examples of railroad track buckling.
Source: Kish and Wesley (2009).

With permafrost thaw, there is also a risk of uneven settling and ground heaves. A section of the rail line on the Copper River Northwestern Railroad has experienced extensive vertical movement of track, so much so that it is referred to as the "roller coaster railroad" (FHWA 2014).

- **Roads:** This category comprises paved and unpaved roads, road drainage, street signs, and traffic lights. Roads can experience surface buckling, softening, and rutting in extreme heat events. Also, roads in Alaska can be severely damaged by permafrost thaw (Figure 4-15). Ice roads, roads across frozen water surfaces, have a shorter season. These roads are essential for travel in Alaska.
- **Solid Waste:** Solid waste management infrastructure includes waste transfer stations, recycling, incineration, composting facilities, and landfills. If flooded, solid waste treatment, storage, and disposal facilities could release contaminants into surface and groundwaters.
- **Stormwater:** Stormwater infrastructure consists of the network of water retention and purification systems designed to slow and direct stormwater flow, prevent flooding, and reduce the contamination of waterways. In heavy rainfall events, stormwater flows can overwhelm systems that combine wastewater and stormwater flows, resulting in combined sewer overflows. Sewage can be released into waterways.
- **Transit:** The US transit infrastructure consists of urban and suburban rail and bus systems that provide local transport and access to jobs, schools, healthcare,

Figure 4-15. Example of roadway collapse on a highway in Alaska because of permafrost melting.
Source: Turchetta (2010).

shopping, and public/private services. Transit systems use local roads and railways and thus are subject to many of the same climate-related issues.

- **Wastewater:** Wastewater infrastructure consists of the network of pipes and treatment systems that collect, treat, and discharge effluents into the environment. Many wastewater treatment systems are located on the coast to make use of ocean outfalls for the disposal of treated wastewater. As such, they are at risk of flooding from rising sea levels and storm surges.

Accelerating: Rate of Change Is Increasing and Nonlinear

Under conditions of nonstationarity, the rate and magnitude of change become important factors in civil infrastructure design. If the rate of change in stressor values is high and accelerating, then adaptation strategies with long lead times could be ineffective. For example, sea levels may rise faster than anticipated because of greater-than-expected ice losses in Antarctica and Greenland.

In response to nonlinearity, adaptation measures should be quick, easy, and low cost: Incremental adjustments are made as new knowledge is acquired and the changes become apparent (FHWA 2017a). For example, for New York City's Governors Island Ventilation Building (Figure 4-16), the New York Metropolitan

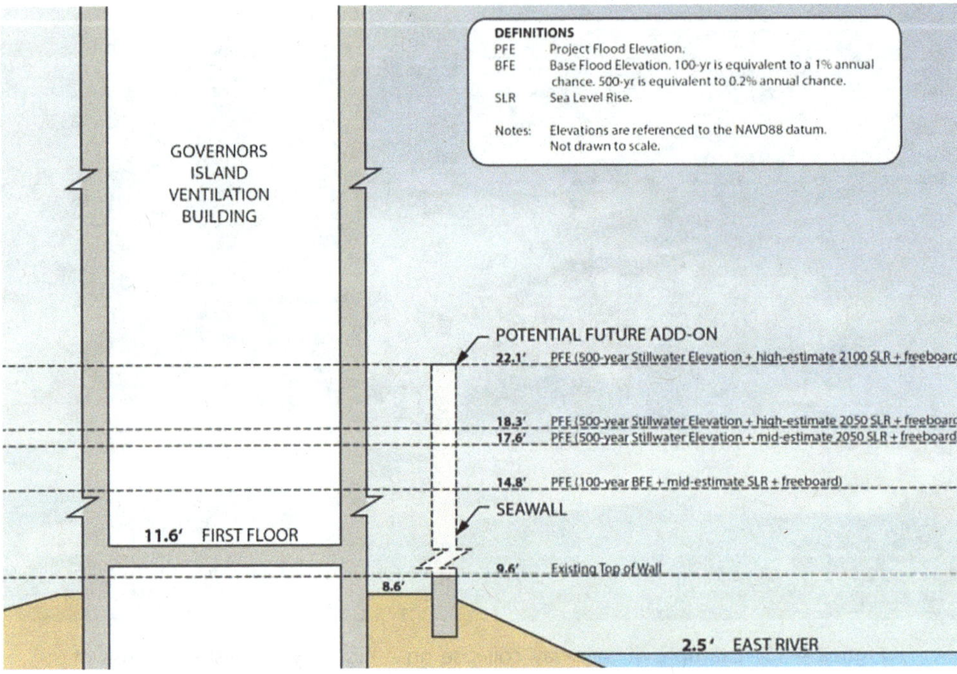

Figure 4-16. Project flood elevation diagram.
Source: FHWA (2017b).

Transportation Authority recommended that the flood wall be designed to be easily raised and strengthened incrementally to accommodate future flooding (FHWA 2017b).

Knock-on Effects

As discussed in Chapter 2, there are potential knock-on effects of climate stressors, secondary, indirect, or cumulative effects that can enhance degradation, produce new stressors, or reinforce other stressors. Under conditions of stationarity, engineers were able to use historical performance information to identify and address knock-on effects in their designs. Now, as climate stressor variable values become more extreme, it is essential for engineers to reconsider potential knock-on effects and address possible impacts.

Compounding: Cumulative Effects of Climate Stressors

Shifts in the probability distributions of climate stressor values may create unanticipated deterioration as variable values, once handled as infrequent occurrences, become considerably more probable and in increasing frequency. For example, paved roadways are built to withstand occasional flooding. However, in the case of multiple flood events occurring at shorter intervals, there are additional chances for flood waters to seep into cracks and weaken the pavement sub-base. This can reduce its load-bearing capacity and support to the surface pavement. Also, an increase in freeze/thaw cycles can bring about unexpected damage to foundations and underground utilities (FHWA 2014).

Compounding, as defined in this work, differs from the concept of compound disasters, defined by the National Academies (NASEM 2022) as “two or more extreme events occurring simultaneously.” For our purposes, compounding focuses on increasing frequencies of a single climate stressor event and its effects. Compound disasters are complex, the result of combinations of multiple hazards (e.g., floods, heat waves, droughts) caused by multiple drivers (e.g., severe storms, cold fronts, high river discharge) occurring simultaneously, in sequence, or in different locations (Zscheischler et al. 2020). Additional research on compound disasters is now underway.

Propagating: Two or More Climate Stressors Combine to Create New Stressors

Shifts in the probability distributions of climate stressor variable values may create instances where two or more stressors combine to reach a tipping point at which new stressors are created. For example, because of the loss of ground cover, soils and vegetation do not absorb stormwater runoff following a wildfire event. After a wildfire, fire-scorched hydrophobic soils let the stormwater run unchecked, bringing with it an assortment of debris, firefighting chemicals, and other pollutants. Debris can clog sewers, create flooding, and disrupt water treatment processes (Patrick 2013).

Reinforcing: Amplification of the Effects of Two or More Climate Stressors

At certain values, climate stressors can combine, amplifying the stressor effects. For example, rising sea levels, high tides, and storm surges can increase flooding. Also, winter storms in Alaska are more or less the norm for coastal villages. However, the loss of sea ice combined with extreme winter storms has caused high levels of coastal erosion. In the past, winter sea ice protected the shorelines from wind and waves from winter storms. As temperatures rise, resulting in less Arctic sea ice, unprotected shorelines are quickly eroded, destroying roads and damaging buildings. For instance, in 1997, the native community of Shishmaref lost 125 ft of beach in a single storm. Kivalina, Alaska, has tested several shoreline erosion measures, but all have failed. A rock revetment wall constructed by the US Army Corps of Engineers in 2010 (Figure 4-17) will give village residents 15 to 20 years of protection from rising seas before needing to relocate (US Climate Resilience Toolkit 2023). The US Army Corps of Engineers has identified more than 180 communities that are threatened by this scale of shoreline erosion (Coffey 2021).

Figure 4-17. Rock revetment wall protecting the village of Kivalina, Alaska, constructed by the US Army Corps of Engineers.
Source: https://www.flickr.com/photos/shorezone/ https://www.flickr.com/photos/shorezone/11471418433, Creative Commons BY 2.0, https://commons.wikimedia.org/w/index.php?curid=64385875.

Infrastructure Design Challenges in a Changing Climate

As a general rule, designers have three choices for dealing with stressor hazards, climate or otherwise. They can avoid the hazard by changing the project location or configuration so that the project is not exposed. They can also accommodate the hazard, designing the project so that the stressor has minimal effect across a broad range of stressor values. Finally, they can adapt, incorporating the ability to modify some of the physical elements of the project and make operational adjustments.

Adaptation Island or Orphaned Resiliency

In making infrastructure projects more resilient to climate stressor hazards, the US Department of Transportation (FHWA 2017a) advises owners and designers to consider the project surroundings carefully. While designed to adapt to changing climate conditions, the project can become an "adaptation island" or be the victim of "orphaned resiliency." In this work I chose to separate the two. In the case of an adaptation island, a bridge may be sufficiently climate resilient.

However, the roads leading to the bridge may be vulnerable to flooding, preventing access to bridge users and maintenance personnel. That same bridge could also be a victim of orphaned resiliency. The bridge may be designed to withstand extreme storms but may be exposed to hazards from other nearby facilities and structures that are not. For example, the bridge could be subject to collisions from upstream barges that were not adequately secured against extreme storm events.

How the choices—avoid, accommodate, and adapt—are applied depends to a large degree on the stressors that need to be addressed. The fact that climate stressors are changing adds to the complexity. Figure 4-18 depicts the infrastructure design challenges faced by engineers organized according to these new uncertainties.

- **Altering:** How are climate stressor values changing, and how will those changes affect my project components or systems? Are there analogous case examples showing the effects of extreme climate stressor values? Can I estimate what those effects might be? What are the risks associated with those extreme values?
- **Accelerating:** How fast are climate stressors changing? Is the rate of change accelerating? If I tried to adapt to these changes, how much time do I have? What adaptation techniques have others applied? How did they perform?
- **Compounding:** Are there climate stressors that can cause compounding failures at some levels of intensity, duration, or frequency (e.g., service disruptions, premature deterioration, or damage)? How has the potential for compounding

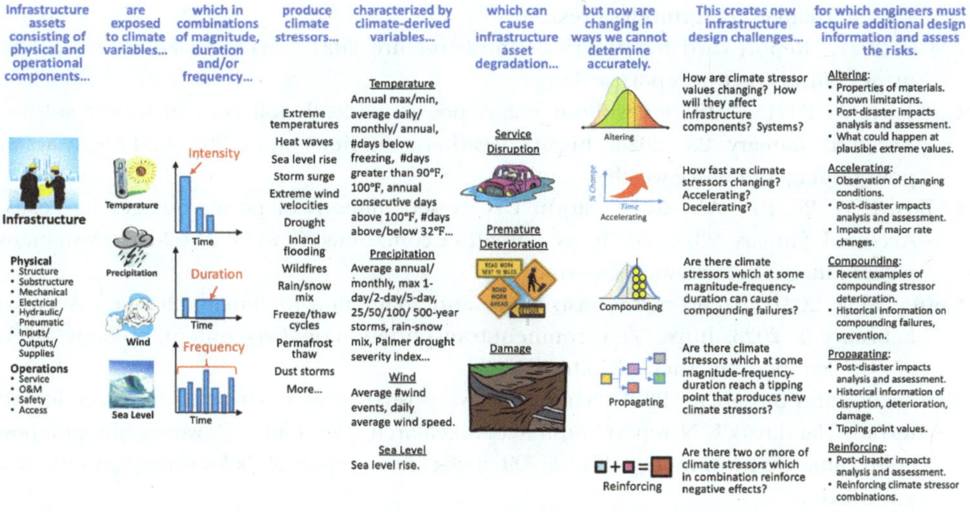

Figure 4-18. New infrastructure design challenges under conditions of nonstationarity.

degradation been handled in the past? Can previous approaches be scaled up or otherwise modified to prevent degradation? Are there examples of successful approaches?

- **Propagating:** Are there climate stressors that, at some levels of intensity, duration, or frequency, reach a tipping point that produces new climate stressors? What are those tipping points? What stressors were produced? What were their effects? How were they handled?
- **Reinforcing:** Are there two or more climate stressors that, in combination, reinforce adverse effects? What were the stressors? What were the effects? How were they handled?

As noted earlier, Appendix A lists the effects of climate stressors organized by infrastructure category and broken out by physical and operational impact. This list can be used as a starting point for identifying possible stressors. When available, altering, accelerating, and knock-on effects were identified.

While these lists of climate stressor effects are useful in identifying potential problems, what is needed is information on infrastructure asset performance under extreme conditions. Knowledge of infrastructure asset performance under extreme conditions is sparse, mostly buried in disaster damage and recovery reports. What is sorely needed is a detailed analysis of these after-action publications, extracting information on the causes and extent of the damage, what failed and what did not fail, and at what stressor values.

References

Anderson, R. C. 1998. *Mid-course correction: Toward a sustainable enterprise: The interface model*, 18–19. Atlanta: Peregrinzella Press.

ASCE. 2021. "Report card for America's infrastructure 2021." Accessed February 4, 2023. https://infrastructurereportcard.org/.

Childs, J. W. 2021a. "Millions without water, power as death toll rises in winter storms." Accessed January 23, 2024. https://weather.com/news/news/2021-02-17-texas-water-power-outages-snow-ice-weather-south.

Childs, J. W. 2021b. "Why winter storm Uri caused millions of power outages in Texas." Accessed January 23, 2024. https://weather.com/news/news/2021-02-16-why-so-many-power-outages-in-texas-winter-storm.

Coffey, M. J. 2021. "Arctic civil infrastructure and adaptation to climate change." Accessed February 6, 2023. https://environment.transportation.org/wp-content/uploads/2021/05/CC_Symposium_MichaelCoffey.pdf.

Dennis, B., and S. Kaplan. 2021. "Humans have pushed the climate into 'unprecedented' territory, landmark U.N. report finds." Accessed March 7, 2024. https://www.washingtonpost.com/climate-environment/2021/08/09/ipcc-climate-report-global-warming-greenhouse-gas-effect/.

Eiseley, L. C. 1954. "Man the fire-maker." *Sci. Am.* **191** (3): 52–57. https://doi.org/10.1038/scientificamerican0954-52.

FHWA (Federal Highway Administration). 2014. "Climate change sensitivity matrix." Accessed January 23, 2024. https://toolkit.climate.gov/tool/transportation-climate-change-sensitivity-matrix.

FHWA. 2017a. *Post Hurricane Sandy transportation resilience study in NY, NJ, and CT.* FHWA-HEP-17-097. Washington, DC: FHWA.

FHWA. 2017b. *Synthesis of approaches for addressing resilience in project development.* FHWA-HEP-17-082. Washington, DC: FHWA.

Godard, O. 2008. "The Stern review on the economics of climate change: Contents, insights, and assessment of the critical debate." Accessed January 23, 2024. http://journals.openedition.org/sapiens/240.

Holdren, J. P. 1993. "'A brief history of 'IPAT' (Impact = Population x Affluence x Technology)." Accessed January 23, 2024. https://mahb.stanford.edu/library-item/a-brief-history-of-ipat-impact-population-x-affluence-x-technology/.

IPCC (Intergovernmental Panel on Climate Change). 2012. "Summary for policymakers." In *Managing the risks of extreme events and disasters to advance climate change adaptation. A special report of Working Groups I and II of the Intergovernmental Panel on Climate Change,* 1–19. Cambridge, UK: Cambridge University Press.

IPCC. 2021. "Climate change 2021—The physical science basis." In *Contribution of Working Group I to the Sixth Assessment report of the Intergovernmental Panel on Climate Change.* Cambridge, UK: Cambridge University Press. https://doi.org/10.1017/9781009157896.

Kish, A., and M. Wesley. 2009. "Truck buckling research." Accessed March 19, 2024. https://rosap.ntl.bts.gov/view/dot/11985.

Levin, K. 2018. "Half a degree and a world apart: The difference in climate impacts between 1.5 °C and 2.0 °C of warming." Accessed March 7, 2024. https://www.wri.org/blog/2018/10/half-degree-and-world-apart-difference-climate-impacts-between-15-c-and-2-c-warming.

Lindsey, R. 2021. "Understanding the Arctic polar vortex." Accessed February 5, 2023. https://www.climate.gov/news-features/understanding-climate/understanding-arctic-polar-vortex.

Machlis, G. E., M. O. Román, and S. T. A. Pickett. 2022. "A framework for research on recurrent acute disasters." *Sci. Adv.* **8** (10): eabk2458. https://doi.org/10.1126/sciadv.abk2458.

McKibben, B. 2010. *Earth: Making a life on a tough new planet,* 1. New York: Henry Holt.

NASEM (National Academies of Sciences, Engineering, and Medicine). 2022. "Resilience for compounding and cascading events." Accessed January 23, 2024. https://doi.org/10.17226/26659.

Patrick, J. 2013. "Wildfires threaten stormwater runoff & water quality." Accessed February 6, 2023. https://www.mswmag.com/online_exclusives/2013/11/wildfires_threaten_stormwater_runoff_water_quality.

Revkin, A. C. 2007. "Stuck on coal, and stuck for words in a high-tech world." Accessed January 23, 2024. https://www.nytimes.com/2007/12/04/science/earth/04comm.html.

Ritchie, H., L. Rodés-Guirao, E. Mathieu, M. Gerber, C. Appel, and C. Giattino. 2023. "Population growth." Accessed January 23, 2024. https://ourworldindata.org/world-population-growth.

Roser, M. 2014. "Human Development Index (HDI)." Accessed September 1, 2021. https://ourworldindata.org/human-development-index.

Roser, M., and H. Ritchie. 2023. "How has world population growth changed over time." OurWorldInData.org. Accessed March 7, 2024. https://ourworldindata.org/population-growth-over-time.

Stern, N. 2007. *The economics of climate change: The stern review.* Cambridge: Cambridge University Press. https://doi.org/10.1017/CBO9780511817434.

Turchetta, D. 2010. "Taking stock: Climate change and transportation." FHWA Public Roads. Accessed March 19, 2024. https://highways.dot.gov/public-roads/marchapril-2010/taking-stock-climate-change-and-transportation.

University of California, Davis. 2021. "Polar vortex." Accessed February 5, 2023. https://www.ucdavis.edu/climate/definitions/what-is-the-polar-vortex.

US Climate Resilience Toolkit. 2023. "Relocating Kivalina." Assessed October 29, 2023. https://toolkit.climate.gov/case-studies/relocating-kivalina.

US Energy Information Administration. 2023. "US energy facts explained." Accessed October 29, 2023. https://www.eia.gov/energyexplained/us-energy-facts/.

USGS (US Geological Survey). 2018. "Hydroelectric power water use." Accessed July 16, 2022. https://www.usgs.gov/special-topics/water-science-school/science/hydroelectric-power-water-use.

WRI (World Resources Institute). 2018. "Unlocking the inclusive growth story of the 21st century: Accelerating climate action in urgent times." Accessed March 24, 2023. https://newclimateeconomy.report/2018/wp-content/uploads/sites/6/2018/09/NCE_2018_FULL-REPORT.pdf.

Zscheischler, J., O. Martius, S. Westra, E. Bevacqua, C. Raymond, and R. M. Horton. 2020. "A typology of compound weather and climate events." *Nat. Rev. Earth Environ.* **1** (7): 333–347. https://doi.org/10.1038/s43017-020-0060-z.

CHAPTER 5

Responding to a Changing Climate

"We basically have three choices – mitigation, adaptation and suffering. We're going to do some of each. The question is what the mix is going to be. The more mitigation we do, the less adaptation will be required, and the less suffering there will be."

John Holdren, former White House Science Advisor

The Work Ahead

As emphasized in the previous chapters, society faces an existential threat: a climate becoming increasingly harsh and producing highly damaging weather events throughout the globe. The global warming tipping point is 1.5 °C (2.7 °F), beyond which damages are projected to increase rapidly (IPCC 2018). Some changes in climate conditions may become irreversible. Table 5-1 shows the succession of responses and strategies available to society for reducing the causes and impacts of a changing climate (NRC 2015b).

To make the necessary changes while protecting and preserving the built environment from more frequent and more severe disasters, society needs to achieve the following in ways that are fair and equitable:

- **Mitigation:** *Reduce greenhouse gas (GHG) emissions, achieving net-zero emissions by 2050.* Limiting global warming to 1.5 °C (2.7 °F) will require extensive carbon mitigation measures: radical improvements in energy efficiency and conservation, capturing carbon emissions, and replacing fossil fuels with energy sources that do not emit carbon. Technologies at various levels of development hold the promise of capturing carbon emissions at the source or removing carbon from the atmosphere. However, at this time, these technologies are very expensive. It may be feasible to reflect more of the sun's energy input by modifying the Earth's albedo, a measure of the ability of the Earth's surface to reflect sunlight. Albedo modification is also costly and has many serious risks: technological, environmental, and sociopolitical. But if other mitigation efforts fail, this may be the world's only option.
- **Adaptation:** *Protect infrastructure assets, existing and planned, from climate-induced deterioration and damage.* While carbon emissions mitigation is underway, society must maintain economic development and human well-being by adapting civil infrastructure to the anticipated changes in climate. Because of the resident

Table 5-1. Portfolio of Responses and Strategies for Reducing Climate Risk: Carbon Emissions Mitigation and Climate Adaptation.

Order	Causal chain of the human-climate system	Strategies for diminishing climate risk
1	Desire for improved well-being	Conservation/less consumption
2	Consumption of goods and services	Efficiency, improved technology
3	Consumption of energy	Low-carbon energy
4	CO_2 emissions reduction	Capture carbon at source
5	CO_2 in atmosphere	CO_2 removal from atmosphere *Possible albedo modification*
6	Changes in climate system	Adaptation to climate impacts, retreat
7	Impacts on humans and ecosystems	Resilience, recovery, retreat

Source: Adapted from the National Research Council (NRC 2015b).

> lifetimes of GHGs in the atmosphere, more frequent extreme weather and climate conditions are expected to continue throughout the 21st century and beyond. New infrastructure assets must be designed to handle these harsh conditions. Existing assets must be refurbished, providing additional protection from extreme conditions. They need to be made more durable and resilient and incorporate means for adaptation. Finally, infrastructure assets must be relocated or abandoned if additional protection, robustness, and adaption efforts are found to be inadequate.

For the United States, along with the rest of the world, the construction and operation of civil infrastructure have been and continue to be the primary source of carbon emissions. And, as shown in previous chapters, carbon is the major component of the US ecological footprint. The same applies to the rest of the world's developed and developing nations. Therefore, reducing carbon, both embodied carbon and carbon emissions, while coping with and adapting to a changing climate must be society's primary goal.

Ideally, achieving that goal should not be at the expense of society's standard of living. Figure 5-1 shows, nations, including the United States, have been able to reduce their ecological footprints while maintaining or improving their Human Development Index (HDI) scores. Although the work required to reduce the US ecological footprint may appear daunting, such reductions are achievable without reducing the US quality of life.

Figure 5-2 offers a close-up view of the US pathway relative to the sustainability quadrant between 1973 and 2014. As discussed throughout this book, increases in the US ecological footprint are driven primarily by carbon emissions, mainly produced by civil infrastructure. These changes are represented by vectors that can be thought of as the contributions of individual infrastructure projects, either increasing or reducing the ecological footprint and the HDI (quality of life). The current position of the United States is an aggregation of multiple vectors, each with

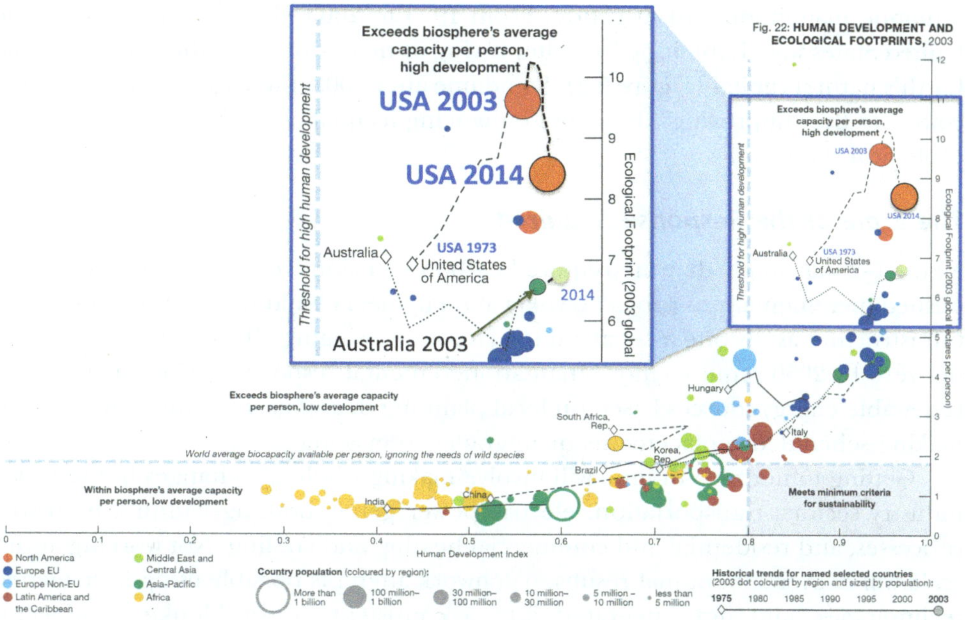

Figure 5-1. Positions of nations relative to the sustainability quadrant; historical trajectories of progress.

Source: Adapted from Global Footprint Network (2006), Creative Commons BY-SA 4.0.

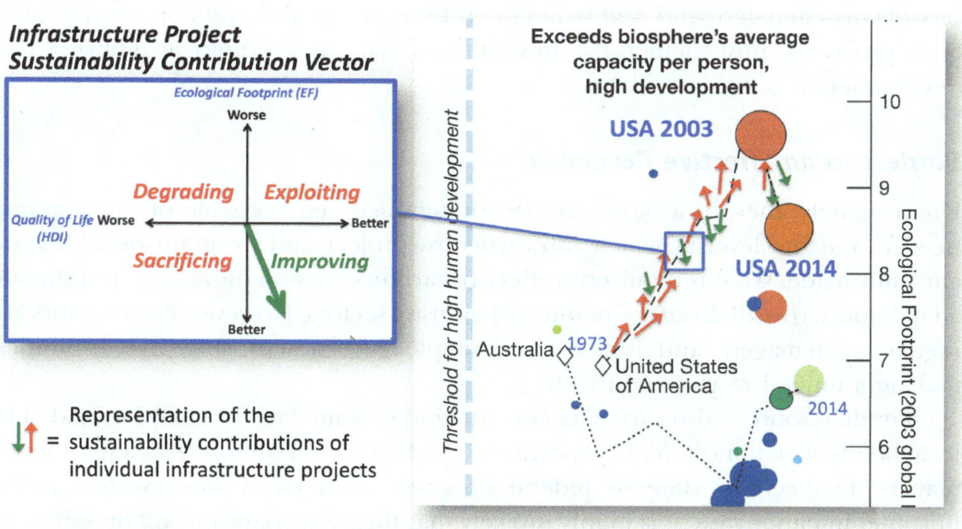

Figure 5-2. Closeup of US progress in development with example depictions of vector contributions by individual infrastructure projects.

Source: Adapted from Global Footprint Network (2006), Creative Commons BY-SA 4.0.

a unique magnitude and direction. From 1973 to 2003, the net direction for the United States was "Exploiting," i.e., improving its quality of life at the expense of the Earth's natural capital (Figure 5-2). Sometime after 2003, the United States changed course to the "Improving" direction, enhancing its quality of life while reducing its ecological footprint.

The Scale of the Response Required

Figure 5-2 may imply that the required carbon emissions reductions can be accomplished incrementally as project owners and engineers build new or retrofit existing infrastructure assets one asset at a time. However, bringing GHG emissions down to net zero by 2050 requires more than an incremental response: a slow migration to renewable energy sources based on local planning decisions, building new assets, or making scheduled replacements or upgrades to existing assets.

Getting to net zero in time will involve making wholesale changes in the major industry sectors: transportation, electric power generation, agriculture, industrial processes, and residential and commercial heating and cooling. Not working at this scale will deliver suboptimal results, patchwork attempts possibly using mismatched technologies, and not integrated into a consistent, forward-looking full-sector response. For example, local efforts to instigate a switch from gas- to electric-powered vehicles may have some success. But adding just a few vehicle-charging stations and dedicated parking slots in a few communities is not enough to precipitate the scale of change needed in the transportation sector. What is required are national programs for migration: design, construction, and operation of vehicle-charging facilities at scale. Fast charging stations with standard connectors are conveniently placed, broadly dissemmated, and well maintained for efficient and cost-effective mobility. Such programs must include the flexibility to adapt as technologies improve and environmental conditions change.

Barriers to an Effective Response

Unfortunately, there is a significant disconnect between the scale of the response needed and the level at which infrastructure project and program decisions are currently made. What is required is effective carbon emission mitigation and climate adaptation across all the aforementioned industry sectors. However, these sectors are regulated, managed, and funded by multiple government and private entities, making a unified response difficult.

Funding sources also vary. The construction of some facilities may be funded by local bonds, with bondholders expecting a good return on their investments. Others may be financed by state or federal programs to achieve some sector-specific purpose. In either case, it is highly unlikely that these mechanisms will be sufficient to reach the overall goal of net zero by 2050 or to make US infrastructure sufficiently climate resilient. A unified national program is required. And, if the US response to

COVID-19 is any example, reaching national unity on a relatively nebulous threat like climate change will be very difficult.

At the global level, the chances of reaching unity in purpose, much less taking effective action, are much the same. As noted in Chapter 3, nations' priorities vary even to the exclusion of climate change. Climate change is not a priority issue if one's survival is day to day. Moreover, there is a global moral and ethical question. The Brundtland Commission Report (UN 1987) assumed that the serious threat to the success of future generations would be sufficient to drive behavioral change in the people and organizations responsible for the current unsustainable conditions. However, that assumption is based on the hypothesis that the current generation cares enough about the success of future generations to alter current behaviors. Based on current commitments by nations to reduce carbon emissions, the current generation does not appear to care enough. According to the Climate Action Tracker (CAT 2023), current national policies and actions will result in approximately 2.7 °C (4.9 °F) warming by 2100, well above the 1.5 °C (2.7 °F) target.

Responding to a Changing Climate: The Details

As John Holdren pointed out in the opening quote for this chapter, the alternatives to extended human suffering from extreme weather events are climate change mitigation and adaptation. These choices and their associated strategies and actions are shown in Figure 5-3.

Climate Change Mitigation

Mitigation involves attacking the causes of climate change: reducing the amount of solar radiation the Earth receives and reducing the quantities of GHGs retained in the atmosphere.

Points of Intervention

For mitigation, there are three points of intervention, as illustrated in Figure 5-4.

1. **Greenhouse Gas Emissions Reduction:** Reducing the quantities of GHGs released at the sources, facilities, and systems burning fossil fuels for heat, power, and mobility.
2. **Greenhouse Gas Removal:** Capturing and storing or destroying GHGs at the point of release and after they are released.
3. **Solar Radiation Management:** Reducing solar energy input by limiting the amount of solar energy received or increasing the amount reflected back into space.

Alternatives that attempt to alter the Earth's atmosphere or modify solar energy input are classified as climate intervention or geoengineering.

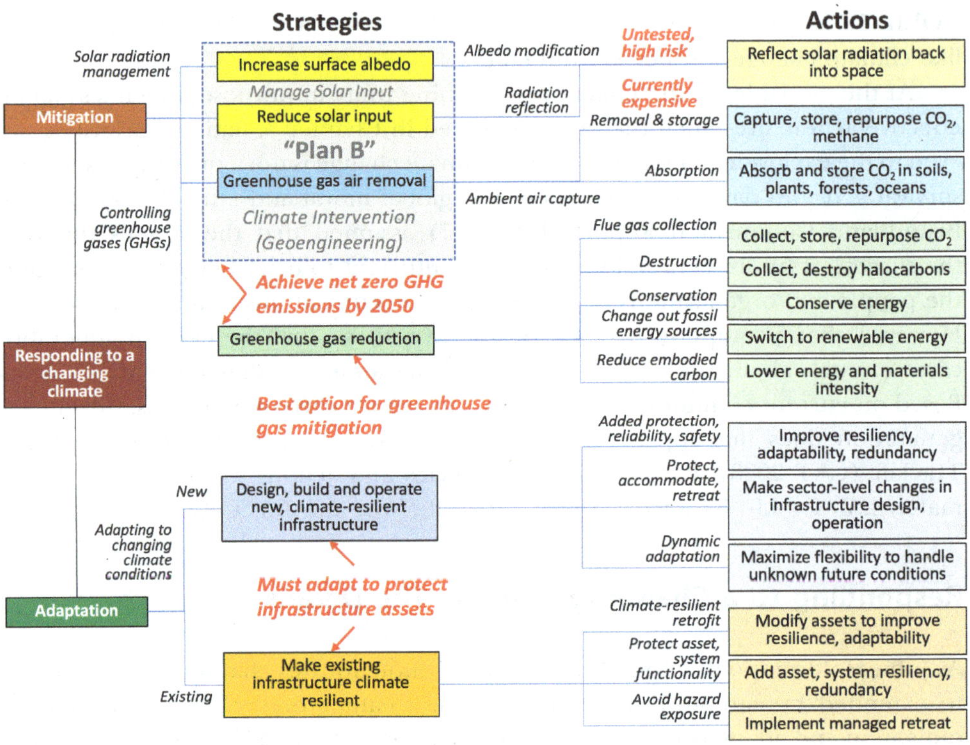

Figure 5-3. Responding to a changing climate: range of responses, strategies, and actions.

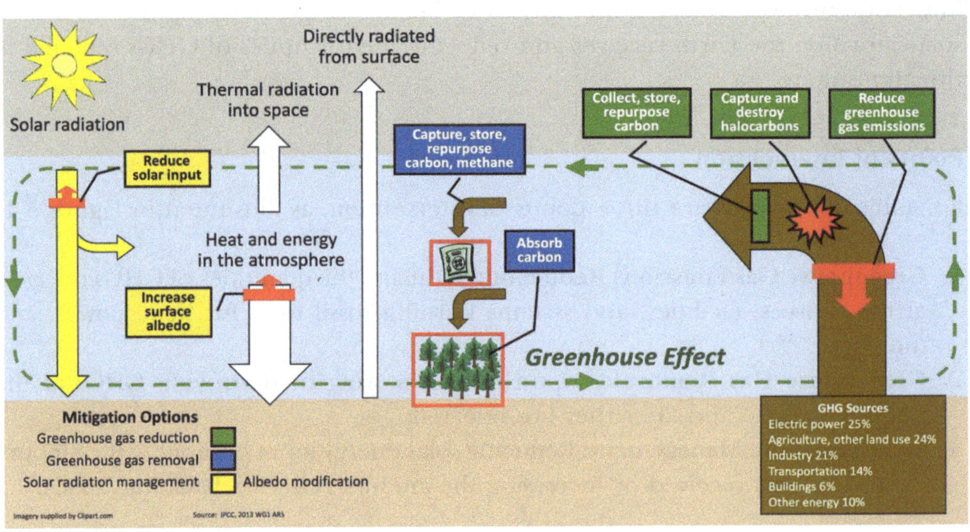

Figure 5-4. Mitigation options. Points of intervention to reduce greenhouse gas (GHG) emissions, reduce solar energy input, and remove GHGs from the atmosphere.

The most effective form of mitigation is through GHG emissions reduction, reducing or eliminating emissions at the source. However, the required source reductions are substantial. None of the other options are attractive. Some may be dangerous because of unintended consequences.

Embodied Carbon and Operational Carbon

When considering carbon emissions reduction, two sources need to be addressed. The most recognized source is operational carbon, that is, GHGs measured in carbon equivalents resulting from facility processes and fossil fuel-based energy generation. The other source is embodied carbon: GHGs released during infrastructure asset life-cycle stages other than actual operation. These stages include materials extraction or harvesting, manufacturing, transportation, construction or installation, maintenance, renovation, and end of life. There are also GHG releases during carbon capture and storage or sequestration. An example of both operational and embodied carbon releases across the life cycle of a building is pictured in Figure 5-5.

Embodied carbon from materials and construction worldwide may be as high as 30% of total GHG emissions. About 40% of that component comes from buildings. Moreover, as the operational energy efficiency in buildings increases, the portion of embodied carbon will also increase.

Not surprisingly, steel and concrete, materials common to most infrastructure projects, are the highest contributors to embodied carbon emissions. These materials are used extensively worldwide and are manufactured through energy-intensive processes, including chemical reactions that produce carbon dioxide (CLF 2020). Carbon dioxide is released in cement manufacture as calcium carbonate ($CaCO_3$) is converted into lime (CaO), the primary ingredient in cement.

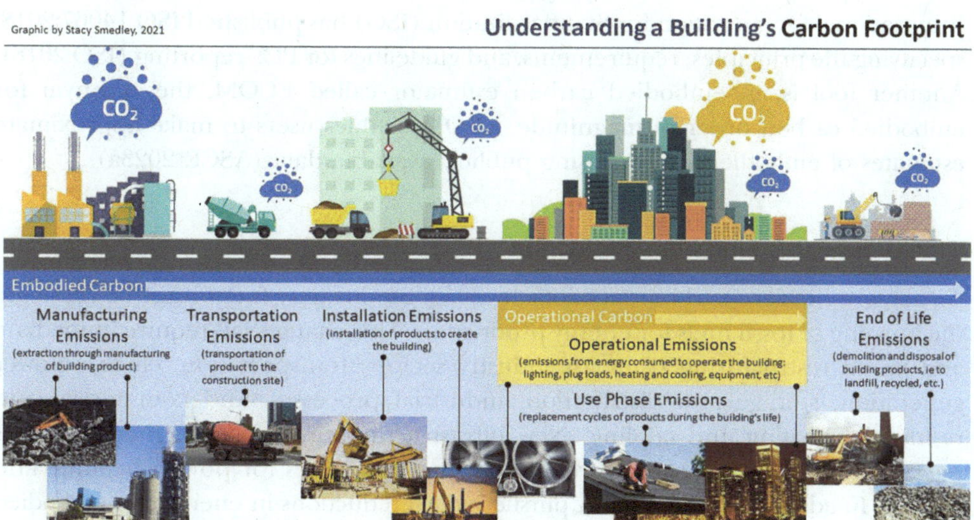

Figure 5-5. Embodied and operational carbon generation across a building's life cycle. *Source*: Courtesy of Stacy Smedley. Reprinted with permission.

The construction industry is now starting to pay attention to this significant source of carbon emissions. This industry has essential roles in embodied carbon reduction: reducing the quantity of high embodied carbon materials used in construction and creating a demand for lower-carbon products (CLF 2020). Accordingly, materials suppliers are keenly interested in having the carbon footprint of their products quantified. Their reasons include reputation, market access, and compliance with expected future regulations (De Schryver and Zampori 2022).

Obtaining reliable and usable information on the embodied carbon of construction materials is becoming easier. In 2011, the Building Services Research and Information Association in conjunction with the University of Bath published *Embodied Carbon: The Inventory of Carbon and Energy (ICE)* (Hammond and Jones 2011) containing approximately 1,800 records of embodied carbon and energy for 34 construction materials. The guide reports both embodied energy and embodied carbon, noting that the quantity of embodied carbon per unit of material depends on the energy sources used to produce the material. According to the authors, the information was sourced from publicly available databases as opposed to proprietary databases (Hammond and Jones 2011).

ASCE is also responding. ASCE's Structural Engineering Institute's Sustainability Committee has developed the SE 2050 Commitment Program, a program "designed to ensure substantive embodied carbon reductions in design and construction of structural systems by the collective structural engineering profession" (ASCE 2023b).

Tools for calculating embodied carbon are available today. One such tool is the Embodied Carbon in Construction Calculator (EC3), a free beta-version of a cloud-based tool that allows benchmarking, assessment, and reductions in embodied carbon (Charles Pankow Foundation 2020). Recognizing the need for sound information, organizations are developing product carbon footprint (PCF) standards. The International Organization for Standardization (ISO) has published ISO 14067:2018, specifying the principles, requirements, and guidelines for PCF reporting (ISO 2018). Another tool is an embodied carbon estimator called ECOM, the acronym for embodied carbon order of magnitude. ECOM enables users to make approximate estimates of embodied carbon using publicly available data (ASCE 2023a).

Greenhouse Gas Emissions Reduction

To achieve net-zero carbon emissions by 2050, the primary focus must be reducing the burning of fossil fuels for energy production. This change will require immediate and broad restructuring of major industry sectors: transportation, electric power generation, storage and distribution, industrial processes, and commercial and residential heating and cooling. New infrastructure projects emerging from this restructuring must incorporate renewable energy sources for power, heating, and cooling. In addition, projects must pursue radical reductions in energy use, embodied energy in materials, and materials intensity. Existing infrastructure assets must be refurbished, converting fossil energy sources to low- or no-carbon sources. Infrastructure

Table 5-2. Global Warming Potential for Greenhouse Gases.

Greenhouse gas	Composition	Lifetime	GWP_{20}	GWP_{100}
Carbon dioxide	81%	100 years	1	1
Methane	10%	12.4 years	86	34
Nitrous oxide	7%	114 years	268	298
Hydrofluorocarbon (HFC-134a)	3%*	13.4 years	3,790	1,550
Chlorofluorocarbon (CFC-11)		45 years	7,020	5,350
Carbon tetrafluoride (CF_4)		50,000 years	4,950	7,350

GWP_{xx} = Global warming potentials after XX years' residence time in the atmosphere: 20 and 100 years.

Global warming potential is a measure of how much energy the emissions of 1 ton of a gas will absorb over a given period of time, relative to the emissions of 1 ton of carbon dioxide (CO_2).

Source: USEPA (2022).
* includes all fluorinated gases.

assets that cannot convert to such energy sources must find cost-effective ways to capture and store carbon emissions or secure carbon emissions credits or offsets.

Although it is the most prevalent, carbon dioxide is not the only greenhouse gas of concern, nor is it the most potent (Table 5-2). Methane emissions account for 10% of the greenhouse gases but have a global warming potential (GWP) many times that of carbon dioxide. Nitrous oxide makes up 7%, but its GWP is many times more than methane. The fluorinated gases (hydrofluorocarbons and chlorofluorocarbons) make up about 3% of the greenhouse gases but have GWPs an order of magnitude above nitrous oxide. These gases are being phased out through international agreements (USEPA 2023).

Carbon Dioxide

Carbon dioxide is naturally removed from the atmosphere through absorption by land- or ocean-based plants and is exchanged between the atmosphere and the ocean as part of the carbon cycle. Carbon dioxide emissions can be reduced by reducing the use of fossil fuels for energy production and by reducing energy demand. For civil infrastructure, reducing carbon dioxide emissions involves the reduction of both embodied and operational carbon. In addition to replacing fossil energy sources with renewable energy sources, designers should consider optimizing material salvage and reuse and the use of low embodied carbon materials (USEPA 2024).

Methane

Methane (CH_4) accumulates in underground coal beds and natural gas deposits. In the Arctic, permafrost and glaciers form a cap over these deposits, trapping methane gas and preventing it from leaking into the atmosphere. As glacial ice retreats and the permafrost melts, methane gas is released from these deposits into the atmosphere (Anthony et al. 2012). Methane is also emitted from livestock and certain agricultural practices, such as rice cultivation using flooded fields. Municipal solid waste landfills also emit methane as organic waste decays. Last, methane is released into the atmosphere in the production and transport of coal, oil, and natural gas. Practices for reducing methane releases include collecting landfill gases, changing agricultural practices, and upgrading oil and gas production, storage, and transport equipment to reduce methane gas leaks.

Nitrous Oxide

Nitrous oxide (N_2O) is a natural component of the atmosphere, part of the Earth's nitrogen cycle. Soil management in agriculture accounts for almost 80% of the nitrous oxide emissions. Activities such as fuel combustion, wastewater treatment, and industrial processes make up the rest. Nitrous oxide emissions can be lowered by reducing the application of, or by more efficient use of nitrogen-based fertilizers. Reductions in fuel use can limit nitrous oxide releases from fuel combustion in transportation and industrial processes.

Fluorinated Gases

Fluorinated gases such as those listed in Table 5-2 are manufactured substances used in refrigerants, aerosol propellants, fire retardants, and certain industrial processes (e.g., aluminum and semiconductor manufacturing). These gases are also ozone-depleting substances, and are being phased out through an international agreement, the Montreal Protocol. Emissions of fluorinated gases can be reduced by better management: reducing leaks and materials substitution.

The Montreal Protocol

The Montreal Protocol is a treaty signed in 1987 by all the world's nations that regulates nearly 100 ozone-depleting substances. The regulated substances include chlorofluorocarbons (CFCs), halons, other fully halogenated CFCs, carbon tetrachloride, methyl chloroform, hydrochlorofluorocarbons (HCFCs), methyl bromide, and hydrofluorocarbons. The production and consumption of these substances are phased out in steps with different timetables for developing and developed nations. The treaty is continually amended and adjusted based on new scientific, technical, and economic developments (UNEP 2023).

Greenhouse Gas Removal

If GHGs cannot be reduced or eliminated, other options are available for removal and destruction. However, at this time, these options are much more costly and energy-consuming than emission reduction. Carbon dioxide and methane can be captured and stored for possible reuse. However, the US National Research Council notes that carbon dioxide removal and sequestration processes are not good substitutes for mitigation and adaptation.

Table 5-3. Capture and Destruction or Reuse of Methane, Nitrous Oxide, and Fluorinated Gases.

Greenhouse gas	Emissions reduction, capture, destruction, and reuse
Methane	Methane gas can be recovered from coal mines. Emission controls can be placed in landfills to capture methane. Once captured and stored, methane can be used as a fuel. However, it does produce carbon dioxide upon combustion. Capturing methane released from permafrost or glacial caps will be very difficult because of the wide dispersion of potential release points
Nitrous oxide	Nitrous oxide emissions can be lowered by (1) reducing the use of nitrogen-based fertilizers, (2) modifying farm manure management practices, (3) reducing fossil fuel combustion, (4) making technological upgrades to industrial processes, and (5) using catalytic converters in combustion exhausts
Fluorinated gases	Fluorinated gas emissions can be reduced through (1) better handling of fluorinated gas refrigerants and (2) finding fluorinated gas alternatives, and fluorinated gas recycling and destruction. The goal is to phase out the manufacture and use of fluorinated gases

Source: USEPA (2022).

Capture and Destruction or Reuse of Methane, Nitrous Oxide, and Fluorinated Gases

Methods for handling methane, nitrous oxide, and fluorinated gas releases are summarized in Table 5-3.

Figure 5-6 shows how carbon dioxide can be captured, stored, and processed. Four processes for the capture, storage, and possible repurposing of carbon dioxide are presented in Figure 5-7 (IPCC 2005). These processes are summarized as follows.

- **Post-combustion:** Flue gases produced from fossil fuel combustion are processed to remove carbon dioxide and release nitrogen and oxygen.
- **Pre-combustion:** Fuels are processed in a reactor using steam and air or oxygen to produce a gas consisting of carbon monoxide and hydrogen. In a second reactor, carbon monoxide is mixed with steam to produce hydrogen and carbon dioxide. These gases are then separated. Carbon dioxide is stored. Hydrogen is burned to produce power and heat.
- **Oxyfuel Combustion:** Oxygen is used instead of air for fossil fuel combustion, producing a flue gas: mostly water vapor and highly concentrated (>80%) carbon dioxide. This process requires oxygen separation from the air, producing high-purity (95%–99%) oxygen.
- **Industrial Processes:** Several industrial applications, such as hydrogen production, already capture carbon dioxide. Carbon dioxide is also separated from raw natural gas (IPCC 2005).

Carbon Storage

Once captured, carbon dioxide can be compressed and transported to locations for long-term storage. Three storage options are

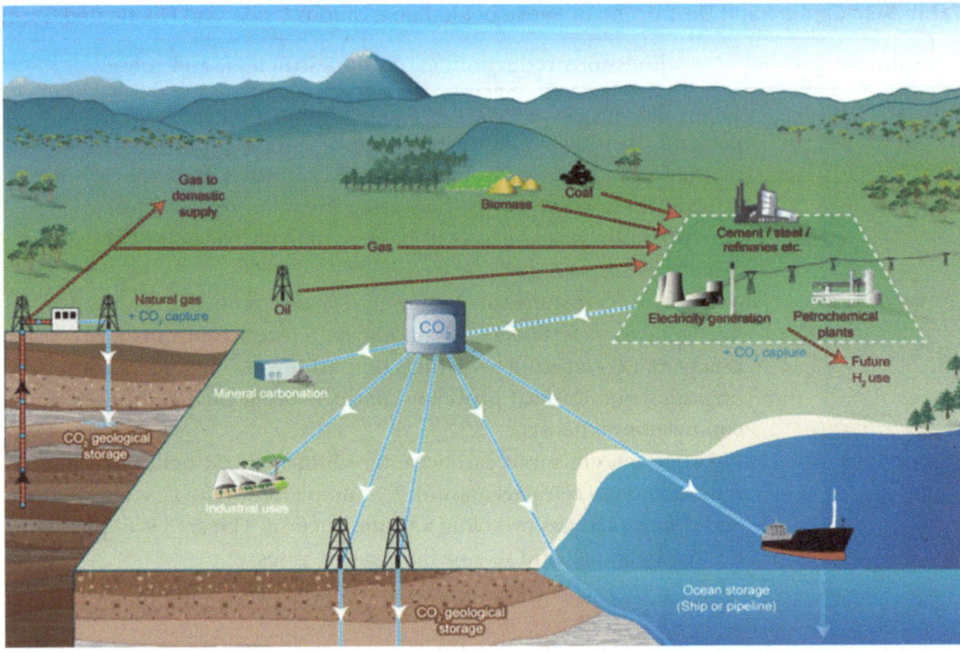

Figure 5-6. Capture, transport, and storage of carbon dioxide.
Source: Figures SPM.1 and TS.1 from IPCC (2005). Reproduced with permission.

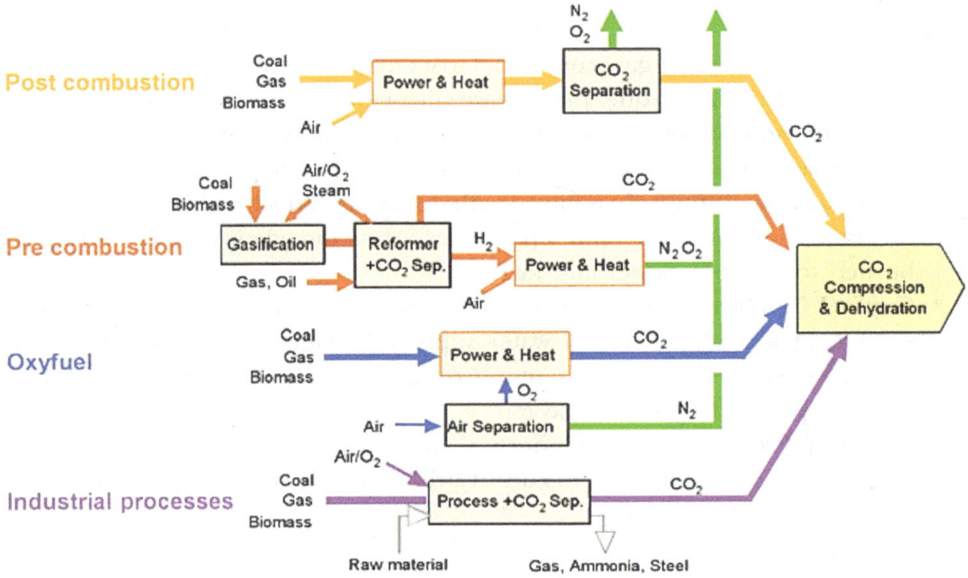

Figure 5-7. Processes for capturing and storing carbon dioxide.
Source: IPCC (2005), Figure TS.3. Reproduced with permission.

- **Geological Storage:** Carbon can be captured and stored in deep geological formations (e.g., oil and gas fields, coal beds, and deep saline formations).
- **Ocean Storage:** This form of storage involves deep discharge of carbon dioxide into the ocean, where it will dissolve, forming heavier than water plumes or hydrates (Pires et al. 2011). Carbon dioxide can also be injected into deep geological formations below the ocean bottom.
- **Industrial Fixation:** A mineral carbonation process creates a carbonate from certain metal oxides (IPCC 2005).

Repurposing Carbon Dioxide

One problem in capturing and repurposing carbon dioxide is that the molecule itself is highly stable, thus requiring significant energy to break it down to create new substances. Mineral carbonation has been offered as a way of sequestering carbon dioxide and reusing the carbonate product in road construction (IPCC 2005). Another suggestion is the use of carbon dioxide as a feedstock in chemical processes requiring carbon.

One form of repurposing is not helpful. Carbon dioxide is being used today to improve gas recovery in depleted gas reservoirs. The practice produces previously sequestered gases that can be converted into fossil fuels (IPCC 2005).

Cost of Carbon Capture

Currently, the cost of carbon capture and storage in electrical power production is expensive relative to the cost of electricity. According to the IPCC, adding carbon capture technology, the costliest element, would add significant costs for electric power.

- **Pulverized Coal Power Plant:** 1.8 to 3.4 US$ ct per kWh (cents per kilowatt-hour).
- **Integrated Gasification Combined Cycle Coal Power Plant:** 0.9 to 2.2 US$ ct per kWh.
- **Natural Gas Combined Cycle Power Plant:** 1.2 to 2.4 US$ ct per kWh (IPCC 2005).

These costs are considerable. The price of electricity in the United States at the time of the referenced report (IPCC 2005) was approximately US$0.08 per kWh.

Carbon Capture from the Air

Although atmospheric carbon dioxide concentrations are low, technologies are available for removing carbon dioxide from the air. Some options studied by the US National Research Council are summarized in Table 5-4. Direct air capture involves drawing ambient air through absorbers to remove carbon dioxide, after which it is stored or reused. At this writing, a carbon capture plant began operating in southwest Iceland. Powered by geothermal energy, the plant can remove 4,000 tons

Table 5-4. Alternatives for Carbon Capture.

Option	Description	Effectiveness	Readiness	Cost	Discussion
Direct air capture	Machines that pull carbon dioxide out of the air	Med	Low	High	Immature technology. Will take decades to deploy. Results easily verifiable. Energy intensive. Low risk of environmental, economic, or social impacts. The land area needed is high
Reforestation, afforestation	Reforesting recently deforested land or afforesting land deforested more 50 years before	High	High	Med	Easily and quickly deployed. Results easily verifiable. Many environmental co-benefits, but also limitations: decreased biodiversity, competition for agricultural land. Large land areas are needed
Ocean fertilization	Adding substances such as iron filings to stimulate plankton that metabolize carbon dioxide	Unknown	Low	Low	Immature technology. Will take decades to deploy. Low carbon capture rate. Difficult to verify. Costs are low. Unknown effects on ecosystems. Currently, there is an international moratorium on ocean fertilization
Bioenergy with carbon capture and storage (BECCS)	Agricultural waste is burned and buried to lock up carbon in the soil	Low	Med	Med	Low effectiveness. Can handle about 1% of fossil fuel resources. Some environmental risks. Large land areas are needed. Transportation of biomass to processors is a limiting factor. Competition for agricultural land also limits application

Source: National Research Council (2015).

of carbon dioxide from the air annually, permanently holding the captured gas in basalt rock formations nearby (IPCC 2005).

The problem with this approach is that because of the low ambient air concentrations of carbon dioxide, significant energy is required for carbon capture and separation. Large surface areas are also necessary to capture meaningful quantities of carbon dioxide (NRC 2015a).

Afforestation or reforestation involves the restoration of forests that have been deforested. The difference between the two is time. Afforestation is the restoration of forests that were deforested for 50 or more years. Reforestation means less than 50 years. While regrowing forests effectively captures and sequesters carbon, preventing deforestation is a much better approach (NRC 2015a).

Ocean fertilization involves stimulating the growth of phytoplankton in specific regions of the ocean. Under the right conditions, phytoplankton growth could increase the supplies of nitrogen and phosphorus. This increase would induce algae and other microscopic plants to metabolize carbon dioxide. After blooming, these plants would die and sink to the ocean bottom, taking carbon dioxide with them. Modeling studies suggest that such a fertilization process could be effective, but its long-term effects on marine ecosystems are unknown and potentially harmful (NRC 2015a).

Bioenergy with carbon capture and storage (BECCS) involves processing biomass, usually agricultural waste, burning it for heat or power, or converting it to fuels. The carbon dioxide created is captured and stored. Theoretically, BECCS could become a significant energy supplier. However, the land area required and the transportation of biomass to processing facilities are limiting factors.

Climate Intervention (Geoengineering)

Climate interventions are strategies and actions designed to change the climate by limiting the sun's energy input. Figure 5-8 illustrates various climate intervention possibilities. Although many federal organizations are studying these options for controlling global warming, the National Academies emphasize that carbon emissions reduction and adaptation should be the priority approaches for addressing climate change (NRC 2015b).

Solar Radiation Management Other approaches involve managing the amount of solar radiation entering the Earth's atmosphere and how much is reflected back into space. Solar radiation management options are listed in Table 5-5.

Space Reflectors Several proposals have been made to limit incoming solar radiation using large mirrors and sunshades. This concept builds on the fact that the Earth currently reflects about 30% of the incoming solar radiation. One idea is to place a large shield in a stationary Earth orbit. The shield would be 2,000 km in diameter and 10 μm thick, all assembled in space. Even at this thickness, the shield

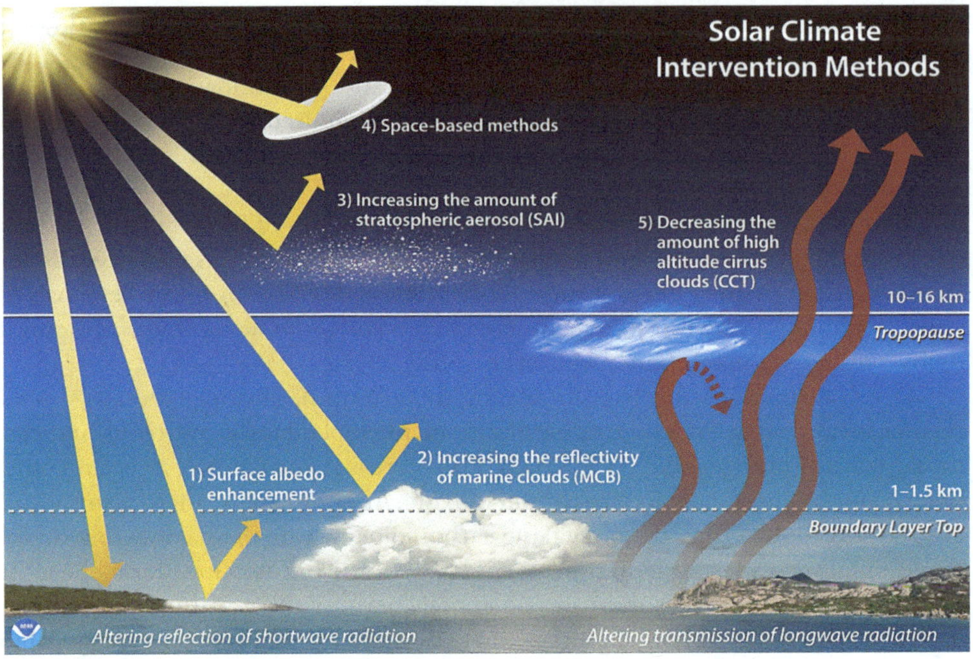

Figure 5-8. Proposed methods for climate intervention that would affect climate by modifying either incoming or outgoing solar radiation.
Source: Chelsea Thompson, NOAA/CIRES.

would weigh 100 Mt. Costs would be in trillions of US dollars. Given the current difficulty of equipment assembly and handling tasks inside and outside of the existing space station, the idea of transporting, assembling, and managing a shield so large seems wildly impractical.

Another plan for reducing solar radiation input is to place trillions of small objects in orbit, each 2 ft in diameter and about 13 μm thick. Together, they would form a cloud and reduce incoming solar radiation by 2%. Presumably, the solar shielding would be enough to offset global warming even with a doubling of the carbon dioxide concentration in the atmosphere. Assembly in space would not be required (Pontin 2007). However, once these objects are placed in orbit, they will be extremely difficult and costly to remove.

Stratospheric Aerosols Recognizing the cooling effect of volcanic eruptions and the resulting releases of aerosols into the stratosphere, proposals have been made to inject SO_2 particles into the air to create a similar cooling effect. Balloons or high-altitude aircraft could make the delivery. Like volcanos, the intentional release of aerosols can be highly effective. However, the potential side effects are many and severe: ozone depletion, unintended changes in precipitation and surface temperatures, ecosystem disruptions, and more.

Table 5-5. Solar Radiation Management Options.

Option	Description	Effectiveness	Readiness	Cost	Discussion
	Reduce solar input				
Space reflectors	Blocking and perhaps reflecting a significant portion of the sun's energy back into space	High	Low	High	Expensive and dangerous. Long-term effects unknown. Ability of state and nonstate actors to deploy
Stratospheric aerosols	Increasing atmospheric reflectivity by introducing reflective aerosols into the stratosphere	High	Low	Low	Can be effective but has potential side effects, unintended consequences. Processes are not well understood
Marine cloud brightening	Delivery of aerosols into the marine boundary layer to brighten clouds and reflect sunlight back into space	Med	Low	High	Mechanisms for cloud brightening not sufficiently understood. Expensive: reduction of 5 watts/m^2 would cost US$1 million a week
	Increase surface albedo				
Surface albedo	Enhance albedo of the land surface by engineered coverings or crops	Low	Low to High	High	Includes white roofs, covering large land areas with reflective coverings, microbubbles spread over large ocean areas, cirrus cloud modification. Mostly low effectiveness and high cost

Source: National Research Council (2015b).

Space Dust Several US astronomers proposed a fanciful idea (Bromley et al. 2023) that solar irradiance bound for Earth could be reduced by 1%–2% by creating a dust sunshade positioned at the L_1 Lagrange point: a point in space at which the gravitational pulls of the Earth and the sun are equal. The quantity of dust required is enormous, more than 10 billion kg or 22 billion lb. Dust could be gathered from the lunar surface and ejected toward L_1. That mass of dust is astronomical (pun intended), 700 times heavier than all the material humans have ever sent into space. This concept has its roots in astronomy. The paper's authors found that planet formation around distant stars generates a lot of dust. The dust forms rings around the stars and blocks a portion of the star's irradiance (Patel 2023).

Marine Cloud Brightening Marine layer clouds, daytime, low-lying stratocumulus clouds, cover 20%–40% of the world's oceans. Increasing the brightness or albedo of those clouds could increase the overall albedo of the Earth and reduce global warming. Introducing aerosols to act as cloud condensation nuclei increases the number of cloud droplets and the cloud reflective properties. Suggested aerosols include seawater, dimethyl sulfide, ships' engine emissions, and military-grade smoke generators. Research is still needed to understand the costs and risks (NRC 2015b).

Reflective Land Areas, Crops Multiple possibilities exist for increasing the reflectivity of the land surface, from painting roofs white to covering large areas of land with reflective materials. Placing reflective particles (microbubbles) on large areas of the ocean surface has also been suggested. Except for ocean microbubbles, all are safe in that they are easily reversible. More research needs to be done on their effectiveness (NRC 2015b).

Geoengineering Summary Getting to the source of global warming (i.e., reducing solar input and removing GHGs from ambient air) is a reasoned approach, space dust being an exception. However, at this juncture, the risks outweigh the benefits.

The arguments for and against climate intervention are listed as follows.

Arguments for

- Solutions may be cheaper to implement than regulations;
- There may be secondary economic benefits;
- Some interventions force investments in valuable, environmentally friendly technologies;
- Equitable cost distribution. Major emitter nations would pay for interventions; and
- Reduced costs. Enables world economies to continue with "business as usual."

Arguments against

- Not knowing what we don't know. If we don't understand how the climate works, we don't want to start something where the cure might be worse than the disease. What if we can't turn it off?
- Little or no ability to control effects;
- It could create new sets of climate stressor extremes in the wrong places;
- The ability of state or nonstate actors to attempt climate intervention; and
- Once started, society can never stop.

"Plan B"

Scientists, supported by leading businesspeople including Bill Gates and Richard Branson, have advocated what has been called "Plan B" (Scharwath 2012). If governments cannot make significant cuts in GHG emissions in time to prevent catastrophic changes to the climate, then climate intervention is society's only alternative. Given that the likelihood of governments taking meaningful and unified action against climate change is not high, "Plan B" is appealing. The concern, however, is that vested interests in either climate modification technologies or fossil fuels (interest in maintaining business as usual) will push this approach despite the risks. In early 2020, the US Congress sent $4 million to NOAA to study two climate intervention methods: stratospheric aerosols and marine cloud brightening (Fialka 2020).

Recognizing the drivers for and challenges of climate intervention, the Carnegie Climate Governance Initiative (C2G) was set up to expand the discussion on climate intervention. The organization seeks effective governance of climate-altering technologies. While the organization states that GHG emissions reduction and climate adaptation are paramount, multiple climate intervention technologies are on display (C2G, n.d.).

Actually, today's "Plan B" should have been called "Plan C." Lester Brown has authored several books advocating a "Plan B," the first published in 2003. His 2009 work, *Plan B 4.0: Mobilizing to Save Civilization*, lays out a four-point roadmap for moving the world toward sustainability, including an 80% cut in global net carbon dioxide emissions by 2020. According to Brown, the cost of his "Plan B" was $187 billion, which at the time was equal to one-eighth the cost of world military spending (Brown 2009).

US Mitigation Activities

Over the years, the US federal government has issued or proposed rules for limiting GHG emissions. They include emission standards for vehicles and engines, limits on CO_2 emissions from future power plants, and GHG reporting programs (GCRP 2022). However, the most far-reaching federal action covering climate change is the Inflation Reduction Act of 2022. Along with tax and healthcare provisions, the act provides $391 billion for energy security and climate change. Funds will be available for renewable energy, grid energy storage, nuclear power, electric vehicles, home

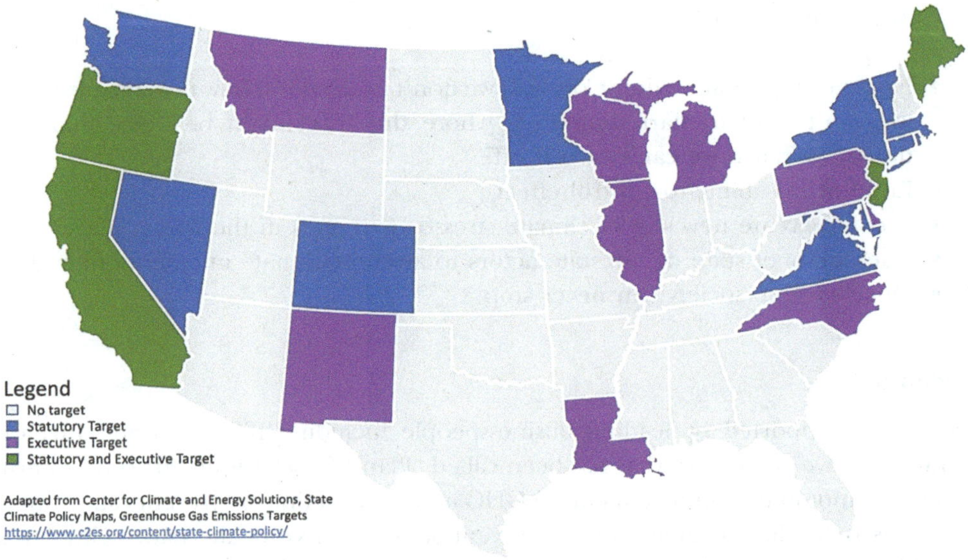

Figure 5-9. State greenhouse gas mitigation targets status as of November 2022. *Source*: C2ES (2023).

energy efficiency, climate-smart agriculture, and more (CRFB 2022, RHG 2022). With all its climate change provisions, the expected US net GHG reductions are 32%–42% below 2005 levels by 2030. Additional efforts by various agencies, cities, and states could bring reductions close to the 45% target levels. Further details are provided in Chapter 9.

Many states have established their own set of targets to reduce GHG emissions (Figure 5-9). These are targets to reduce GHG emissions, either sector-specific or for the entire state economy, by some specified quantity and by a certain date. As of September 2023, economy-wide targets were set in 23 states and the District of Columbia. Three states have recommended targets. Five states, California, Connecticut, Oregon, Rhode Island, and New Jersey, had 2020 emissions targets. California was the only state that achieved its target by 2020 (C2ES 2023).

Climate Change Adaptation

Regardless of mitigation actions, most infrastructure assets, existing and new, will likely be subjected to extreme climate and weather conditions throughout their design life. Billion-dollar disasters will continue, moderated to some extent by mitigation successes. However, the effects of previous GHG emissions will continue through the 21st century. And unless global warming is limited to 1.5 °C (2.7 °F), those effects will become markedly worse.

All of this makes infrastructure adaptation a critical element of infrastructure asset management. Infrastructure assets and systems must be made resilient (i.e., designed or reinforced to withstand extreme climate stressor values). However, adding resilience to infrastructure assets has its limits. Creating infrastructure projects to withstand the most extreme stressor values is wasteful. Adaptation, having the ability to make physical or operational changes to adjust to changed conditions, is a more cost-effective approach.

Adaptation Alternatives

To maintain infrastructure asset functionality in the face of climate variable extremes, project owners and engineers have three alternatives.

1. **Avoid/Retreat:** Locate or move the infrastructure asset away from exposure to climate hazards. For example, locating structures far from an urban–wildland interface will reduce the chances of wildfire damage. Locating assets out of flood zones will avoid flood damage. However, determining which areas have a low risk of flooding is becoming difficult. Properties previously designated by the Federal Emergency Management Agency (FEMA) as having a low risk of flooding (1% chance in any given year) are being flooded with increasing frequency. As a result, FEMA is revising its risk-rating methodology to cover more flood risk variables (FEMA 2022).

   Managed retreat, voluntarily and equitably moving people, buildings, and infrastructure out of harm's way, can be more economical than trying to fortify the asset at its current location or endure the impacts of climate extremes. For example, instead of trying to cope with permafrost melting, rising sea levels, erosion, and flooding, the Alaskan village of Newtok has voted to relocate to a new location, Mertarvik, 9 miles away. While the Newtok move is seen as a success, there can be complications. Moving people from their homes can create social and psychological problems (Hino et al. 2017). Also, Hino et al. (2017) describe the "levee effect" feedback loop. Development tends to expand behind previously constructed protective measures, making later retreat efforts more difficult.
2. **Accommodate:** Design the infrastructure asset to maintain structural, mechanical, and electrical integrity throughout extreme weather events. Full functionality may or may not be maintained throughout the event. However, the asset can be designed for resiliency, the ability to return to full functionality quickly and at a low cost. An example of accommodation would be a home on the coast in which the living area on the first level is raised well above ground level. If flooding occurs, access to the home may be limited for a time, but floodwaters are allowed to pass freely through an undeveloped and water-resistant lower level, leaving the living area undamaged.
3. **Protect:** Protect the infrastructure asset from a changing climate and extreme weather events by applying combinations of robustness, resiliency, redundancy, and adaptability to the infrastructure asset components and systems.

- *Robustness and Resilience:* Robustness in infrastructure design is the ability to function effectively across a broad range of conditions. A resilient design allows for fast recovery from extreme events for a relatively low cost. These design strategies are linked. Because the statistical properties of climate stressor variables are no longer constant, it is impossible to accurately determine what the demarcation line between function and failure ought to be. To deal with this problem, designers should incorporate a range of resiliency: a range of climate variable values for which infrastructure asset recovery from failure is quick and low cost. This will provide a buffer against sudden or catastrophic failures, very costly failures that take a long time to fix.
- *Redundancy:* Incorporating redundancy into an infrastructure asset enables that asset to maintain system functionality when one or more components fail. Designers can include additional components to bypass failed components until repairs are made. Redundancy at the systems level means other assets can take on services that an infrastructure asset provides if that asset fails.
- *Adaptability:* For infrastructure designed for adaptation, designers incorporate the ability of components to be physically altered in anticipation of changes in climate stressor variables. For infrastructure assets designed to operate under conditions of nonstationarity, it may be too expensive to add robustness to the components to handle an expanding range of climate conditions. Alternatively, engineers can design infrastructure components for an initial range of climate conditions and incorporate features that enable operators to alter the components as conditions change or new knowledge is acquired. During the operations phase, specific climate stressor variables are monitored. Preplanned changes are made when predetermined values of these variables are reached.

An example of a protective strategy is a flood wall designed to protect homes and businesses from flooding. The wall could be designed to a certain height (robustness), calculated as sufficient for a 500-year flood event. Floodwaters overtopping the wall could be channeled into a holding pond (resiliency). Accessibility to homes and businesses by highways that may be flooded in an extreme flood event is provided by smaller but usable sideroads out of flood range (redundancy). Finally, the wall can be designed so that additional height can be added relatively quickly and economically (adaptability).

US Adaptation Activities

In 2018, the US National Climate Assessment stated that while the public and private sectors are working on adaptation planning, only some adaptation measures have been implemented (GCC 2022). While much has changed over the intervening years, adaptation measures are still lagging. Some state, regional, and local agencies have completed or have planning efforts underway (Figure 5-10). Multiple local and regional plans are being prepared. The Georgetown Climate Center maintains a record of state and local adaptation planning progress organized by state (GCC 2022).

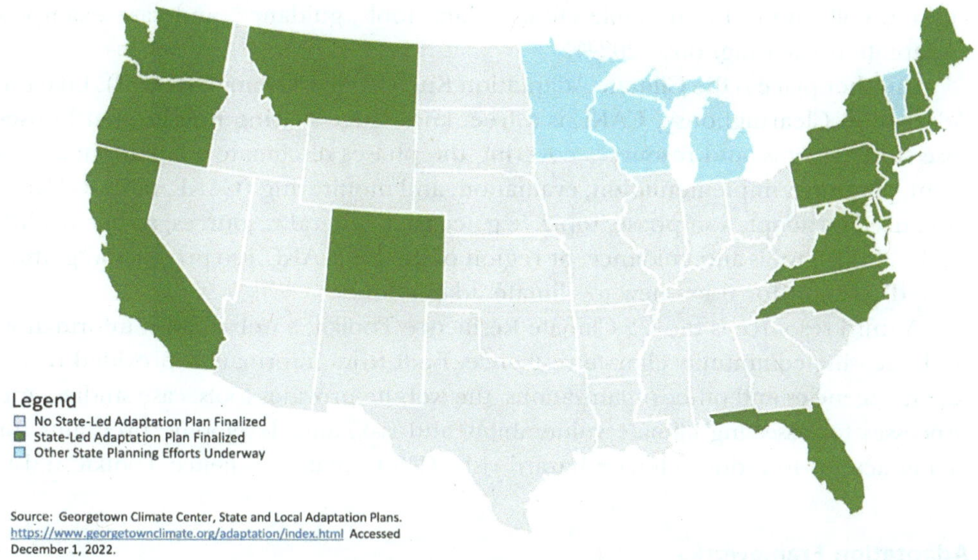

Figure 5-10. Status of state adaptation plans as of November 2022. Alaska and Hawaii also have planning efforts underway.
Source: Georgetown Climate Center, state and local adaptation plans (2022).

California: Paying It Forward

California's adaptation efforts are significant for engineers in that the state recognizes that climate adaptation is essential and that barriers exist in incorporating climate impacts into state infrastructure designs. In 2016, the California state legislature passed AB 2800, a bill to ensure that current and future climate change impacts are considered in state infrastructure projects (Moser and Finzi Hart 2018). The bill called for a panel of scientists, engineers, and architects to determine how to make this happen. In 2018, this panel, called the Climate-Safe Infrastructure Working Group, issued its report, *Paying it Forward: The Path Toward Climate-Safe Infrastructure in California.* The report recognized the inadequacies of existing codes and standards based on past climate conditions. Among the recommendations are calls for more interaction and collaboration between climate scientists and engineers, updating climate-related state architectural and engineering standards, and developing additional climate-related guidelines where current codes and standards are insufficient (CSIWG 2018).

Available Guidance on Adaptation

To increase adaptation measures, many agencies and organizations have issued guidance on assessing the risk of infrastructure damage from a changing climate. For broad adaptation guidance, one place to stop is the Adaptation Clearinghouse, a comprehensive resource and networking site for information on climate adaptation. Developed by the Georgetown Climate Center, this free online database houses a curated collection of searchable climate data, tools, guidance, and case examples (Adaptation Clearinghouse 2024).

Another place is the Climate Adaptation Knowledge Exchange (CAKE). Like the Adaptation Clearinghouse, CAKE is a free, knowledge-sharing website that houses case studies, tools, and resources covering the phases of climate adaptation: assessment, planning, implementation, evaluation, and monitoring (CAKE, n.d.). CAKE is organized by adaptation phase, topic (e.g., coastal, federal resources, public health, sea level rise), tools and guidance, or region of the US. CAKE also provides a "getting started" section for users new to climate adaptation.

A third resource is the US Climate Resilience Toolkit, a website with information on increasing community climate resilience. Built from information provided by US federal agencies and other organizations, the website provides tools, case studies, and processes for assessing climate vulnerability and risk, and developing programs for taking actions to reduce climate hazard risks (US Climate Resilience Toolkit, n.d.).

Adaptation Frameworks

Several federal and state agencies have developed adaptation frameworks which are detailed approaches for assessing climate risks and vulnerabilities specific to their areas of responsibility. The intent is to provide decision makers with systematic approaches for evaluating conditions and devising approaches for reducing climate risks. Three such frameworks are presented as follows.

Federal Highway Administration: Vulnerability Assessment and Adaptation Framework

The US Federal Highway Administration has been active in helping state, regional, and local transportation agencies assess their transportation system vulnerabilities to the effects of climate change. They provide a detailed process, the Vulnerability Assessment and Adaptation Framework, that enables users to study, assess, and make decisions on transportation projects to reduce climate risks. The process is described in detail in a published document (FHWA 2017b).

In addition, the framework is placed on a FHWA website accompanied by overview videos and a framework process diagram (Figure 5-11). The diagram has hotlinks to information for each process step (FHWA 2017b). Each of the framework's elements is further defined and developed in outline form. Each section of the outline is hot-linked to subsections containing explanations, video descriptions, examples of applications, additional resources, and references. Beyond the framework, the website provides links to research activities, case studies, tools, publications, webinars, and workshops.

US Army Corps of Engineers: Coastal Storm Risk Management Framework

The US Army Corps of Engineers (USACE) has developed and published guidance for coastal communities to address climate change. The guidance is the product of its North Atlantic Coast Comprehensive Study, a 2-year effort following Hurricane

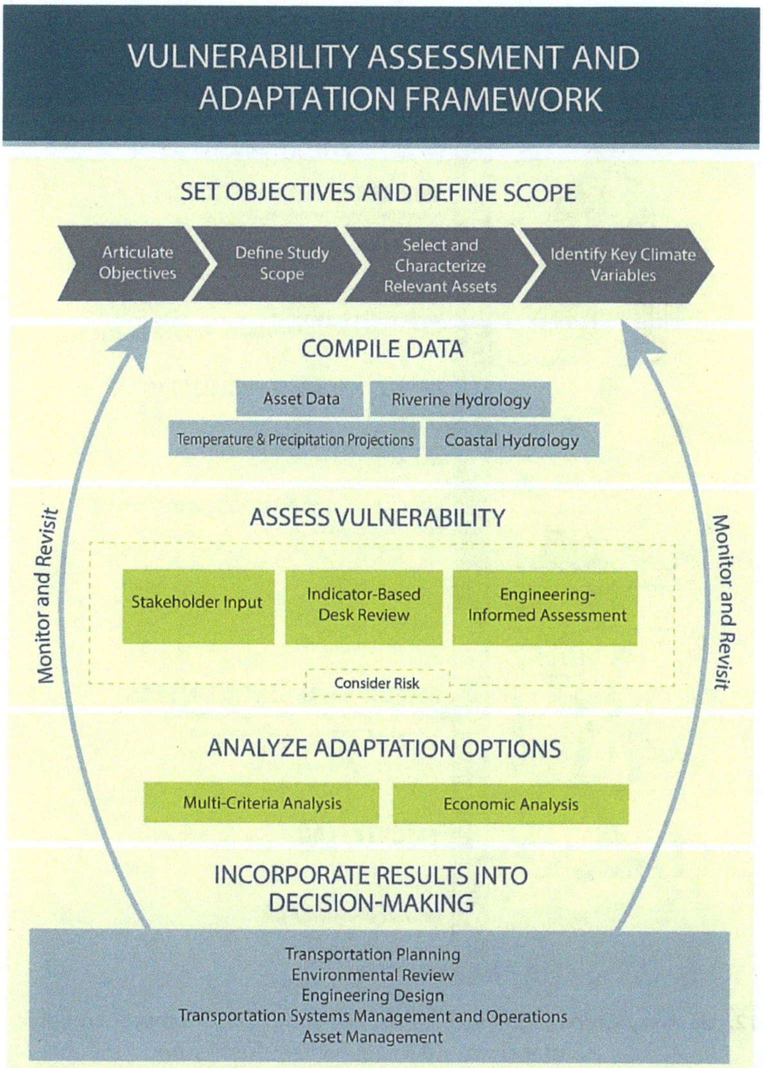

Figure 5-11. FHWA vulnerability assessment framework conceptual diagram. Users can click on any section in the diagram to navigate to that section or use the browser's back button to go back to the previous section. *Source*: FHWA (2017b).

Sandy to address the climate-related vulnerabilities and risks to communities along the North Atlantic seaboard. One of the products was the Coastal Storm Risk Management Framework, which is similar to the FHWA Framework. The US Army Corps of Engineers' nine-step process (Figure 5-12) allows coastal communities to analyze risks and vulnerabilities, determine and implement solutions, and measure

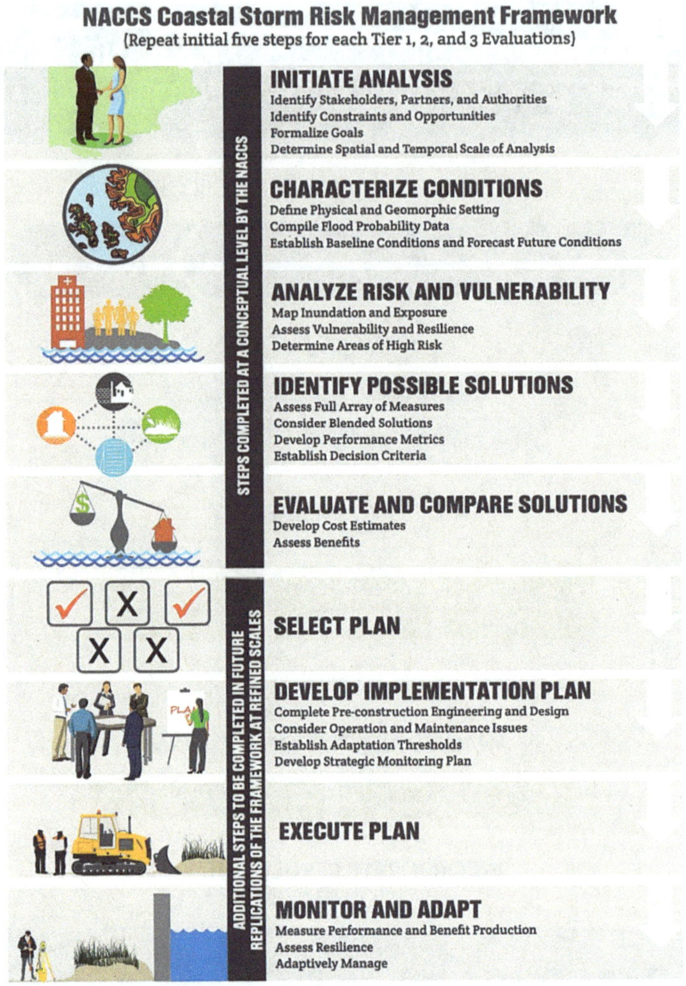

Figure 5-12. US Army Corps of Engineers North Atlantic Coast Comprehensive Study (NACCS) coastal storm risk management framework.

performance. The process is scalable, ranging from the entire North Atlantic Coast Study area to state and large watersheds and finally to local and small watershed areas (USACE 2015). In addition to the framework, USACE offers an assortment of flood risk management strategies for coastal communities. The strategies range from structural to natural and nature-based.

Delta Water Stewardship Council: Adaptive Management and the Delta Plan

An adaptive management framework has been incorporated into California's Delta Plan, a comprehensive program for managing water resources in the Sacramento–San Joaquin River Delta. In 2009, the California legislature passed the Delta Reform

Act, setting co-equal goals of providing California with a more reliable water supply in the long term while protecting, restoring, and enhancing the Delta ecosystem (Delta Stewardship Council 2013). The Delta refers to the 1,300 sq. mi. of wetlands formed by the San Joaquin, Sacramento rivers, and the Suisun Marsh. Today, this area encompasses a vast network of water supply conduits, electric powerlines, natural gas pipelines, highways, railways, and shipping channels. It is also home to more than a half-million people.

For more than 150 years, the Delta has suffered from the effects of hydraulic mining, attempts at flood control, and agricultural and urban development. The Delta Reform Act established and directed the Delta Water Stewardship Council to develop an enforceable Delta Plan to achieve the co-equal goals. The council recognized that the Delta in 2100 will be much different than today. Climate change will bring a rise in sea level, along with an increased risk of droughts, flooding, and reductions of snowpack. In addressing these changes, the act directed the council to apply the best available science and adaptive management principles. Adaptive management is defined as "a framework and flexible decision-making process for ongoing knowledge acquisition, monitoring, and evaluation leading to continuous improvements in management planning and implementation of a project to achieve objectives" (California.Public.Law 2023). Adaptive management can be applied at the program, plan, or project level.

In applying adaptive management, the Delta Stewardship Council uses a three-phase framework, similar to the Plan-Do-Check-Act cycle of continuous improvement, which is part of the lean manufacturing philosophy. The council's version is Plan (Development of the Delta Plan), Do (Implementation and Oversight of the Delta Plan), and Evaluate and Respond (Update and Revision of the Delta Plan). Figure 5-13 shows the three phases subdivided into steps (Delta Stewardship Council 2013).

Plan

1. *Define/Redefine the Problem.* Define the problem to be addressed and show how it is linked to program goals. Define problem boundaries: time scale, geography.
2. *Establish Goals and Objectives.* Set goals and objectives based on the best available science. Goals are defined as broad statements of solutions. Objectives are specific, often quantitative, allowing evaluation of the extent to which the objectives are being achieved.
3. *Model Linkages between Objectives and Proposed Action(s).* Models (conceptual, statistical, physical, decision support, or simulation) are links from the objectives to the proposed actions.
4. *Select Action(s) (Research, Pilot, or Full-scale) and Develop Performance Measures.* Evaluate and select action(s) based on (a) level of action(s) to be taken, (b) geographic and temporal scale, (c) degree of confidence in benefits, and (d) consequences of being wrong.

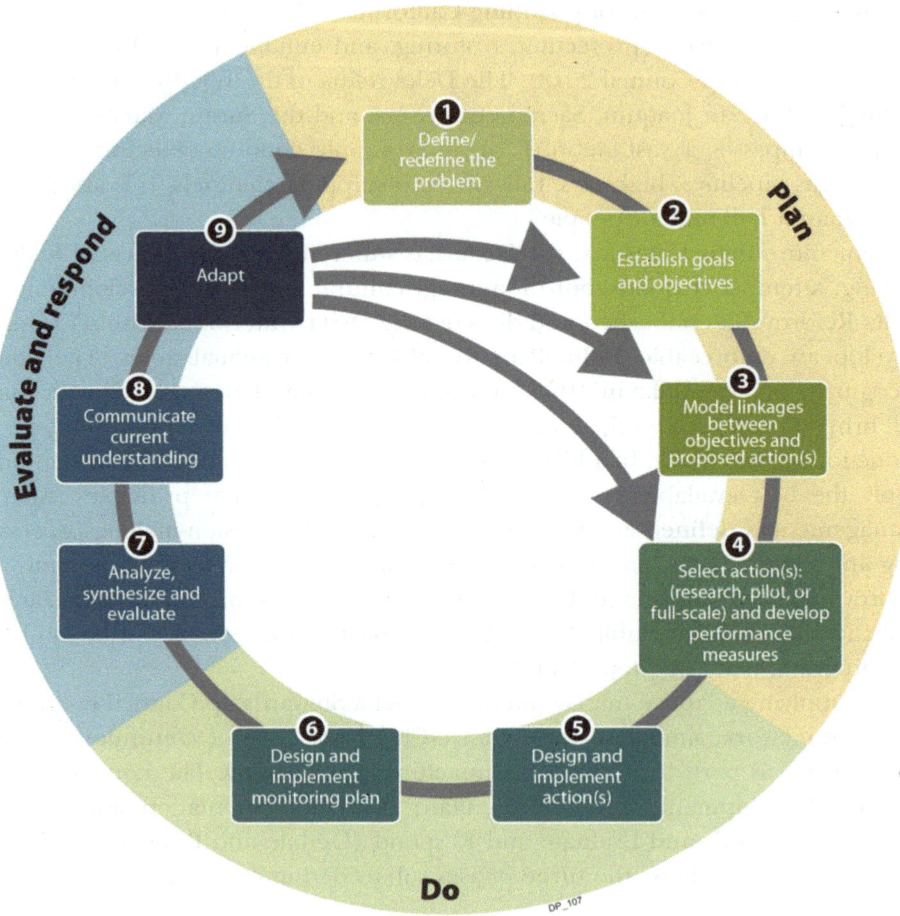

Figure 5-13. Delta plan nine-step adaptive management framework.
Source: The Delta Plan, Appendix C (Delta Stewardship Council 2013). Reproduced with permission.

Do

5. *Design and Implement Action(s).* Create a plan to implement the selected action(s) and monitor responses. Clearly describe the activities of the selected action(s) and how they will link to the monitoring plan. Describe existing uncertainties. Action(s) should be linked to meeting goals and objectives.
6. *Design and Implement a Monitoring Plan.* Monitoring plans consist of (a) compliance monitoring required by permits, (b) performance monitoring for measuring the achievement of targets, (c) mechanistic monitoring with concurrent targeted research for testing and understanding of linkages to the conceptual models, and (d) system-level monitoring.

Evaluate and Respond

7. *Analyze, Synthesize, and Evaluate.* Evaluate whether performance measures indicate if objectives are being met. If not met, state the reasons why. This includes how conditions have changed, expectedly or unexpectedly. Evaluations should consider if the action(s) prevented deterioration that would have occurred if no actions were taken.
8. *Communicate Current Understanding.* Communicate what has been learned to policymakers, managers, stakeholders, and the public.
9. *Adapt.* With updated knowledge of current conditions and an understanding of action(s) performance, decision makers can determine what changes should be made.

Adaptive Management for Infrastructure Projects

ASCE (2015) recognized early that a changing climate would significantly impact the civil engineering profession. Researchers noted that a wide gap exists between engineering practice and climate science. Bridging that gap requires a new paradigm for engineering practice (Olsen et al. 2015), one that can reduce future climate uncertainties. One approach suggested was the Observational Method, a technique used in geotechnical engineering. Here, project designs are based on the most probable conditions but incorporate preplanned design modifications to accommodate changed conditions. Conditions are monitored during operation, and preplanned modifications are made as conditions reach predetermined thresholds. The Observational Method is limited to applications in which changes in climate stressor variables are gradual (e.g., sea level rise or permafrost melting). In those cases, constructed modifications can be scheduled based on stressor variable measurements. In contrast, the Observational Method is ineffective for sudden extreme and damaging climate and weather events (Olsen et al. 2015, ASCE 2018).

Wallace et al. (2017) proposed an Observational Method-based approach for designing, constructing, and operating infrastructure projects, accomplishing three objectives.

1. **Deliver the Project:** Meet the project owner's requirements and specifications on time and within budget.
2. **Improve Sustainable Performance:** Make a significant contribution to improving the overall sustainable performance of infrastructure. Reducing the project's carbon footprint in both embodied and operational carbon is a critical component of performance. Guidance for improving sustainable performance was based on the Institute for Sustainable Infrastructure's Envision sustainable infrastructure rating system.

3. **Address Nonstationarity:** Account for changing environmental operating conditions through combinations of increased robustness and resilience, redundancy, and adaptation.

The paper compares the elements of traditional infrastructure project delivery under conditions of stationarity to those required under conditions of nonstationarity.

ASCE expanded its work on climate adaptation with a manual of practice, providing guidance for adaptive design and risk management, bringing in the analytical methods necessary for implementing the Observational Method or other adaptive management techniques (ASCE 2018).

In an ASCE journal, the article "Adapting to Risk from Climate Change" (Roth 2022) describes an adaptive management process (Figure 5-14) for civil infrastructure projects, a simplified version of the nine-step process of the Delta Plan Nine-Step Adaptive Management Framework.

Barriers to Adaptation

Despite these frameworks and the volume of resources available for adapting infrastructure projects to a changing climate, implementation has been slow. Although there is increasing awareness and assessment of climate change threats, actual adaptation actions generally respond to recent extreme events, focusing on recovery rather than protection against future climate threats. As noted in the Fourth National Climate Assessment (GCRP 2018), most decision makers in public and private organizations still work under the assumption of an unchanging climate. Adapting to the threats posed by future climate conditions requires a sea change in current practices which are based on the assumption of stationarity.

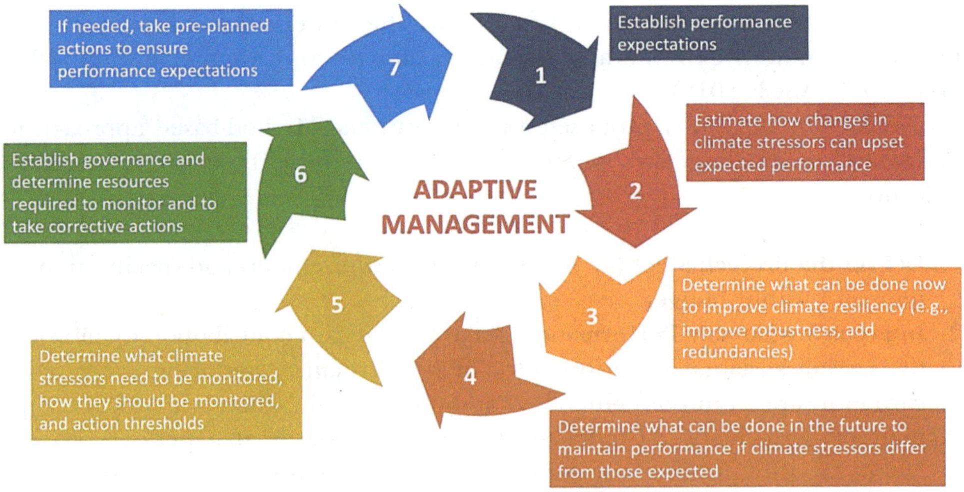

Figure 5-14. Simplified adaptive management process.
Source: Roth (2022). Reprinted with permission.

The Joint Federal and California Program (CALFED) has struggled to implement adaptive management strategies for its Ecosystem Restoration Program. The barriers to implementation identified included (1) decision makers not understanding the need for adaptation measures, (2) inability to lead the more complex adaptation management processes, and (3) lack of funding for monitoring requirements (Delta Stewardship Council 2013). As noted in an earlier chapter, the US Department of Transportation (FHWA 2017b) describes the problem this way:

> To date, few transportation agencies in the United States have explicitly required design changes in anticipation of future climate change due, in part, to the lack of understanding of how best to incorporate projections into designs.

The Post Hurricane Sandy Transportation Resilience Study in New York, New Jersey, and Connecticut (FHWA 2017a) expanded the discussion of barriers to adaptation, explaining the common themes that arose during the study.

- **Multiple Climate Projections:** Efforts to apply climate models to project future conditions, such as rising sea levels, have produced numerous estimates and probabilities. To decision makers, the lack of certainty can lead to indecision and delays.
- **Spotty Historical Data on Impacts of Discrete Weather Events:** A lack of good information about the effects of past events limits a decision maker's ability to quantify the impacts and make good decisions for adaptation to future events.
- **Information on Specific Transportation Facilities Can Be Hard to Find or Is in a Non-Digital Format:** There is a gap in understanding current asset conditions, level of deterioration, remaining useful life, user demand, and other information needed for planning.
- **Cross-Agency Coordination and Jurisdictional Issues:** Multiple agencies have jurisdiction over vulnerable areas affected by climate change, creating response problems.
- **Legal and Regulatory Hurdles:** Barriers to right-of-way acquisition, lawsuits from impacted landowners, and required environmental and community impact studies can stop adaptation projects.
- **Limited Funding:** Funding for adaptation projects is limited and highly competitive. Some funding sources can only be accessed after a disaster happens.

From the engineer's vantage point, the situation is similar. In a survey of climate change-related client services (Laxo et al. 2023), the authors found a significant gap between climate science and architect and engineering (A&E) services. Only 34% of the participating companies offering sustainable design services said they use climate change projection data in their building and infrastructure systems designs. The rest reported that they were unfamiliar with climate change projections, or, if they were familiar, they did not use them. One-quarter of the companies stated that their clients were not asking for climate projections. On this point, the survey noted:

> Clients generally assume that A&E professionals will apply the right data and decision-making to projects, trusting that the clients' own lack of knowledge will be filled by the training and expertise of the licensed A&E professionals with whom they are collaborating.

The survey also highlighted other problems. For climate projection data to be useful, it must be at a level of resolution that fits the project location. It also must be in a format compatible with engineering design tools. Also, given the range of climate scenarios and modeling methodologies, it is hard to determine the most appropriate projections to use.

No matter how many frameworks, methodologies, or guidance manuals are published, implementation of climate adaptation measures will likely continue to stall. Chester and Allenby (2019) point out that people have been trying to solve climate change problems in civil infrastructure design as if they were normal, albeit complicated, engineering problems, problems that could be handled through the traditional approaches that produce optimization and efficiency. Codes and standards have been developed that state what to do and how to do it. Engineers are trained and licensed to operate in this problem domain, sufficiently knowledgeable of the applicable codes and standards, and well equipped for applying the appropriate methodologies to deliver a cost-effective solution. For example, engineers can provide a community with possible layouts for a stormwater management system on local terrain with the capacity to handle the expected local rainfall intensity. Engineers will size the system for a specified maximum storm intensity to protect the health, safety, and welfare of the public residing in the affected area. They will also adhere to treatment standards for suspended solids, oils, and other pollutants.

However, in this era of climate change, civil infrastructure design problems have been pushed into the complex domain, a domain that is changing and unpredictable. As the climate changes, most climate-related codes and standards designed to protect the public are no longer effective. In addition, the institutions that govern these codes and standards find it extremely difficult to respond: devising new rules for civil infrastructure design and operation when future climate conditions are unpredictable. Chester and Allenby (2019) note that these institutions have been configured to address complicated rather than complex systems. Each is partitioned into single disciplines (e.g., water, transportation, power generation). Each discipline has its own bureaucracy, each with a unique set of rules, responsibilities, structures, knowledge, and competencies. Making matters worse, these institutions are not prepared for increasingly frequent and more intense climate and weather disasters, disasters regarded as extremely rare or unimaginable in intensity and frequency.

Conclusion

As stated in Chapter 1, the purpose of this book is to instigate change in how civil infrastructure is planned, designed, constructed, and operated, made ready for changing climate conditions while reducing carbon emissions and correcting other unsustainable practices. Although climate change is a complex problem and barriers to climate adaptation exist, engineers and other practitioners have devised ways to handle them. Chapter 1 noted that our assumptions about climate variability have shifted, moving from historical consistency to multiple plausible futures. They have shifted, all or in part, into the realm scientists call *deep uncertainty* in which many outcomes are possible. Ways to deal with deep uncertainty in a changing climate are discussed in Chapter 6. Chapters 7 and 8 offer methodologies to address deep uncertainty in infrastructure projects and systems. As it turns out, we have always handled situations with deep uncertainty in our work and lives. We must now add climate uncertainty into the mix.

References

Adaptation Clearinghouse. 2024. Accessed March 1, 2024. https://www.adaptationclearinghouse.org/

Anthony, K. W., P. Anthony, G. Grosse, and J. Chanton. 2012. "Geologic methane seeps along boundaries of Arctic permafrost thaw and melting glaciers." *Nat. Geosci.* **5** (6): 419–426.

ASCE. 2015. *Adapting infrastructure and civil engineering practice to a changing climate.* Reston, VA: ASCE. https://doi.org/10.1061/9780784479193.

ASCE. 2018. *Climate-resilient infrastructure: Adaptive design and risk management.* Reston, VA: ASCE. https://doi.org/10.1061/9780784415191.

ASCE. 2023a. "ECOM – Embodied carbon estimator." Accessed July 20, 2022. https://se2050.org/ecom-tool/.

ASCE. 2023b. "Program goals." Accessed January 26, 2024. https://se2050.org/program_goals/.

Bromley, B. C., S. H. Khan, and S. J. Kenyon. 2023. "Dust as a solar shield." *PLOS Clim.* **2** (2): e0000133. https://doi.org/10.1371/journal.pclm.0000133.

Brown, L. 2009. "Plan B 4.0: Mobilizing to save civilization." Accessed February 9, 2023. http://www.earth-policy.org/books/pb4.

C2ES (Center for Climate and Energy Solutions). 2023. "U.S. state greenhouse gas emissions targets." September 2023. Accessed March 1, 2024. https://www.c2es.org/document/greenhouse-gas-emissions-targets/.

C2G (Carnegie Climate Governance Initiative). n.d. Accessed February 9, 2023. https://www.c2g2.net/.

CAKE (Climate Adaptation Knowledge Exchange). n.d. Accessed December 1, 2022. https://www.cakex.org/.

California.Public.Law. 2023. "California water code 85052." Accessed January 26, 2024. https://california.public.law/codes/ca_water_code_section_85052.

CAT (Climate Action Tracker). 2023. "CAT warming projections: Global temperature increase to 2100." Accessed January 25, 2024. https://climateactiontracker.org/global/cat-thermometer/.

Charles Pankow Foundation. 2020. "Embodied carbon in construction calculator." Accessed April 17, 2022. https://www.pankowfoundation.org/our-work/research-grants/sustainability/carbon-reduction/06-18-embodied-carbon-in-construction-calculator/.

Chester, M. V., and B. Allenby. 2019. "Infrastructure as a wicked complex process." *Elem. Sci. Anth.* **7** (April): 21. https://doi.org/10.1525/elementa.360.

CLF (Carbon Leadership Forum). 2020. "Embodied carbon 101." Accessed April 17, 2022. https://carbonleadershipforum.org/embodied-carbon-101/.

CRFB (Committee for a Responsible Federal Budget). 2022. "CBO scores IRA with $238 billion of deficit reduction." Accessed November 30, 2022. https://www.crfb.org/blogs/cbo-scores-ira-238-billion-deficit-reduction.

CSIWG (Climate-Safe Infrastructure Working Group). 2018. "Paying it forward: The path toward climate-safe infrastructure in California." Accessed January 26, 2024. https://resources.ca.gov/CNRALegacyFiles/docs/climate/ab2800/AB2800_ES_FINAL.pdf.

De Schryver, A., and L. Zampori. 2022. "Product carbon footprint standards: Which one to choose?" Accessed April 16, 2022. https://pre-sustainability.com/articles/product-carbon-footprint-standards-which-standard-to-choose/.

Delta Stewardship Council. 2013. "About the delta plan." Accessed January 26, 2024. https://deltacouncil.ca.gov/delta-plan/.

FEMA (Federal Emergency Management Agency). 2022. "NFIP's pricing approach." Accessed January 26, 2024. https://www.fema.gov/flood-insurance/risk-rating.

FHWA (Federal Highway Administration). 2017a. *Post Hurricane Sandy transportation resilience study in NY, NJ, and CT.* FHWA-HEP-17-097. Washington, DC: FHWA.

FHWA. 2017b. *Vulnerability assessment and adaptation framework.* FHWA-HEP-18-020. Washington, DC: FHWA.

Fialka, J. 2020. "NOAA gets go-ahead to study controversial climate Plan B." Accessed February 9, 2023. https://www.scientificamerican.com/article/noaa-gets-go-ahead-to-study-controversial-climate-plan-b/.

GCC (Georgetown Climate Center). 2022. "State adaptation progress tracker." Accessed March 1, 2024. https://www.georgetownclimate.org/adaptation/plans.html.

GCRP (US Global Change Research Program). 2018. *Impacts, risks, and adaptation in the United States: Fourth national climate assessment.* Vol II, edited by D. R. Reidmiller, C. W. Avery, D. R. Easterling, K. E. Kunkel, et al. Washington, DC: GCRP. DOI 10.7930/NCA4.2018.

GCRP. 2022. "Mitigation." Accessed November 30, 2022. https://nca2014.globalchange.gov/report/response-strategies/mitigation.

GFN (Global Footprint Network). 2006. "Living planet report 2006." Accessed February 29, 2024. https://wwfint.awsassets.panda.org/downloads/lpr_living_planet_report_2006.pdf.

Hammond, G., and C. Jones. 2011. *Embodied carbon: The inventory of carbon and energy (ICE).* Accessed January 26, 2024. https://www.emccement.com/pdf/Full-BSRIA-ICE-guide.pdf.

Hino, M., C. Field, and K. Mach. 2017. "Managed retreat as a response to natural hazard risk." *Nat. Clim. Change* **7** (5): 364–370. https://doi.org/10.1038/nclimate3252.

IPCC (Intergovernmental Panel on Climate Change). 2005. "Carbon dioxide capture and storage." Accessed February 29, 2024. https://www.ipcc.ch/site/assets/uploads/2018/03/srccs_wholereport-1.pdf.

IPCC. 2018. "Summary for policymakers." Accessed January 26, 2024. https://www.ipcc.ch/sr15/7.

ISO (International Organization for Standardization). 2018. *Greenhouse gases—Carbon footprint of products—Requirements and guidelines for quantification.* ISO 14067:2018. Geneva: ISO.

Laxo, A., B. Hoppe, H. Roop, and P. Cipriano. 2023. "Climate forward? How architects and engineers are(not) using climate projections to inform design." Accessed January 26, 2024. https://hga.com/climate-forward.

Moser, S., and J. Finzi Hart. 2018. "AB 2800 and California's climate safe infrastructure working group." Accessed February 20, 2023. https://acwi.gov/climate_wkg/ab_2800_and_california%E2%80%99s_csiwg_water_resources_adaptation_to_climate_change_wg_03_29_2018.pdf.

NRC (National Research Council). 2015a. *Climate intervention: Carbon dioxide removal and reliable sequestration.* Washington, DC: National Academies Press. https://doi.org/10.17226/18805.

NRC. 2015b. *Climate intervention: Reflecting sunlight to cool earth,* 19. Washington, DC: National Academies Press. Accessed February 29, 2024. https://nap.nationalacademies.org/catalog/18988/climate-intervention-reflecting-sunlight-to-cool-earth.

Olsen, R., B. Ayyub, D. Walker, A. Barros, W. Lei, and F. Lombardo. 2015. *Adapting infrastructure and civil engineering practice to a changing climate.* Reston, VA: ASCE. https://doi.org/10.1061/9780784479193.

Patel, K. 2023. "Researchers want to create a dust shield in space to fight climate change." *The Washington Post,* February 8, 2023.

Pires, J. C. M., F. G. Martins, M. C. M. Alvim-Ferraz, and M. Simões. 2011. "Recent developments on carbon capture and storage: An overview." *Chem. Eng. Res. Des.* **89** (9): 1446–1460. https://doi.org/10.1016/j.cherd.2011.01.028.

Pontin, M. W. 2007. "Cooling the planet." Accessed February 9, 2023. https://www.technologyreview.com/2007/02/13/130428/cooling-the-planet/.

RHG (Rhodium Group). 2022. "A turning point for us climate progress: Assessing the climate and clean energy provisions in the inflation reduction act." Accessed November 30, 2022. https://rhg.com/research/climate-clean-energy-inflation-reduction-act/.

Roth, L. H. 2022. "Adapting to risk from climate change." *GEOSTRATA* **26** (3): 20–27. https://ascelibrary.org/doi/abs/10.1061/geosek.0000418.

Scharwath, K. 2012. "Bill Gates supports controversial climate change plan." Accessed February 9, 2023. https://www.triplepundit.com/story/2012/bill-gates-supports-controversial-climate-change-plan/68516.

UN. 1987. *Our common future: Report of the world commission on environment and development.* New York: Oxford University Press. Accessed February 29, 2024. https://sustainabledevelopment.un.org/content/documents/5987our-common-future.pdf.

UNEP (United Nations Environment Programme). 2023. "About Montreal protocol." Accessed March 24, 2023. https://www.unep.org/ozonaction/who-we-are/about-montreal-protocol.

USACE (US Army Corps of Engineers). 2015. "North Atlantic coast comprehensive study report, risk management strategies for coastal communities." Accessed December 2, 2022. https://www.nad.usace.army.mil/CompStudy/.

US Climate Resilience Toolkit. n.d. Accessed December 2, 2022. https://toolkit.climate.gov/.

USEPA (US Environmental Protection Agency). 2022. "Overview of greenhouse gases." Accessed April 4, 2022. https://www.epa.gov/ghgemissions/overview-greenhouse-gases.

USEPA. 2023. "Greenhouse gas emissions: Understanding global warming potentials." Accessed February 9, 2023. https://www.epa.gov/ghgemissions/understanding-global-warming-potentials.

USEPA. 2024. "What is embodied carbon?" Accessed March 1, 2024. https://www.epa.gov/greenerproducts/what-embodied-carbon.

Wallace, B., D. Ellison, and R. Daugherty. 2017. "Delivering safe, cost-effective, sustainable civil infrastructure projects under conditions of non-stationarity." In *Proc., Int. Conf. on Sustainable Infrastructure 2017: Technology*. New York: ASCE. https://doi.org/10.1061/9780784481219.

Additional References (Not Cited)

Ali, A., and J. Xiao. 2017. "Estimation and minimization of embodied carbon of buildings: A review." Accessed April 12, 2022. https://climateactiontracker.org/global/cat-thermometer/. https://doi.org/10.1038/ngeo1480.

CLF. 2023. "Embodied carbon 101." Accessed February 9, 2023. https://carbonleadershipforum.org/embodied-carbon-101/.

Webster, M. D., ed. 2020. "Achieving net zero embodied carbon in structural materials by 2050." Accessed April 18, 2022. https://carbonleadershipforum.org/sei-sustainability-committee-publishes-white-paper-achieving-net-zero-embodied-carbon-in-structural-materials-by-2050/.

CHAPTER 6

Civil Infrastructure and Deep Uncertainty

The Third Ape's Problem

> A fruit-laden tree grows on a prehistoric savanna. A bipedal ape-human comes to the edge of the brush and ponders the risk of crossing the grasslands to gather a meal. Is a lion lurking in the high cover? She notices the swish of a tail: Swishing tail => Active lion => Danger. She withdraws. Later, a second ape long familiar with this patch comes by. No swishing tail, yet she knows that every so often a troop mate has been lost gathering fruit here. How often? She considers the risks, balances her need, and determines whether to cross. Now a third ape approaches this stretch of savanna for the first time. She can draw upon neither concrete information nor familiarity with this patch as a basis for logical deduction. She is, however, quite hungry and her survival hangs in the balance. She searches for familiar patterns and weighs them against her experience. She takes a few steps forward and then looks for changes in the patterns. How far can she proceed and still scramble back to safety? Where might a lion hide in this brush? Is that movement over there solely due to the wind? She ventures forth step-by-step, updating information, planning for contingencies, perhaps tossing a rock or two to probe for any lurking predators. In this manner she proceeds into a potentially terrifying unknown.
>
> *Robert J. Lempert and Steven W. Popper (2005)*

A Framework for Infrastructure Planning and Design Decision Making

As noted in Chapter 1, the years with the greatest number of billion-dollar weather and climate events occurred in the last decade. In 2021, the United States had 20 separate billion-dollar weather and climate disasters (NOAA NCEI 2022), just two events less than the record set in 2020. 2022 had 18 separate billion-dollar events, tying 2017 and 2011 for the third-highest number. At US$165 billion, 2022 was also the third highest in costs, behind 2017 and 2005 (Smith 2023). 2023 had 28 separate billion-dollar events, breaking the record of 22 set in 2020 (Smith 2024). 2023 was also the warmest year on record, with 1.2 °C (2.2 °F) over the twentieth century average of 13.9 °C (57.0 °F) (NOAA NCEI 2024).

Reducing the number and cost of these disasters requires new approaches to civil infrastructure planning, design, and operation, which can handle future climate conditions, conditions that climate scientists have labeled deeply uncertain. On its face, this challenge is enormous, requiring a change out of standards and practices that relied on static climate conditions, revising or replacing them with ones that handle radical climate fluctuations and potential knock-on effects. Handling this challenge requires an understanding of how infrastructure design decisions are made and how climate uncertainties can be managed effectively over the infrastructure project lifecycle.

A generalized framework for infrastructure planning and design decision support (Marchau et al. 2019) is presented in Figure 6-1. Decision makers determine policies (in the case of infrastructure projects, performance requirements, and specifications) designed to influence the behavior of a system (infrastructure asset with physical and operational aspects) to achieve the outcomes of interest to the stakeholders. Those outcomes become criteria relevant to the decision makers and are weighted according to stakeholder preferences. Those preferences can be significant. There can also be conflicts among stakeholders, preferences, and concerns that can change over time.

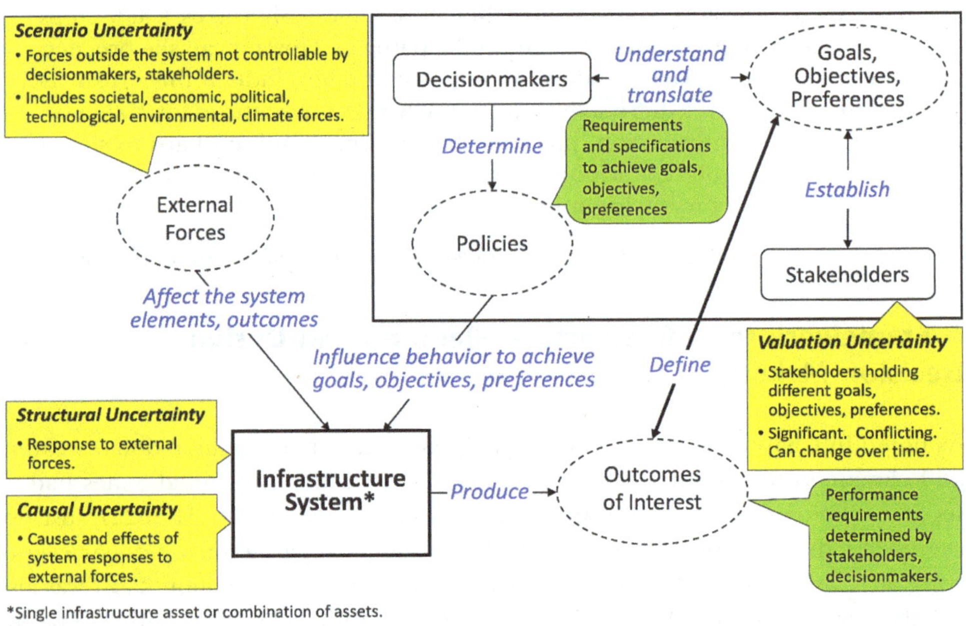

Figure 6-1. Framework for infrastructure planning and design decision making, including areas of uncertainty. Policies can be infrastructure asset performance requirements and specifications for achieving outcomes of interest (i.e., stakeholders' desired outcomes).
Source: Adapted from Marchau et al. (2019), Creative Commons Attribution 4.0 International License.

Decision makers translate stakeholder goals, objectives, and preferences into the infrastructure asset requirements and specifications for producing the desired outcomes, which are translated into infrastructure asset performance criteria. The infrastructure asset or system can be a combination of natural, engineered, and human elements, some of which may not be controllable. External forces exist outside the system that are not controllable by the decision makers or stakeholders. These include societal, economic, political, technological, and environmental forces, any or all of which can affect the system elements and outcomes.

Marchau et al. (2019) categorize several areas of uncertainty inside and outside of the system boundaries.

- **Scenario Uncertainty:** Significant external developments outside of the system influence system performance. These developments may change in scope, direction, and impact. Climate is one of those forces.
- **Structural Uncertainty:** Uncertainty about how social, environmental, and economic systems respond to external developments. There is also uncertainty about the internal workings of the system itself and how it will respond to external forces.
- **Causal Uncertainty:** Uncertainty about the causes of the system's responses because of insufficient knowledge about interactions within the system.
- **Valuation Uncertainty:** Uncertainty about how the stakeholders value different goals, objectives, and preferences. The differences can be significant, conflicting, and change over time.

Levels of Uncertainty

Marchau et al. (2019) define the uncertainty spectrum ranging from complete certainty to total ignorance. In between are multiple levels of uncertainty, as depicted in Figure 6-2. These are described in the context of civil infrastructure planning, design, and delivery as follows.

- **Complete Determinism:** Everything is known. This level is not achievable.
- **Level 1: Clear Enough:** Uncertainty is low enough for the intended purpose. For example, according to ASTM standards, 4340 VAR AMS 6414 is a low-alloy chromium-nickel-molybdenum premium aircraft-quality steel (SSA). Its carbon content is limited to 0.38% and 0.43% by weight (Bringas 2016). The uncertainty of carbon composition in this type of steel is 0.05%, but that uncertainty is still sufficient for its use in high-stressed aircraft parts.
- **Level 2: Know the Probabilities:** The uncertainty of variables can be defined in a probability distribution or as a known set of predicted values to which probabilities can be assigned. An example related to infrastructure is the Earth's previous climate conditions, conditions that were statistically stable (condition of

	Complete Determinism	Level 1	Level 2	Level 3	Level 4		Total Ignorance
					Level 4a	Level 4b	
Context	Accepted fact	Clear enough future	Alternative futures with probabilities	A few plausible futures	Many plausible futures	Unknown future	Don't know what we don't know
		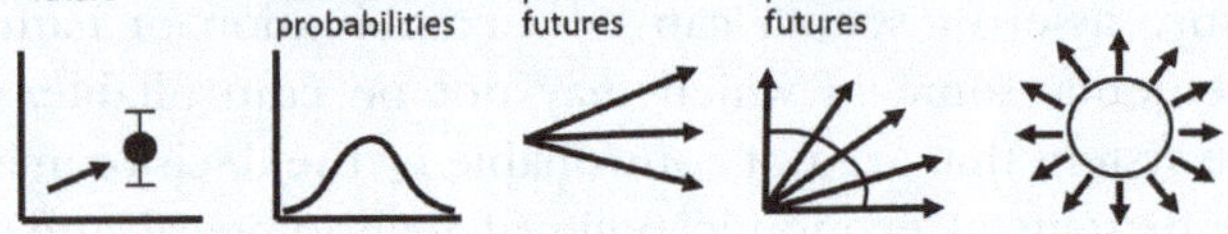					
		Clear enough	Know the probabilities	Bracket the outcomes	Bracket the alternatives	Know what we don't know	
System model		A single deterministic system model	A single stochastic system model	A few alternative system models	Many alternative system models	Unknown system model	
System outcomes		A point estimate for each outcome	A confidence interval for each outcome	A limited range of outcomes	A wide range of outcomes	Unknown outcomes	
Stationarity		Properties of materials, supplies	Soil structure, properties Climate variables				
Non-Stationarity		Properties of materials, supplies	Soil structure, properties Climate variables, short term		Climate variables, long term Climate knock-on or cascading effects Human/climate interventions		

Figure 6-2. Framework for infrastructure planning and design decision-making, including areas of uncertainty. Policies can be infrastructure asset performance requirements and specifications for achieving outcomes of interest (i.e., stakeholders' desired outcomes). (This figure is a duplicate of Figure 1-14).
Source: Adapted from Marchau et al. (2019), Creative Commons Attribution 4.0 International License.

stationarity). Most of today's climate-related engineering standards are based on the assumption of a statistically stable climate.

- **Level 3: Bracket the Outcomes:** Similar to Level 2, there are a known and limited number of potential variable values in Level 3, but probabilities cannot be assigned with sufficient accuracy. For example, in the case of hazardous waste site remediation, subsurface contamination is known to exist, but the waste quantities and their exact location are unknown. Using the observational or learn-as-you-go method, engineers can select a course of action based on the most probable conditions while anticipating and planning for other reasonably possible conditions. As the remediation process proceeds, engineers monitor actual site conditions and modify their course of action based on what they discover during on-site remediation (Brown et al. 1990).

The observational method is applicable in designing for a changing climate in situations where the range of impact of specific climate stressors is known (including the most probable and the most unfavorable) and the likelihood of knock-on effects is low. In some cases, protection from flooding because of rising sea levels may be handled as Level 3 uncertainty. Engineers could design sea walls so that the height of the walls could be raised incrementally as sea levels rise.

The observational method has been applied in multiple locations to handle the effects of climate change. For example, rail trestles for the Los Angeles to San Diego (LOSSAN) Rail Corridor are designed to allow the insertion of additional pier segments to raise the trestles as the flood hazard increases (Ayyub 2018).

- **Level 4: Deep Uncertainty:** This is the deepest level of uncertainty with two sublevels.
  - *Level 4a: Bracket the alternatives*: At this level, there are multiple possible system models. However, the operative model is not known.
  - *Level 4b: Know what we do not know:* Like Level 4a, this level has multiple possible system models. However, in addition to multiple models, other unknown interactions may produce other extreme events. These events can be unexpectedly high climate stressor values, rapid acceleration of the rate of value change, or knock-on effects (i.e., compounding, propagating, or reinforcing). These are sometimes labeled as "black swans," events that were never expected to happen based on past experience (Taleb 2007).
- **Total Ignorance:** Nothing is known about how the system works or how outside forces will affect its operation.

Levels of Uncertainty in Business Management

These four levels of uncertainty are not unique to infrastructure design under changing climate conditions. In their *Harvard Business Review* article, authors Courtney, Kirkland, and Viguerie discuss four levels of business strategy uncertainty using a similar construct (Courtney et al. 1997)

- *Level 1: A Clear-Enough Future*. A single forecast of the future can be developed. While not exact, its residual uncertainty is small enough to be ignored.
- *Level 2: Alternate Futures*. The future is limited to a few alternatives.
- *Level 3: A Range of Futures*. A limited number of variables define the range of possible futures, but the actual future could be anywhere within the defined range.
- *Level 4: True Ambiguity*. A large number of variables with many possible interactions exist, making it impossible to predict future conditions.

Engineers Have Always Managed Deep Uncertainties

As it turns out, engineers and other practitioners have always had to handle high levels of uncertainty, unexpected situations, and events in their work and life. Engineers, throughout their careers, manage civil infrastructure projects, each with known scopes, objectives, and specifications. Conditions for success in managing those projects are a matter of achieving the project owner's objectives: achieving the project requirements and specifications on an agreed-on schedule and within budget.

Successfully managing these projects then becomes a matter of managing change. Experienced project managers are focused not only on the owner's objectives but also on the conditions for project success (i.e., what must go right over the course of project planning, design, and delivery for the project to succeed). As projects proceed, project managers will likely face multiple unforeseen situations and events that can create deviations from the conditions for success. For example, a key person accepts another job and leaves the project. Materials delivered to the project site arrive late or are not what was ordered. A heavy rainstorm delays construction.

In project assignments, the project manager determines the project's conditions for success along with potential deviations. Recognizing that losing key project personnel is possible, the project manager will designate and train other engineers to take on a wider range of roles and responsibilities. Material delivery and storm delays are handled by incorporating extra time in the project schedule or by short-term increases in project staff to compensate for lost time.

Deep Uncertainty in Our Lives

We face a wide assortment of deep uncertainties in our daily life and work. Events such as illness, accidents, unemployment or loss of a loved one are part of life and not possible to predict with a high degree of certainty. But we have found ways to cope and manage. In the face of these uncertainties, we have defined personal success as achieving our own sense of a good life and well-being. We seek a reasonable level of security, sufficient food and shelter, sound health, and good relations with others. Based on our personal appetite for risk, we determine the level of risk we are willing to take on and what we will do to reduce that risk. We expect and plan for deviations by placing extra money in savings accounts, buying insurance, having regular health checkups, and maintaining supporting relationships. For certain extreme circumstances, we may seek government assistance. We also scan for opportunities to improve our personal situation and that of our friends and family. By managing life's challenges logically and with reasonable expectations, we develop personal objectives for how we will thrive and achieve success throughout our lifetime.

Global Climate Change and Deep Uncertainty

Marchau et al. (2019) state and the IPCC confirms that the debate over whether the world's climate is changing through human causes has been settled. However, as society begins to address climate change, there is considerable uncertainty over the magnitude and speed of change, the effects at various locations, and how to deal with the consequences. This point was expanded in Chapter 4 in the discussion of climate knock-on effects. Not only is the magnitude and speed changing, but the continuous shift in the distribution of climate stressor values opens up new problems, extreme climate variable values in uncharted territory, and values at which engineers have little or no experience. In this territory, infrastructure project components may be subject to unexpected compounding failures: repeated values outside the optimal performance zone causing premature degradation of performance and possibly failure. Threshold values propagating new climate stressors could be reached (tipping points). Also, in this unexplored range, two or more climate variables may combine to reinforce adverse effects.

Despite these unknowns, infrastructure assets must still be planned, designed, built, operated, and repaired to meet growing and changing societal demands. New approaches for delivering and maintaining infrastructure assets are needed for successful operation under these new and often deeply uncertain climate conditions.

Approaches to Decision Making under Deep Uncertainty

Marchau et al. (2019) describe five approaches for handling deep uncertainty:

- **Robust Decision Making (RDM):** Test strategies over many plausible future scenarios to identify vulnerabilities and to find the most robust strategy (i.e., the one that is least sensitive to assumptions made, works over the widest range of futures, and keeps most options open).
- **Dynamic Adaptive Planning (DAP):** Design and implement an initial plan and provide for monitoring and adaptation over time as conditions change or new knowledge is acquired.
- **Dynamic Adaptive Policy Pathways (DAPP):** Develop a series of actions to be taken in a predetermined sequence as conditions change. Each action, a policy, operational change, or physical alteration, is taken at the point when the previous action can no longer perform as intended based on changed conditions.
- **Info-Gap Decision Theory (IG):** Focus on what is known and what needs to be known about a particular problem to make a reliable and responsible decision. Instead of optimizing the infrastructure system design to highly uncertain future conditions, Info-Gap determines which designs best handle errors in assumptions, modeling, or other project inputs.

- **Engineering Options Analysis (EOA):** Assess the value of options (i.e., the benefits of flexibility in the timing, size, and location of changes in engineered systems). The user then has the ability, but not the obligation, to make changes to an infrastructure asset over its design life based on new information.

Robust Decision Making

Under conditions of deep uncertainty, making predictions about the future is not helpful. Rather than trying to improve predictive models, robust decision making (RDM) "stress tests" the proposed strategies against a broad set of plausible scenarios and identifies the factors for which their decisions meet or miss the intended results. From those results, decision makers can determine the most robust strategies, those that deliver the best performance over that broad range of scenarios.

The RDM process steps are depicted in Figure 6-3. The four steps are summarized as follows.

Step 1: Decision Framing

Stakeholders define key factors in the analysis: objectives and criteria, alternatives for achieving those objectives, uncertainties that can affect the outcome, and relationships among actions, uncertainties, and alternatives. These factors are

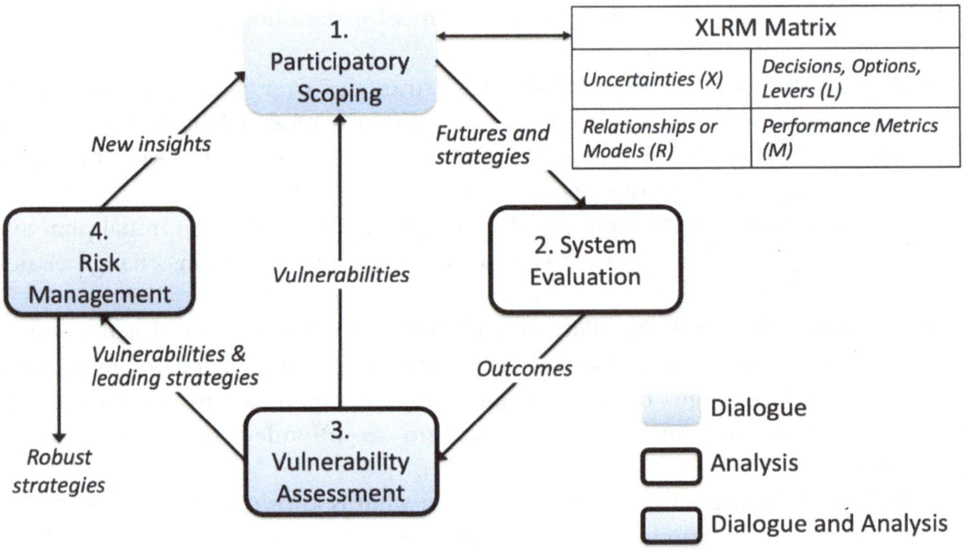

Figure 6-3. Robust decision-making (RDM) framework.
Source: Adapted from Marchau et al. (2019), p. 31; Creative Commons Attribution 4.0 International License.

organized in an "XLRM" matrix: future uncertainties (X), management actions, strategies or policy levers (L), relationships, models (R), and measures of performance (M). Initially, inputs to this matrix may be qualitative. However, as this information is developed, ranges of plausible variable values, numerical performance values, and thresholds would be established, enabling the stakeholders to determine if their objectives are being met (Groves et al. 2013b).

Step 2: Evaluate Strategy Across Futures

The performance of the various actions, strategies, and policies in the identified plausible futures is evaluated against stakeholder performance objectives. In this step, no judgments are made about the likelihood of one future over another. As a result, many combinations of plausible futures and strategy combinations or management actions (strategies or policy levers) are possible. Here, computer models are used to calculate and compare performance.

Step 3: Vulnerability Analysis

In this step, the outcomes of applying the various strategies across plausible futures are evaluated and characterized for vulnerabilities. Of interest are the decision-relevant scenarios: scenarios in which the actions, strategies, or policies meet or miss the performance objectives. Those that meet the performance objectives under many plausible futures are grouped for further evaluation in Step 4. Those that miss the objectives can be returned to Step 1, providing additional insight for devising new approaches.

Step 4: Tradeoff Analysis

Leading strategies that are most robust in uncertain futures are further evaluated. Decision makers and stakeholders bring together their values and assumptions regarding the objectives and future conditions. In this step, tradeoffs are discussed, seeking to converge on the most robust actions, strategies, and policies (i.e., those that work for the largest reasonable set of futures). If convergence is reached, actions can then be taken. If decision makers and stakeholders cannot reach convergence, they can return to Step 1 with new insights on the performance of the initial strategy set.

Example: RAND Corporation Application of RDM for the Colorado River Basin

The Colorado River and its tributaries provide water and hydropower for approximately 40 million people in the Southwestern United States. Today, the federal and

Table 6-1. Simplified XLRM Framing Matrix from the Colorado River RDM Analysis.

Uncertainties or scenario factors (X) • Demand for Colorado River water • Future streamflow or water-supply climate drivers • Reservoir operations post-2026	**Management options and strategies (L)** • Current management • Four portfolios composed of individual options: demand reduction, supply augmentation
Relationships or systems model (R) Colorado River Simulation System (CRSS)	**Performance metrics (M)** • 10-year average streamflow at Lee Ferry (Upper Basin reliability) • Lake Mead Pool Elevation (Lower Basin reliability) • Cost of option implementation

Source: Groves et al. (2013a).

state agencies overseeing this water system are concerned about its viability. Initial water allocations were made at a time when river water flow was higher than normal. Unfortunately, years of high water demand plus recent drought conditions have brought water reservoir storage in the Colorado River Basin down to half capacity. In response, these agencies undertook the Basin Study to evaluate the viability of this critical resource over the next 50 years and to develop effective water management strategies. Basin boundaries are shown in Figure 6-4.

RAND Corporation was a contributor to this study, employing RDM to identify future vulnerabilities in the Colorado River Basin and develop water management strategies that are robust across a range of water resource and use scenarios (Groves et al. 2013a). A simplified XLRM Matrix for the RDM Basin Study is presented in Table 6-1.

The Basin Study Team developed multiple scenarios—four supply, six demand, and two operations—covering a broad range of possibilities starting in 2012 and extending out to 2060. The supply scenarios were based on historical stream flows, as well as future flows derived from climate models downscaled to the Colorado Basin level. The demand scenarios incorporated variations in population growth and environmental awareness. The operations scenarios were either to extend existing guidelines beyond their 2026 expiration date for storage and shortage allocation or "No Action." These combinations of climate-driven supply, demand, and operational futures produced more than 23,000 possibilities (called *traces*).

The RAND team developed four strategy combinations (portfolios) and ran computer simulations to see how the strategies performed across the scenarios. The portfolios were described as (A) inclusive, (B) reliability focus, (C) environmental performance focus, and (D) options common to B and C.

The team looked at future vulnerabilities: (1) under which futures are water delivery objectives not met, and (2) what future conditions create significant

Figure 6-4. Colorado River Basin.
Source: US Geological Survey (2016).

vulnerabilities. Then, the team assessed cost effectiveness, comparing the annual costs of each portfolio to its effectiveness in handling system vulnerabilities. System vulnerabilities were measured in the percentage of years in which two key performance metrics of system reliability, the Lee Ferry Deficit and Lake Mead Pool Elevation, were below set numbers. The Lee Ferry Deficit is a measure of required versus actual water flow at Lee Ferry, a point about 15 mi downstream of the Glen

Canyon Dam. The Lake Mead Pool Elevation is for Lower Basin reliability, a reservoir level below 1,000 ft, the level below which water can no longer be withdrawn using its current lowest level water intake.

When considering cost versus vulnerability reduction, using RDM allowed the team to compare portfolios for their cost-effectiveness. Considering all 23,000+ possibilities (i.e., all the traces together), there was little difference between portfolios. When the traces associated with declining water supply were isolated from the rest, Portfolio A resulted in the highest reduction in vulnerability for both the Upper Basin (Lee Ferry Deficit) and the Lower Basin (Lake Mead Pool Elevation). However, the cost of Portfolio A was considerably higher than Portfolio B.

Dynamic Adaptive Planning

For long-lived infrastructure projects in a changing climate, dynamic adaptive planning (DAP) recognizes that the longer the design life, the greater the uncertainty of climate impacts (Marchau et al. 2019). The approach is to explicitly recognize and include provisions for adaptation as climate conditions change and new knowledge is acquired. Dynamic adaptive planning creates plans that combine immediate actions with commitments to shape and preserve flexibility for the future. Both infrastructure vulnerabilities and opportunities are recognized and accounted for. Dynamic adaptive planning employs monitoring systems designed to take preplanned actions when predetermined thresholds are reached. The five steps in the DAP process are presented in Figure 6-5.

Step I: Stage Setting

The process starts by setting objectives for decision makers and stakeholders along with significant constraints. Success is defined, and options for achieving success are established.

Step II: Assembling the Initial Plan

Here, objectives and definitions for success from Step I are used to define the necessary conditions for success. One or more related policy actions and their plans for implementation are also determined.

Step III: Increasing the Robustness of the Initial Plan

The overall robustness of the plan is then increased by first identifying plausible future conditions that could create vulnerabilities and opportunities. For those vulnerabilities and opportunities, there are four types of actions based on relative certainty or likelihood of exposure to a hazard (vulnerability) or developments that improve the chances for a successful outcome (opportunities). These are defined in

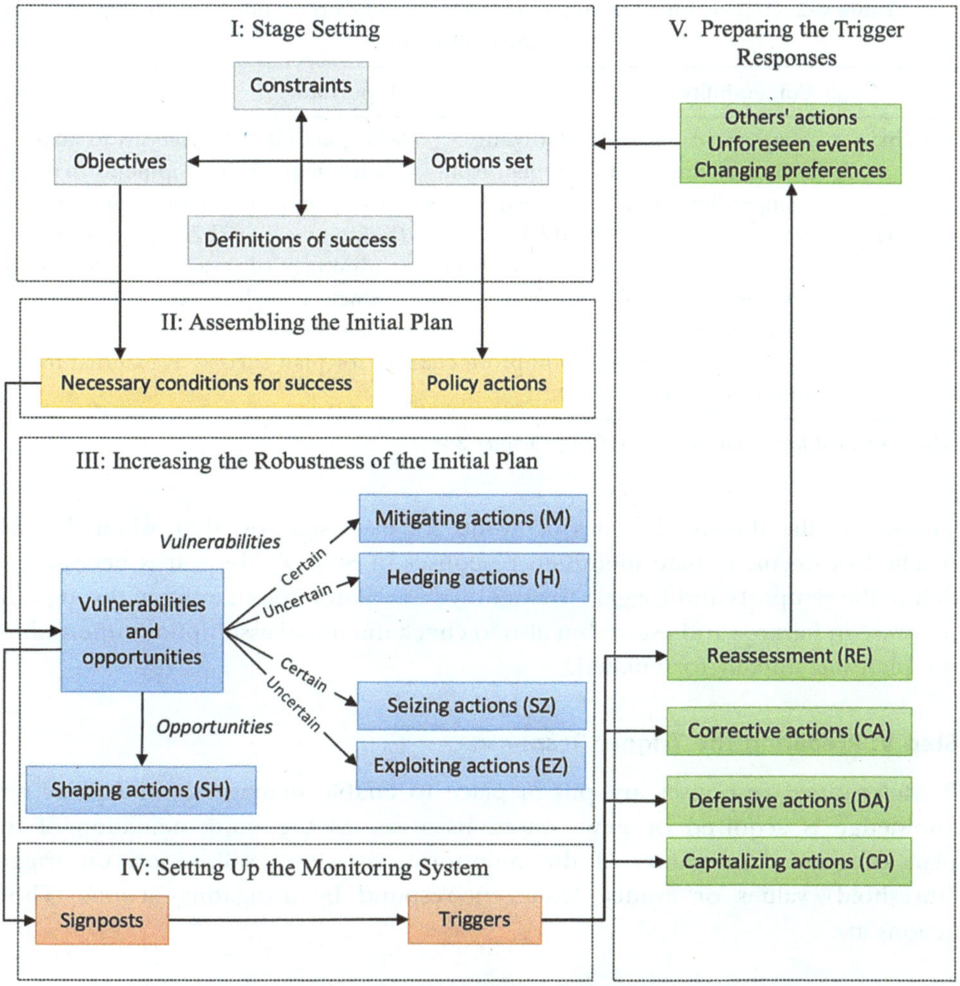

Figure 6-5. Dynamic adaptive planning process steps.
Source: Adapted from Marchau et al. (2019), p. 55; Creative Commons Attribution 4.0 International License.

Table 6-2. In addition, DAP recognizes shaping actions (SH), which are actions taken to improve chances for plan success as external trends, events, or conditions emerge.

Step IV: Setting Up the Monitoring System

To take advantage of opportunities or reduce vulnerabilities, a monitoring system is established to detect signals of change and trigger the appropriate responses. *Signposts* are the predetermined indicators of change that are to be monitored.

Table 6-2. Actions for Handling Vulnerabilities and Opportunities under Dynamic Adaptive Planning.

	Vulnerability	Opportunity
Certain	*Mitigating actions* (M). Reducing adverse impacts to the plan from impending hazards and exposures	*Seizing actions* (SZ). Actions to take advantage of developments that arise which improve chances of success
Uncertain	*Hedging actions* (H). Reducing adverse impacts to the plan from known but uncertain hazards and exposures	*Exploiting actions* (EZ). Actions to take advantage of possible developments, which, if realized, could improve chances of success
Shaping Actions (SH). Actions taken to improve chances for plan success as external trends, events, or conditions emerge		

Source: Adapted from Walker et al. (2019), Chapter 3.

Triggers are the threshold values or events for each signpost that, when they are reached or occur, initiate identified responses in Step V. Here, it is necessary to define the signposts and triggers to not only steer plan implementation through the unforeseen hazards and events but also to check the initial assumptions upon which the plan was initially formulated.

Step V: Preparing the Trigger Response

Predetermined responses are put in place to enable managers to adapt as new knowledge is acquired or new conditions arise. As the implementation of the plan progresses, operators of the monitoring program will report on trigger (threshold) values or events. Managers respond by instigating actions. Those actions are

- **Defensive actions (DA):** Preplanned actions to address exposures to hazards or other events that jeopardize the success of the plan, leaving the initial plan essentially unchanged.
- **Corrective actions (CA):** Actions to adjust the initial plan during operation to avoid plan failure in response to new knowledge or conditions.
- **Capitalizing actions (CP):** Actions to adjust the initial plan to take advantage of new opportunities in response to new knowledge or conditions.
- **Reassessment (RE):** A complete reevaluation and possible revision of the current plan if new knowledge or significant unforeseen conditions undermine the initial plan assumptions, objectives, preferences, and constraints.

Example: Impact of the Panama Canal Expansion on Caribbean Ports

In 2016, a third set of locks in the Panama Canal were opened, enabling larger (Neo-Panamax) vessels to transit through the canal. This change in lock capacity is

Figure 6-6. View of the port in Barranquilla, Colombia.
Source: Wikimedia Commons, Attribution-ShareAlike 3.0 Unported (CC BY-SA 3.0).

expected to have a significant impact on ports within the region. Old Panamax vessels will be replaced by newer and more efficient Neo-Panamax vessels, creating new opportunities for Caribbean ports that can handle larger vessels and increasing the vulnerabilities of ports without them.

A new port terminal is to be added to the current port complex in Barranquilla, Colombia (Figure 6-6). In light of the Panama Canal expansion and other related developments, port planners in Barranquilla are evaluating multiple options for the new terminal. The port's current business consists of handling petroleum liquids, grains, and containers. Long-term projections for each of these cargoes transiting through the port are subject not only to the advent of larger vessels but also to actions by other competing ports, a new and larger capacity canal, and the vagaries of international trade.

Using the DAP framework (Reyes et al. 2018), planners worked through Steps I and II by establishing the objectives and constraints for the port expansion and definitions of success. In the case of the port expansion, success is defined as the best net present value gained from the expansion across several options.

For Step III, the team used brainstorming sessions to identify plausible future conditions that could impact the plan. First, they created a set of developments and potential impacts in four major categories: technology, market and economy, politics and legislation, and society and environment. They also identified external sources of uncertainty (e.g., a slowdown in China's economy and the prospect of liquified natural gas becoming recognized as an environmentally cleaner fuel). Planners also identified external wildcards—highly improbable trends or events that have a high impact and thus present a significant risk.

Figure 6-7. Container vessel of the Neo-Panamax class squeezing through the Miraflores locks of the Panama Canal.
Source: Ralf Broskvar/Shutterstock.com.

Based on these possibilities and their impacts, the planners worked to increase plan robustness by reconfiguring future developments and impacts into potential vulnerabilities and opportunities. Per Step III, vulnerabilities and opportunities can be further refined by their level of certainty. The team then identified the actions for the likely developments (certain). For example, in the case of the opportunity created by the Panama Canal expansion and the increase in Neo-Panamax container vessels (Figure 6-7), the planning team considered a seizing action response: designing and building port infrastructures that can handle this larger scale of vessels and gain market share.

Shaping actions were also identified. The planning team identified another reasonable but uncertain development: the creation of a super deep-water port that would compete with the planned port operations. To "shape" this potential development, the planners would work to establish cooperation agreements with the other terminal operators to focus on other complementary niche markets.

This example focused on the container market segment. For the evaluation, the planning team had identified a base case plus five flexibility options for the port expansion. Flexibility strategies were developed by varying the quay wall length, the size of the container handling yard, the port depth, and the extent to which the port expansion could handle Neo-Panamax vessels.

For Step IV, monitoring thresholds and triggers were established for reassessment, corrective, defensive, and capitalizing actions. For example, based on the vulnerability that the smaller Panamax vessels would be quickly replaced with Neo-Panamax vessels, port planners would reassess plans for adding Neo-Panamax port capacity. In making this decision, planners would monitor vessel scrapping and new vessel orders and reassess their decision based on predetermined thresholds for the scrapping of old Panamax vessels and new orders for Neo-Panamax vessels, even if the current Neo-Panamax traffic is low.

These options were simulated for different strategies out to the year 2040 and expected net present values were calculated. The results showed that despite considerable uncertainties, two of the ten strategies produced the highest estimated net present value.

Example: The Oakland Touchdown

Wall et al. (2015) applied DAP to planning the Oakland approach to the San Francisco–Oakland Bay Bridge, also known as the Oakland Touchdown (Figure 6-8). The Oakland approach is a critical transportation link between the two cities. Whereas the bridge itself is not likely to be significantly affected by changes in sea levels or intense storms, the Oakland approach is vulnerable to many flood possibilities. Sea level rise and storm surges could flood the bridge approach, damaging toll plaza booths, buildings, and associated equipment. If bridge access is disrupted, the traffic on Interstate 80 crossing the bridge will be disrupted.

Climate uncertainties regarding the bridge approach are not the only concern. There are significant uncertainties in long-term travel and land-use patterns in the East Bay area, the result of climate change, as well as social, economic, and technological developments. For example, the effects of the COVID-19 pandemic transformed how

Figure 6-8. The dynamic adaptive planning study area for the San Francisco–Oakland Bay Bridge. *Source*: Wall et al. (2015).

and where people work and altered commuter traffic to and from cities. Before 2020, the start of the pandemic, San Francisco's office occupancy rate was close to 100%. But in March 2020, restrictions put in place to prevent the spread of the disease reduced the occupancy rate to 10%. COVID-19's resurgence in 2021 kept the rates low. A year later, high vaccination rates, a relaxation of mask requirements, and economic incentives brought many workers back to the city. By October 2022, the occupancy rate had risen to about 40% but still well below the rate before the pandemic. Office occupancy in cities is unlikely to return to what it used to be. High office space costs, commuter frustrations, office worker values, and remote work technologies have changed how companies operate (Symon 2022). Some office spaces are being turned into apartments.

The process steps for the Oakland approach are summarized in Figure 6-9 and described as follows.

Step I: Setting the Stage

The stated objective (Wall et al. 2015) for this plan was to "ensure adequate transportation links that serve the (current and future) demand to move people and goods between San Francisco and Alameda County." Success is then defined as the assurance that "a certain volume of people and goods can travel between San Francisco and Alameda County during the identified time frame." Constraints included the cost to alter the bridge approach, public and political acceptance, and spatial restrictions for any substantial physical alterations.

Five options were considered. The first option was to do nothing, meaning no flood resilience features would be added to the bridge approach. The second option was to keep the current approach configuration but devise various management strategies to handle flood disruptions as they occur. The third option was to improve flood protection and resilience by improving drainage, raising the approach roadway surface, and raising or water proofing the toll plaza. The fourth option involved the construction of berms, levees, or flood walls to prevent flooding, modified as needed to address rising sea levels and increasing storm intensity and duration. The fifth option was to relocate the entire bridge approach and eliminate the threats.

Step II: Assembling the Basic Plan

In Wall et al. (2015), the policy actions declared for the basic plan are Option 2, Current Configuration + Management Activities. However, as the robustness of the basic plan is increased in Step III, other elements are applied.

The policy actions in this example are shown as a blend of Options 2, 3, and 4. The conditions for success, what must happen for the project to be successful, are

1. **Travel demand develops as originally forecast:** The volume of users of the Oakland Touchdown is within expected bounds.
2. **Current bridge approach design/location is sufficient:** The bridge can withstand projected flooding from sea level rise and storm surges. Users consider traffic disruptions from flooding as rare.

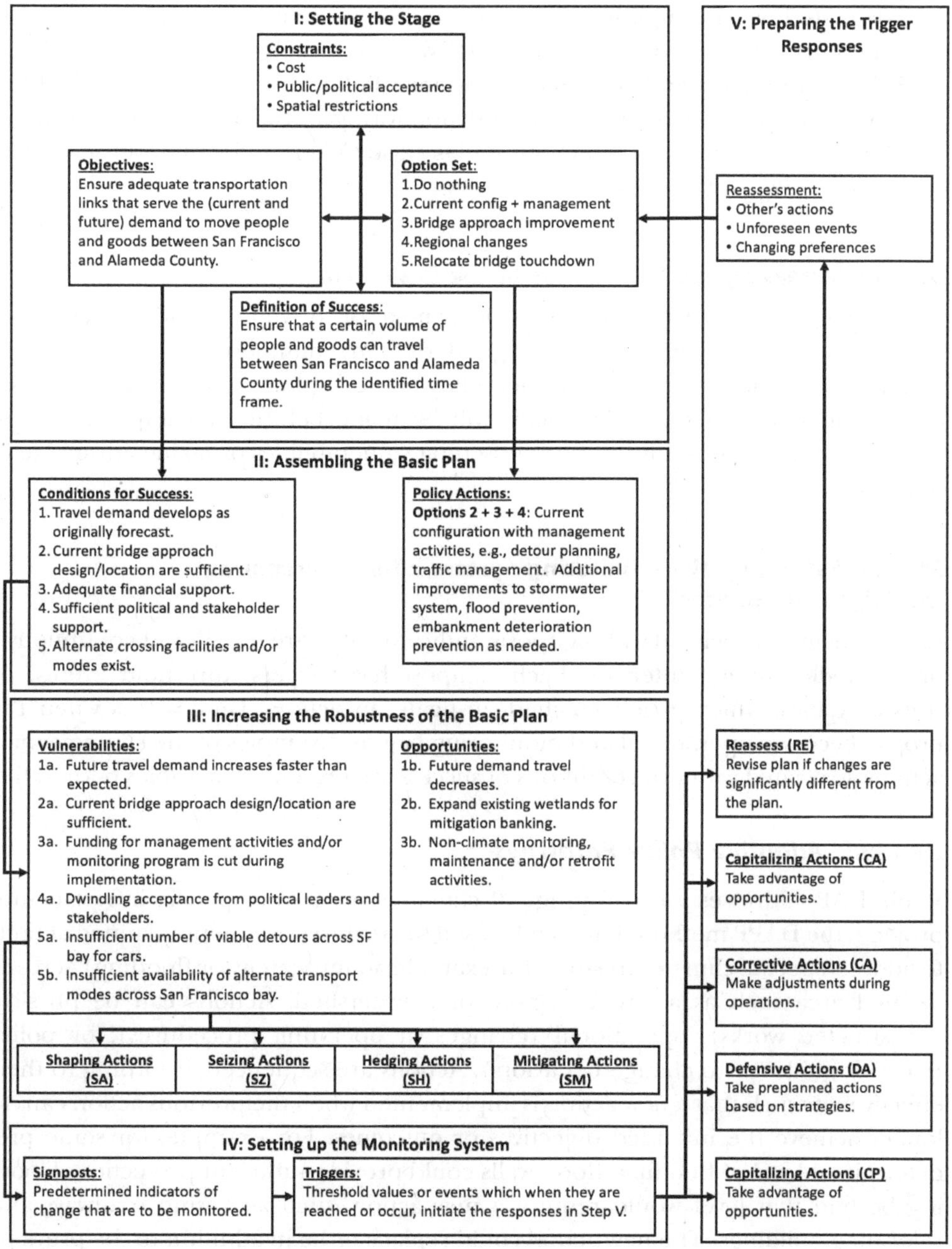

Figure 6-9. Dynamic adaptive planning process for the San Francisco–Oakland Bay Bridge approach.
Source: Adapted from Wall et al. (2015).

3. **Adequate financial support:** Sufficient funding is available to support the basic plan, plus any needed monitoring and adaptation modifications.
4. **Sufficient political and stakeholder support:** Political leaders and stakeholders support the plan along with the needed monitoring and adaptation modifications.
5. **Alternate crossing facilities and/or modes exist:** Other modes of travel across San Francisco Bay are available.

Step III: Increasing the Robustness of the Basic Plan

Increasing plan robustness involves a determination of vulnerabilities and opportunities (i.e., identifying deviations, good or bad) numbered to coincide with conditions for success. For each vulnerability and opportunity, strategies are developed. The strategies state what plans will be made before plan implementation (time < 0) for taking various actions: lessening vulnerabilities or taking advantage of opportunities.

Step IV: Setting up the Monitoring System, Step V: Preparing the Trigger Responses

Each strategy is associated with signposts, indicators that are monitored continuously or at predetermined intervals. Each signpost has triggers, threshold events, or stressor values, which, when reached, instigate an action. Time = 0 is when the project becomes operational and monitoring begins. Examples of the strategies and actions associated with two conditions of success are presented in Tables 6-3 and 6-4.

Dynamic Adaptive Policy Pathways

While DAP addresses a broad range of climate stressors on specific infrastructure projects, the DAPP methodology can be used to create a series of preplanned actions to address specific climate stressors (for example, water scarcity or flooding) across a defined area such as a city, a region, or a watershed. Actions can be physical (constructed works), operational (changes in operating procedures), or policy (rules or guidance to change behaviors). Actions are sequenced according to their effectiveness such that a new action is implemented when the previous action can no longer achieve the intended objectives or outcomes. For example, for some predetermined level of flooding, flood walls could provide sufficient protection. Flooding beyond that level would require a new action, such as raising roadways and relocating buildings. The new action could replace or be in addition to the previous action.

Conditions are monitored for predetermined variable threshold values. Reaching a specific threshold constitutes an adaptation signal, triggering a shift to one or more preplanned policy actions (pathways). This planning process consists of seven steps, as shown in Figure 6-10 (Marchau et al. 2019).

Table 6-3. Strategies and Actions for Condition for Success. Travel Demand Develops as Originally Forecast.

Condition for success: Travel demand develops as originally forecast			
Vulnerability or opportunity	**Strategies developed at time < 0** (strategies to increase robustness)	**Signpost monitoring** (starts at time = 0)	**Adaptive actions** (actions taken at time > 0)
Strategies: (**SM**) Mitigation, (**SH**) Hedging, (**SZ**) Seizing, (**SA**) Shaping. **Actions:** (**DA**) Defensive, (**CA**) Corrective, (**CP**) Capitalizing, (**SA**) Shaping, (**RE**) Reassessment			
a. Vulnerability: Future travel demand increases faster than expected *Uncertain*	(**SH**) Develop alternate traffic management plans for increased demand scenarios (e.g., tolling, public transit, or telecommuting incentives) above those planned for in the basic plan	**Monitor:** bridge traffic volume against established thresholds **Trigger:** if average daily volume exceeds upper threshold, take **CR-1** and **CR-2** actions **Monitor:** updated land-use and travel demand forecasts, regional visioning scenarios for new developments that incorporate climate-related factors **Trigger:** if available and forecasts differ significantly from those available at basic plan implementation, **RE**	(**CR-1**) Implement additional traffic management plans (e.g., increased tolling) in case those developed as part of the basic plan become insufficient (**CR-2**) Retrofit toll plaza to increase throughput capacity (**RE**) Reassess the plan
b. Opportunity: Future travel demand decreases *Uncertain*	(**SZ**) Conduct analysis of alternate uses of bridge roadway facilities, e.g., transit facilities, bicycle facilities, touristic facilities. Develop strategies to handle decreased traffic volume, alternate uses	**Monitor:** bridge traffic volume **Trigger 1:** if volume drops below lower threshold, take **CP-1** action **Trigger 2:** if volume remains below threshold for predetermined amount of time, e.g., several years, take **CP-2** action	(**CP-1**) Implement revised traffic management plans, e.g., adjust tolling to decreased volumes (**CP-2**) Implement alternate uses of portion of the bridge, e.g., bicycle facilities, touristic facilities

Source: Adapted from Wall et al. (2015).

Table 6-4. Strategies and Actions for Condition for Success: Bridge Approach Design and Location Sufficient for Expected Changes in Climate Conditions.

Condition for success: Bridge approach design and location are sufficient for expected changes in climate conditions			
Vulnerability or opportunity	**Strategies developed at time < 0** (strategies to increase robustness)	**Signpost monitoring** (starts at time = 0)	**Adaptive actions** (actions taken at time > 0)
Strategies: (**SM**) Mitigation, (**SH**) Hedging, (**SZ**) Seizing, (**SA**) Shaping. **Actions:** (**DA**) Defensive, (**CA**) Corrective, (**CP**) Capitalizing, (**SA**) Shaping, (**RE**) Reassessment			
a. Vulnerability: Significant climate-related impacts disrupt the bridge approach, and therefore access to the bridge *Uncertain*	(**SM-1**) Develop plans to increase storm drainage capacity (**SM-2**) Develop plans to protect and improve resilience of toll plaza systems, toll booths, and related buildings (**SM-3**) Develop plans to construct berms, levees, and elevation of approach roadways to protect from flooding (**SM-4**) Develop plans for installing armoring material (e.g., riprap) to protect embankment from erosion or deterioration	**Monitor:** storm drainage volume relative to system capacity **Trigger 1:** if storm drainage volume exceeds capacity for a specified number of incidents per year, take **CR-1** action **Monitor:** frequency of roadway operations disruption **Trigger 2:** if disruption frequency exceeds <u>low</u> threshold, take **CR-2** action **Trigger 3:** if disruption frequency exceeds <u>mid</u> threshold, take **CR-3** action **Trigger 4:** if disruption frequency exceeds <u>high</u> threshold, take **CR-4** action **Monitor:** bridge approach embankment for erosion or deterioration because of storm surge and wave action **Trigger 5:** if any erosion or deterioration has occurred beyond a specified level, take **CR-3**, **CR-5** actions	(**CR-1**) Improve/retrofit stormwater drainage system to increase capacity and allow for more rapid drainage of floodwaters (**CR-2**) Retrofit/waterproof toll plaza electrical systems, toll booths, and related buildings to increase resilience to inundation (**CR-3**) Construct berm to protect roadway and toll plaza from mid-century flooding (**CR-4**) Construct levee to elevate the approach roadway surface above end-of-century projections (**CR-5**) Install armoring material (e.g., rip-rap) to protect embankment from further erosion

Source: Adapted from Wall et al. (2015).

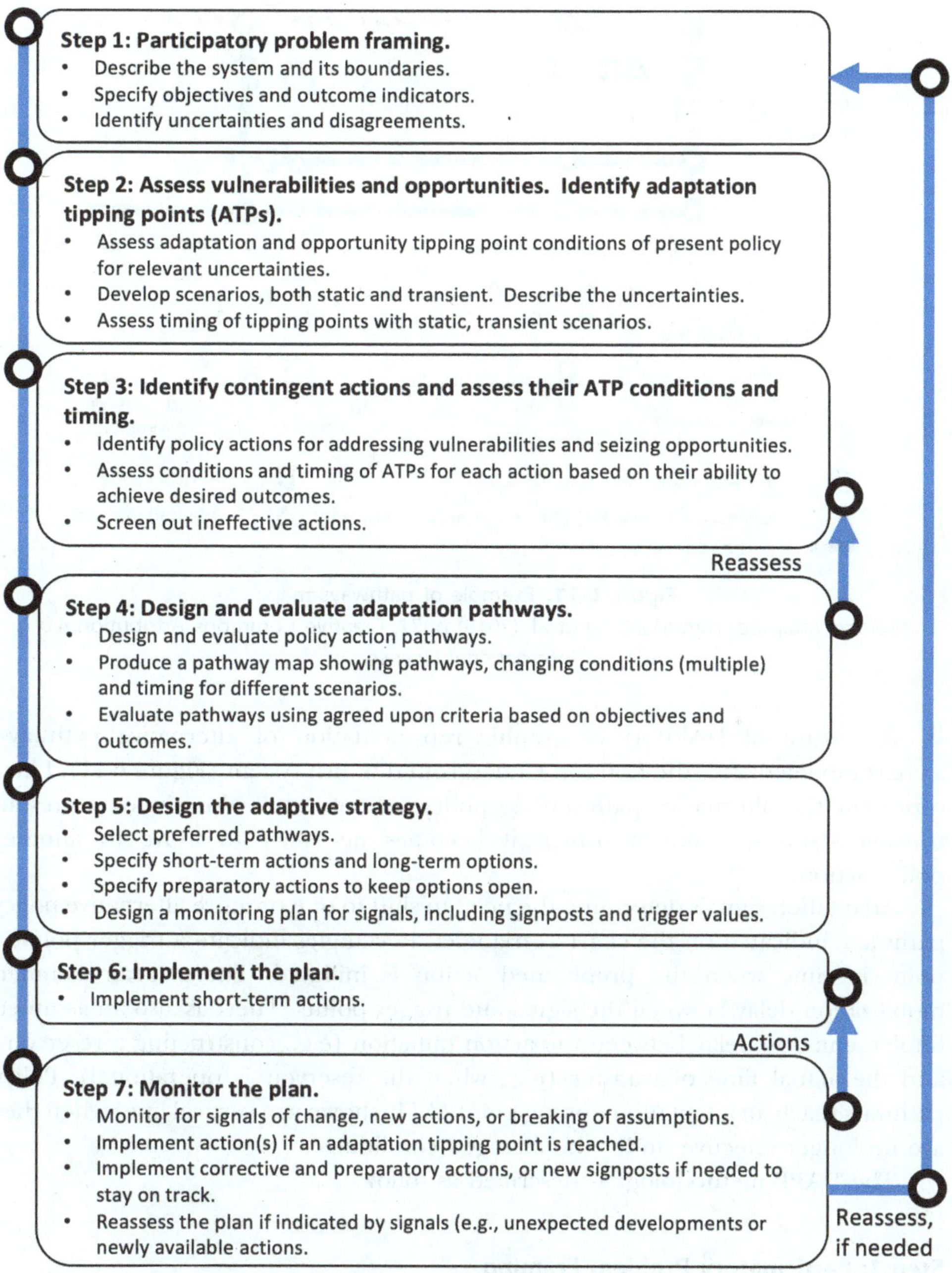

Figure 6-10. DAPP approach.
Source: Adapted from Marchau et al. (2019), p. 74; Creative Commons Attribution 4.0 International License.

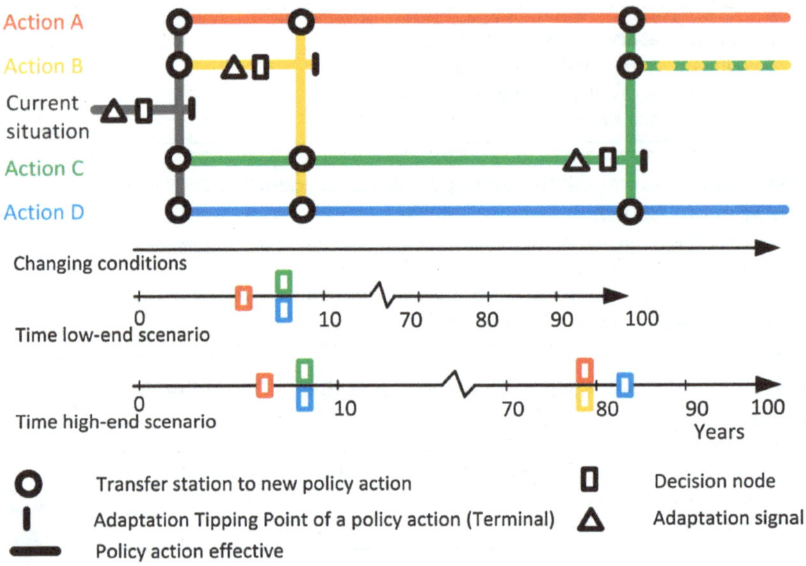

Figure 6-11. Example of pathways map.
Source: Adapted from Marchau et al. (2019) p. 77; Creative Commons Attribution 4.0 International License.

A feature of DAPP is its graphic representation of alternative pathways, a chart modeled after the London Underground transit system (Figure 6-11). Lines represent the alternative pathways or policy actions, while the circles represent transfer stations, points at which it becomes necessary to move to another policy action.

Adaptation signals determine the need to shift to one or more alternative policy pathways, indicated on the chart as triangles. Rectangles indicate a trigger point, a point in time when the preplanned action is initiated. There is an assumed mobilization delay between the signal and trigger points. There is also an assumed implementation delay between the action initiation (e.g., constructing a reservoir) and the actual time of transfer (e.g., when the reservoir is operational). Policy pathways reach an adaptation tipping point (ATP, shown as a vertical line) when they are no longer effective under the changed conditions.

The DAPP methodology is described as follows:

Step 1: Participatory Problem Framing

In this initial step, project owners, engineers, and other stakeholders establish the setting, system characteristics, objectives, current and future constraints, and definitions of success. Desired outcomes are stated in terms of measurable indicators and performance targets. Significant uncertainties, disagreements, and preferences among the stakeholders are also identified and described.

Step 2: Assess Vulnerabilities and Opportunities. Identify Adaptation Tipping Points

Based on an understanding of the system setting, characteristics, objectives, and constraints, the stakeholders identify the possible actions that could be used to achieve the desired outcomes. For each action, the stakeholders identify the *adaptation tipping points* (ATPs), also called *sell-by dates*, for different climate scenarios. The ATPs are the points at which the action will fail to perform in a way that will achieve the desired outcomes (i.e., does not meet its performance targets). The purpose of this analysis is to determine what actions could be used to achieve the desired outcomes across what range of changing conditions.

Step 3: Identify Contingent Actions and Assess Their Adaptation Tipping Points, Conditions, and Timing

Once the ATPs are determined, the stakeholders assess the actions for their ability to address the objectives and constraints. Actions considered ineffective are eliminated. The timing of action is important, as some actions (e.g., operational changes and policies) can generally be taken immediately and economically. Physical alterations typically have a longer lead time.

Thinking about the water shortages in Cape Town discussed in Chapter 1, the alternative actions were to reduce water demand through multiple conservation measures and increase water supply by using groundwater resources and constructing desalination plants. Conservation measures could be implemented immediately, while increasing the water supply required a much longer lead time. For Cape Town, one of the ATPs was the point when the reservoirs dropped to 13.5% of full capacity, the level so low that current conservation measures would allow the reservoirs to run dry before the fall rains came. At 13.5% capacity, the city would institute drastic water conservation measures. Municipal taps would be shut off, and water would only be supplied to hospitals and other critical services. Water would be supplied to Cape Town residents only at designated distribution centers (Lin 2019).

Step 4: Design and Evaluate Adaptation Pathways

With an array of possible actions, participants would evaluate these actions arranged in a logical sequence to achieve the objectives and desired outcomes. The timings for when specific actions need to be taken can be shown as a range of scenarios presented as different scales on the time axis. Multiple sequences of possible actions can then be assessed qualitatively or quantitatively based on feasibility, cost, public acceptance, or other factors. Pathway maps, such as the example in Figure 6-11, can be drawn manually or using software (e.g., Pathway Generator, available from Deltares 2017).

Step 5: Design the Adaptive Strategy

Based on the evaluations performed in Step 4, a preferred plan is selected consisting of the sequence of actions that produce the most net benefit and plan robustness

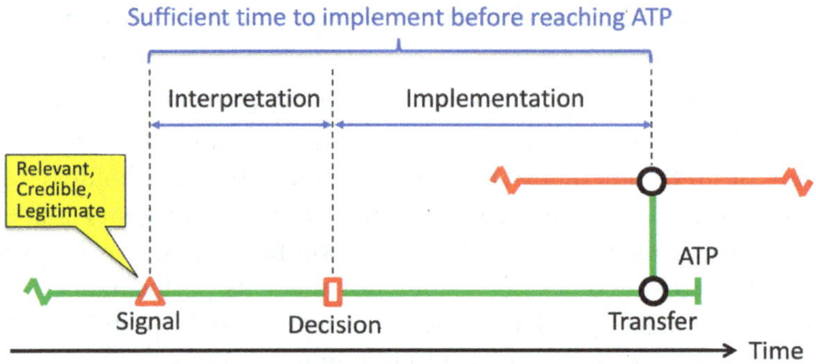

Figure 6-12. Development of a signal monitoring system.
Source: Adapted from Haasnoot et al. (2018), p. 278.

among the potential futures. Adaptation pathways will require monitoring systems that provide early warnings (signals) for upcoming ATPs, early enough to initiate and complete the preplanned actions needed to transfer to the new policy action. In such a monitoring system, the signals must be relevant (measurable, timely, and reliable), credible (technically sound and convincing), and legitimate (acceptable to the stakeholders, garners stakeholder support).

One of the difficulties in applying DAPP is determining when to transfer to the next action. Operationally, DAPP is based on the decision maker's ability to postpone investments in future actions until it is clear that shifting to the next action is warranted by changed conditions. As shown in Figure 6-12, climate monitoring programs must produce signals that the detected changes are relevant, credible, and legitimate, sufficient to instigate the next action. Once initiated, the transfer must occur before the current action reaches its ATP. This assumes that the DAPP planners have a reasonable estimate of the rate of change and have set up lead times sufficient for implementation.

Step 6: Implement the Plan

Once the strategies are set, the plan is implemented, following the path selected by the stakeholders and decision makers.

Step 7: Monitor the Plan

Monitoring programs will also begin with plan implementation. Plan managers will look for the agreed-on signals and decide whether the signal is strong enough to warrant a change to a new policy action. As new knowledge is developed, the plan may have to be reassessed.

Example: Water Resource Adaptation in Southern Portugal

Water is becoming scarce in Algarve's agricultural and resort communities, the southernmost region of Portugal, as shown in Figure 6-13. Two-thirds of its water

Figure 6-13. Algarve region of southern Portugal.
Source: Porcupen/Shutterstock.com.

resources are used for agricultural irrigation, while the remaining third supports its population and tourism. Agricultural irrigation is primarily for oranges (Figure 6-14) that demand a lot of water. Golf course irrigation (Figure 6-15) uses 8% of the region's water resources (Nunes et al. 2020).

Figure 6-14. Oranges on trees ready for harvesting in the Algarve region of southern Portugal.
Source: ADV images/Shutterstock.com.

Figure 6-15. Golf course in the Algarve region of southern Portugal.
Source: Saintee/Shutterstock.

Table 6-5. Climate Change Adaptation Options.

Adaptation solution	WEI* in 2010 (RCP 8.5)	Assessment by stakeholders and researchers
Target	43%	Keep WEI at 43%
No action	78%	Severe water stress—unacceptable
1. Improve water efficiency	74%	Region is already drought aware. Irrigation is efficient
2. 1 + Decrease irrigation needs	53%	Already tried in the past. Not socially acceptable. DISCARD!
3. 1 + Wastewater recycling	66%	Not much wastewater to recycle. Costly to distribute
4. 3 + Build new reservoirs	60%	The best places for dams already have them
5. 4 + Landscape water retention	Scalable until ~ 45%	Improve traditional water conservation. Many small-scale works
6. Desalination	Scalable until < 20%	Costly and energy-intensive now. May become less expensive over time

* WEI = Water Exploitation Index. <20%: No stress; 20%–40%: Moderate Stress; 40%–70%: Severe Stress; >70%: Scarcity

Source: Adapted from Nunes et al. (2020).

The Algarve climate is dry. Water resources consist of intermittent rivers and coastal aquifers. Current demands coupled with recurring droughts are creating water scarcity and use conflicts. The effects of climate change in this region are expected to worsen the situation.

To address these problems, the Algarve Intermunicipal Community took a multistakeholder approach to creating a climate adaptation plan (Nunes et al. 2020). The community used DAPP methodology to arrive at a sequence of adaptation measures that satisfied the stakeholders and was most cost-efficient. The stakeholders devised and assessed six adaptation solutions, comparing them against the RCP 8.5 climate scenario (business as usual) and their potential effectiveness and cost (Table 6-5).

Based on this assessment, the community developed an adaptation plan. The region's adaptation objective was to keep the Water Exploitation Index (WEI) at its current level of 43%. The WEI is the ratio of the mean annual total demand for freshwater resources divided by the long-term mean annual freshwater resources available, expressed as a percent.

A stakeholder group comprised of the affected municipalities, regional water authorities, agricultural and business associations, researchers, and environmentalists held a series of meetings to solve their water resource problems. The initial set of adaptation pathways is shown in Figure 6-16.

The researchers, composed of technical experts in climate and water resources, presented the impacts of climate change along with the region's vulnerabilities. The stakeholders and researchers proposed solutions for climate adaptation and used

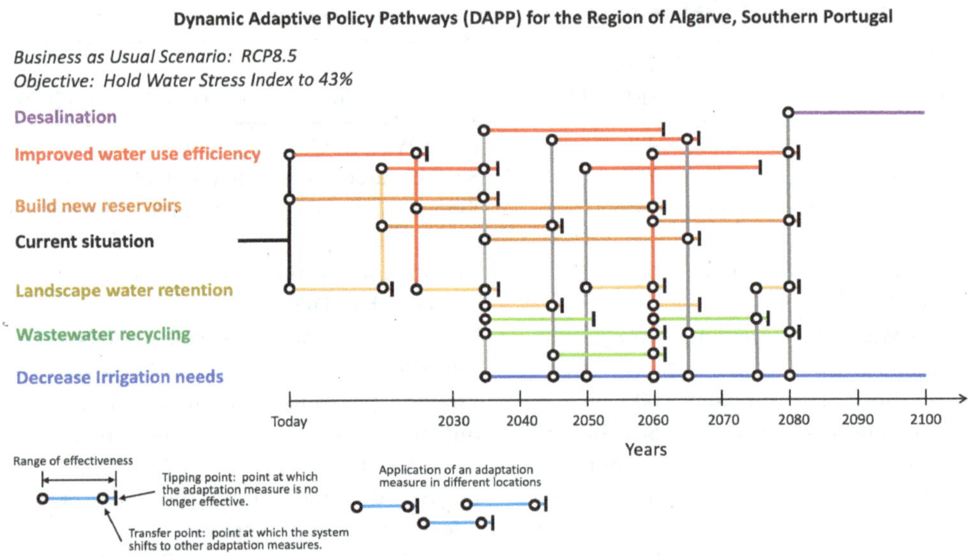

Figure 6-16. Dynamic adaptive policy pathways for the region of Algarve, southern Portugal. The initial set of pathways.
Source: Adapted from Nunes et al. (2020), Creative Commons Attribution 4.0 License.

DAPP to determine the pathway offering the best solution. This was judged based on effectiveness and lowest economic and social cost.

Six approaches were developed, consisting of activities that would keep the region's WEI at or below 43%. Tipping points were determined for each activity (i.e., when that activity would no longer be able to meet the 43% WEI objective). At that level, a new activity, within or outside the adaptation measure, would need to be initiated.

After reviewing the actions developed by the stakeholders and researchers, the group agreed on a pathway (Figure 6-17). Near-term efforts included improvements in water use efficiencies coupled with landscape water retention. When these efficiency improvements were no longer effective, wastewater recycling was seen as the next

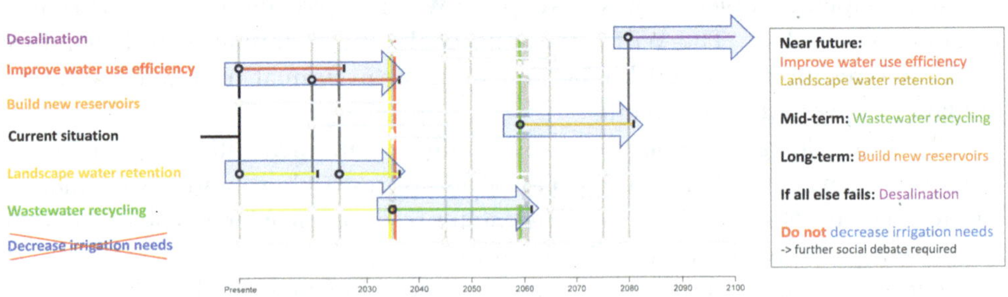

Figure 6-17. Selected adaptation approaches for the Algarve region.
Source: Adapted from Nunes et al. (2020), Creative Commons Attribution 4.0 License.

activity. Beyond wastewater recycling, the region would turn toward building new reservoirs, leaving the construction and operation of desalination plants as a last resort.

The researchers in the group offered the possibility of decreasing irrigation needs. The orange trees planted in this region are not adapted to drought and consume a large amount of water. Researchers suggested that farmers replace their orange crops with other less water-consuming crops such as almonds and olives. However, that suggestion was resisted. First, that solution has already been discussed in the past. The crops proposed would be less profitable. Furthermore, redirecting water resources from agriculture to tourism is not a socially acceptable answer to the water scarcity problem.

Information-Gap Decision Theory (IG or Info-Gap)

Info-Gap decision theory is an approach for determining what management options are good enough over an unbounded range of uncertainty in multiple parameters. It can be applied in cases where the parameter uncertainty is severe.

In the traditional infrastructure design process, the designer uses the expected future conditions to design a system that meets the project owner's requirements and specifications. However, in instances where future conditions are not just high but severely uncertain, the traditional approach is unlikely to produce a satisfactory result.

Information Gap Decision Theory (Info-Gap) alters the traditional design perspective. Instead of trying to optimize the infrastructure system design to severely uncertain future conditions, Info-Gap determines which designs are best able to handle the uncertainties of future conditions.

This approach examines two decisions:

1. **Robustness:** To what extent can a design alternative handle the error in project assumptions, modeling, and other inputs before it is unable to meet the project owner's requirements and specifications?
2. **Opportuneness:** To what extent does error in project assumptions, modeling, and other inputs produce the greatest increase in outcomes that are better than expected? Design options that produce the greatest gain in performance for small degrees of error are preferred over those that deliver gains at larger degrees of error (Smet 2017).

This approach organizes the variables into two groups:

1. Those groups that, with increasing or decreasing uncertainty, push the system most quickly toward failure. For example, in the case of water resources, uncertainty in water demand increases, and uncertainty in the intensity and duration of droughts would push water supply to critically low levels.
2. Those groups that, with increasing or decreasing uncertainty, push the system most quickly toward success. Using water resources again as an example, decreasing served populations or success in the application of residential and commercial water conservation measures would increase water supplies (Korteling et al. 2013).

The general steps in the Info-Gap analysis are described as follows using a water resources example.

A. **Build and calibrate the simulation model of the system**. The simulation model of water storage includes the capacity of the reservoir, water inflow, evaporation, and water demand.

B. **Characterize the uncertainty**. Elements of uncertainty include changes in water inflow, population, efficiency measures, demand reduction through rainwater harvesting, greywater use, and cost. These variables are further characterized by their central tendency, the central positions of those data sets. The average or mean is the most popular measure of central tendency, but others like median (middle value of a data set) or mode (most frequent value observed) could be used.

C. **Define the performance criteria**. Criteria include the minimum reservoir storage levels over the course of a year, chances of water use restrictions, streamflow below the reservoir outfall, and energy costs.

D. **Quantify the uncertainty**. An analysis is performed over random combinations of variable values using an appropriate planning horizon.

E. **Identify robust management strategies using multi-criteria decision analysis**. Generate robustness and opportuneness curves for increasing uncertainties.

Examples of an Info-Gap Analysis Application in Water Infrastructure Design

Info-Gap analysis has been applied theoretically to flood risk management in a rural location in the United Kingdom. The objective was to determine the most cost-effective management options—installation of a levee, channel widening, a combination of both, or managed retreat—all compared against a "do nothing" option. The decision criterion used was net present value. Using digital modeling of river flows and cost–benefit analysis, a 2 m high levee option had the greatest opportuneness (i.e., most chances of realizing a high net present value in flood damage avoidance for the investment made). In contrast, investing in channel widening had the least (Hine and Hall 2010).

The application of Info-Gap analysis was also assessed for high water-stressed areas in the city of Bristol (Figure 6-18) in the United Kingdom. The Bristol Water Resource Zone is considered high stress because of an increasing population and climate variability. The purpose of this analysis was to determine the best strategies for upgrading the multiple existing water resources, a combination of surface and groundwater sources, each with highly variable outputs and climate change impacts. The objective was to determine the best sequence of upgrades that would yield the most robustness in water supplies at the lowest cost.

Water supply scenarios were derived from the UK Climate Impacts Programme. River flows incorporating climate change were projected out to the year 2100. Six

Figure 6-18. Bristol: boats and buildings seen from Redcliffe Bridge, United Kingdom.
Source: Tupungato/Shutterstock.com.

water demand scenarios were also produced, assuming a low to high range of population growth. A list of 31 intervention strategy options was created for a 50-year planning horizon. Options included reductions in water consumption, system losses, and provisions for additional water resources. Each option included an estimate of costs (capital and operating) and water yield.

Computer simulations were run on approximately 2,000 supply–demand combinations. The results, summarized as percent robustness compared to net present value, were compared to Bristol Water's chosen plan. The Bristol Water plan produced a robustness value of 54% compared to 85% derived from the Info-Gap-derived strategies. Comparing the cost of Bristol Water's chosen plan to the strategy set produced by the Info-Gap analysis, the Info-Gap strategies were more cost-effective by a wide margin (Roach et al. 2015).

Engineering Options Analysis

Engineering options analysis (EOA) is a methodology that addresses uncertainty by incorporating flexibility into infrastructure project designs (Marchau et al. 2019). Flexibility means having the ability but not the obligation to make changes to an infrastructure asset over its design life based on new information. Uncertainties about future conditions, trends, or events are considered risks and opportunities. Additional flexibility incorporated into the design is valued based on its ability to mitigate risks and take advantage of opportunities as they arise. Given that infrastructure designs have many options, EOA helps decision makers determine what options have the most value.

For infrastructure projects, for example, a commercial building, the owner, faced with an uncertain market for commercial space, may decide to limit the initial building height. However, the owner can reinforce the building foundation beyond what was necessary for the initial number of stories to preserve the option to add stories should the market for commercial space increase. Additional foundation reinforcement adds cost to the initial project. Still, it brings the value of money saved on constructing additional building floors to take advantage of increased demand for commercial space (de Neufville 2017).

The steps in an EOA are summarized as follows (Marchau et al. 2019):

Step 1. **Formulate the Problem:** State the problem to be solved.

Step 2. **Establish the Objectives and Outcomes:** State what must be achieved to consider the problem to be solved. What is the definition of success?

Step 3. **Develop a Model of the System:** Devise an analytical model that accurately describes the system. What variables affect the system outcomes? For buildings, the market demand for commercial space is based on economic conditions, customer base, competition from other commercial building developers, price, future construction costs, and more. The model must be capable of running different options and generating results that reflect the objectives and outcomes established in Step 2.

Step 4. **Establish the Option:** The options should be chosen based on their effectiveness in handling uncertainties and have significant effects on system performance.

Step 5. **Determine the Relevant Uncertainties and Generate Scenarios:** Run the model over a wide range of variable values for each variable. The range and distribution of values should be based on past experience and an understanding of plausible range limits.

Step 6. **Calculate System Performance across a Range of Scenarios:** Produce results showing how the various options affected system performance across multiple scenarios. Looking again at the commercial building example, the information of interest is which options produced the best results under what circumstances.

Step 7. **Reduce the Information in Step 6 for Management/Stakeholder Consideration:** Present the information in a form that compares options against objectives and outcomes.

Step 8. **Select the Preferred Strategy:** Based on the information presented in Step 7 and considering tradeoffs among objectives and outcomes, management and stakeholders select the best strategy. The chosen strategy should be the one that delivers the best results and retains the most flexibility.

Step 9. **Implement the Selected Strategy and Monitor for Changes in Key Variables:** Management proceeds with the chosen strategy and monitors for changes. Management exercises options as conditions change and appear advantageous.

Example: An Engineering Options Analysis (EOA) Approach for the Replacement of the Netherlands IJmuiden Pumping Station

The pumping station in the city of IJmuiden (Figure 6-19) is one of many in the Netherlands. It is a component of the inland waterways system used for transportation, flood control, drought mitigation, water quality control, and more (Marchau et al. 2019). First constructed in 1975 and expanded in 2004, the pumping station is the largest in Europe, capable of pumping water at 260 m^3/s.

As it approached the end of its design life, the Rijkswaterstaat (the managing government agency) was faced with a critical decision: how to design the next pumping station in a way that would meet its primary functions: flood control and water level regulation (Van Vuren et al. 2015). Its ability to perform these functions satisfactorily is difficult because of climate uncertainties: sea level rise and precipitation. Other factors include economic development (increased precipitation runoff from impervious surfaces) and land subsidence. Engineering options analysis techniques were applied to evaluate various designs for the pumping station replacement.

Several design alternatives were considered (Figure 6-20).

- **Fixed design** using an extended factor of safety.
- **Reactive–adaptive design**, allowing for changes as future conditions warrant.

Figure 6-19. The port city of IJmuiden, Netherlands straddling the North Sea Canal. *Source*: Aerovista Luchtfotografie/Shutterstock.com.

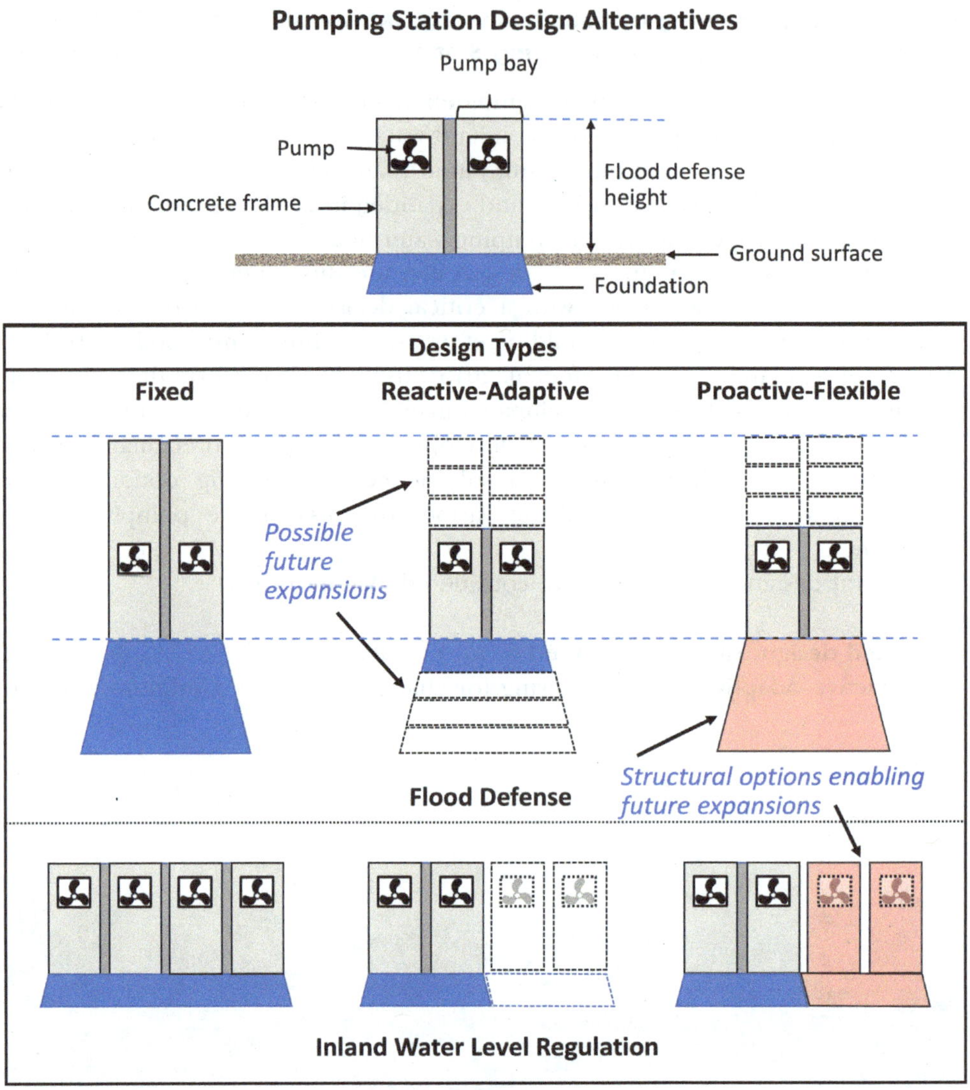

Figure 6-20. EOA design alternatives for the IJmuiden pumping station.
Source: Adapted from Marchau et al. (2019), Table 11.6, p. 242; Creative Commons Attribution 4.0 International License.

- **Proactive–flexible design** that allows for short-term adaption while anticipating and incorporating structures needed for possible future adaptation requirements.

Here, it is important to note that Dutch law requires that flood defenses are capable of protecting the area from flooding from the North Sea, limiting the probability of flooding to once every 10,000 years (Higgins 2012). In summary, the

Table 6-6. Design Alternative Performance against Objectives.

Design alternatives	Description	Water level regulation	Flood prevention
Fixed	Traditional. Minimum level of service plus safety margin	Reject. All pumping capacity installed at start	Both have comparable lifetime economic performance
Proactive–flexible	Flexible design plus incorporate options for adaptation	Better prepared for large future environmental changes	
Reactive–adaptive	Design with best available information. Make changes as needed	Lower first costs but higher long-term costs	Reject. Short-term savings do not outweigh future risks

Source: Adapted from de Neufville et al. (2019).

primary functions of the pump station are flood defense coupled with regulating inland water sea levels.

In this example, two sea level rise scenarios and four precipitation scenarios were applied across a 2015 to 2100 time frame. The scenarios were Netherlands-specific and provided by the Royal Netherlands Meteorological Institute. In the best-case scenario, the project was designed and built in such a way as to facilitate the dismantling of the pumping station and repurposing of the components and materials for other uses.

Engineering optional analysis used a physical performance model to project future climate-related conditions to the performance indicators of interest (e.g., water levels) to determine the conditions at which the physical system no longer provides the required level of service. An economic evaluation model used those outputs to compare the design alternatives in terms of life-cycle costs using discount rates of 0% (no discounting of future costs) and 5.5%.

Engineering optional analysis provided useful insights for the decision makers, summarized in Table 6-6. For regulating water levels, the reactive–adaptive and the proactive–flexible design were preferred over the fixed design. Between these two, the reactive–adaptive design offers short-term savings by requiring a smaller structure, but future costs may be incurred because of environmental changes. In contrast, the proactive–flexible design is better prepared for these changes. For flood prevention, both the fixed and the proactive–flexible have comparable lifetime economic performance.

Conclusion

Engineers have always managed uncertainty in both work and life. Today, with the advent of climate change, engineers working in the built environment must tackle new levels of uncertainty. This chapter presented five methodologies for handling changed conditions. Chapters 7 and 8 present two of the five in more detail, modified versions of DAP and DAPP. Dynamic adaptive planning is useful for handling climate stressors and other uncertainties for a single infrastructure asset. Dynamic adaptive policy pathways address one or two climate stressors (e.g., water shortage, flooding) for a defined area such as a city, region, or watershed.

References

Ayyub, B. M., ed. 2018. "Climate resilient infrastructure: Adaptive design and risk management." Accessed February 5, 2024. https://sp360.asce.org/PersonifyEbusiness/Merchandise/Product-Details/productId/244232276.

Bringas, J. E., ed. 2016. *Handbook of comparative world steel standards.* ASTM DS67D. West Conshohocken, PA: ASTM.

Brown, S. M., D. R. Lincoln, and W. A. Wallace. 1990. "Application of observational method to hazardous waste engineering." *J. Manage. Eng.* **6** (4): 479–500. https://doi.org/10.1061/(ASCE)9742-597X(1990)6:4(479).

Courtney, H., J. Kirkland, and P. Viguerie. 1997. "Strategy under uncertainty." Accessed February 5, 2024. https://hbr.org/1997/11/strategy-under-uncertainty.

de Neufville, R. 2017. "Engineering options analysis for infrastructure decision-making." Accessed February 5, 2024. https://www.deepuncertainty.org/wp-content/uploads/2018/01/15Nov-0930-Neufville-Engineering-options-analysis.pdf.

de Neufville, R., K. Smet, M.-A. Cardin, and M. Ranjbar-Bourani. 2019. "Engineering options analysis (EOA): Applications." In *Decision making under deep uncertainty: From theory to practice,* edited by V. A. W. J. Marchau, W. E. Walker, P. J. T. M. Bloemen, and S. W. Popper. Cham, Switzerland: Springer. https://doi.org/10.1007/978-3-030-05252-2.

Deltares. 2017. "Pathway generator." Accessed March 8, 2024. https://publicwiki.deltares.nl/display/AP/Pathways+Generator.

Groves, D. G., J. R. Fischbach, E. Bloom, and D. Knopman. 2013a. "Adapting to a changing Colorado river: Making future water deliveries more reliable through robust management strategies." Accessed February 5, 2024. https://www.rand.org/pubs/research_reports/RR242.html.

Groves, D. G., J. R. Fischbach, N. Kalra, E. Molina-Perez, and D. Yates. 2013b. "Developing robust strategies for climate change and other risks: A water utility framework." Accessed February 5, 2024. https://www.rand.org/pubs/research_reports/RR977.html.

Haasnoot, M, S. van't Klooster, and J. van Alphen. 2018. "Designing a monitoring system to detect signals to adapt to uncertain climate change." *Global Environ. Change* **52** (2018) 273–285. https://doi.org/10.1016/j.gloenvcha.2018.08.003.

Higgins, A. 2012. "Lessons for US from a flood-prone land." Accessed February 5, 2024. https://www.nytimes.com/2012/11/15/world/europe/netherlands-sets-model-of-flood-prevention.html.

Hine, D., and J. W. Hall. 2010. "Information gap analysis of flood model uncertainties and regional frequency analysis." *Water Resour. Res.* **46** (1): W01514. https://doi.org/10.1029/2008WR007620.

Korteling, B., S. Dessai, and Z. Kapelan. 2013. "Using information-gap decision theory for water resources planning under severe uncertainty." *Water Resour. Manage.* **27** (4): 1149–1172. https://doi.org/10.1007/s11269-012-0164-4.

Lempert, R. J., and S. W. Popper. 2005. "High-performance government in an uncertain world." In *High performance government: Structure, leadership, incentives*, edited by R. Klitgaard and P. C. Light, 117–118. Santa Monica, CA: Rand Corporation.

Lin, J. 2019. "Avoiding a water crisis: How Cape Town avoided 'Day Zero'." Accessed February 5, 2024. https://globalresilience.northeastern.edu/avoiding-a-water-crisis-how-capetown-avoided-day-zero/.

Marchau, V. A. W. J., W. E. Walker, P. J. T. M. Bloemen, and S. W. Popper. 2019. *Decision making under deep uncertainty: From theory to practice.* Cham, Switzerland: Springer Nature. https://doi.org/10.1007/978-3-030-05252-2.

NOAA NCEI (National Oceanic and Atmospheric Administration and National Centers for Environmental Information). 2022. "US billion-dollar weather and climate disasters." Accessed February 5, 2024. https://www.ncdc.noaa.gov/billions/.

NOAA NCEI. 2024. "2023 was the warmest year in the modern temperature record." Accessed March 7, 2024. https://www.climate.gov/news-features/featured-images/2023-was-warmest-year-modern-temperature-record.

Nunes, J. P., L. F. Dias, B. A. Aparício, I. Morais, et al. 2020. "Using dynamic adaptive policy pathways and hydrological modelling to co-create water resource adaptation policies for climate change: A practical example for southern Portugal." Accessed February 5, 2024. https://doi.org/10.5194/egusphere-egu2020-11015.

Reyes, O. S., P. Taneja, B. A. Pielage, and M. van Schuylenburg. 2018. "Seizing opportunities from Panama Canal expansion through adaptive port planning: A case study`PF the Caribbean Port of Barranquilla." Accessed February 5, 2024. https://coms.events/pianc-panama/data/x_abstracts/x_abstract_193.pdf.

Roach, T., Z. Kapelan, and R. Ledbetter. 2015. "Comparison of info-gap and robust optimisation methods for integrated water resource management under severe uncertainty." Accessed February 5, 2024. https://www.sciencedirect.com/science/article/pii/S1877705815026259.

Smet, K. S. M. 2017. "Engineering options: A proactive planning approach for aging water resource infrastructure under uncertainty." Ph.D. thesis, Graduate School of Arts and Sciences, Harvard University. https://edepot.wur.nl/446335.

Smith, A. 2023. "2022 U.S. billion-dollar weather and climate disasters in historical context." Accessed April 18, 2024. https://www.climate.gov/news-features/blogs/beyond-data/2022-us-billion-dollar-weather-and-climate-disasters-historical.

Smith, A. 2024. "2023: A historic year of U.S. billion-dollar weather and climate disasters." Accessed April 18, 2024. https://www.climate.gov/news-features/blogs/beyond-data/2023-historic-year-us-billion-dollar-weather-and-climate-disasters.

Symon, E. 2022. "San Francisco office occupancy still under 40% despite ending mask mandate." Accessed February 5, 2024. https://californiaglobe.com/articles/san-francisco-office-occupancy-still-under-40-despite-ending-mask-mandate/.

Taleb, N. N. 2007. *The black swan: The impact of the highly improbable.* New York: Random House.

Van Campenhout, C. 2023. "2023 on track to become hottest year on record, says EU climate service." Accessed October 31, 2023. https://www.reuters.com/business/environment/2023-track-become-another-record-breaking-year-temperature-hits-new-high-2023-10-05/.

Van Vuren, S., V. Kinings, T. Jansen, M. Van der Vlist, and K. Smet. 2015. "Dealing with aging of hydraulic infrastructure: An approach for redesign water infrastructure networks." In *Proc., 36th IAHR Congress.* The Hague, Netherlands.

Wall, T. A., W. E. Walker, V. A. Marchau, and L. Bertolini. 2015. "Dynamic adaptive approach to transportation-infrastructure planning for climate change: San-Francisco-bay-area case study." *J. Infrastruct. Syst.* **21** (4): 05015004. https://doi.org/10.1061/(ASCE)IS.1943-555X.0000257.

US Geological Survey (USGS). 2016. "Colorado river basin map." Accessed April 18, 2024. https://www.usgs.gov/media/images/colorado-river-basin-map.

Walker W. E., W. A. W. J. Marchau, and J. H. Kwakkel. 2019. "Dynamic adaptive planning." In *Decision making under deep uncertainty: From theory to practice.* Cham, Switzerland: Springer Nature.

Additional References (Not Cited)

Agence France-Presse. 2021. "Cost of world's 10 most expensive natural disasters rises to $170bn, aid group finds." Accessed January 19, 2022. https://www.theguardian.com/world/2021/dec/27/cost-of-worlds-10-most-expensive-natural-disasters-rises-to-170bn-aid-group-finds.

Lempert, R. J., and M. T. Collins. 2007. "Managing the risk of uncertain threshold responses: Comparison of robust, optimum, and precautionary approaches." *Risk Anal.* **27** (4): 1009–1026. https://doi.org/10.1111/j.1539-6924.2007.00940.x.

New Zealand Ministry for the Environment. 2017. "Preparing for coastal change: A summary of coastal hazards and climate guidance for local government." Accessed February 5, 2024. https://environment.govt.nz/publications/preparing-for-coastal-change-a-summary-of-coastal-hazards-and-climate-change-guidance-for-local-government/.

NOAA NCEI (National Oceanic and Atmospheric Administration and National Centers for Environmental Information). 2022. "US billion-dollar weather and climate disasters." Accessed February 5, 2024. https://www.ncdc.noaa.gov/billions/.

CHAPTER 7

A Proposed Dynamic Adaptive Planning Methodology

"We cannot direct the wind, but we can adjust the sails."

Dolly Parton

Introduction

Chapter 6 covered multiple approaches for handling the uncertainties, particularly those associated with climate and weather conditions. Two stood out as relatively straightforward and practical approaches for project owners and engineers. Dynamic adaptive planning (DAP) is a methodology for handling a broad set of uncertain conditions for single infrastructure projects. As a complement, dynamic adaptive policy pathways (DAPP) is a methodology for dealing with one or more changing conditions over a defined area, e.g., protecting a city from flooding or handling water scarcity for a particular region. In this case, the risk of damage from these specific changing conditions is reduced for the populations, businesses, and infrastructure assets located within the defined geographic area.

This chapter will address the challenge of making infrastructure projects sustainable and climate resilient using DAP. A modified version of DAP is presented here, which will help clarify the DAP steps and better align with traditional infrastructure design and delivery practices.

Additional steps were added to cover the entire life cycle of a project: planning, design, construction, and operation. A new Step IV was added to distinguish the process of evaluating and selecting the best project alternative. The monitoring step was expanded to recognize a scanning element designed to identify external trends and events that the project owner or operator could use to shape the project outcome: mitigate vulnerabilities or take advantage of opportunities. A new Step VII (Deliver the Project) was added to make project design and construction an explicit part of the process. Successful project delivery is integral for overall project success.

Resiliency strategies were included in Step III to enhance the existing robustness and adaptation measures. It is highly likely that some extreme climate stressors in the future may be too much for the current measures to handle. Therefore, it is important to incorporate some resiliency measures that enable quick and cost-effective recovery from such extreme events. Resiliency strategies align with the recovery actions outlined in Step VI.

Finally, some of the terminology has been changed, replacing the term *actions* to improve robustness in Step III with *strategies.* This was done to differentiate them from the actions described in Step VI.

Modified Dynamic Adaptive Planning: A Summary

The steps involved in this modified DAP process are depicted in Figure 7-1 and summarized as follows.

- **Step I: Set the Stage:** Working with the project stakeholders, the project owner sets down the project objectives, taking into consideration the project's importance as well as its constraints. The objectives are a set of definable results or outcomes expected from the completed project. In general, success is defined as achieving the owner's requirements and specifications reliably, safely, and cost effectively throughout the project life cycle. From this, a set of options can be devised. The results and outcomes should be expressed as targets and acceptable ranges. This will allow designers to balance risk against performance in the later steps.
- **Step II: Assemble the Initial Plan:** From the options set, project alternatives can be produced. The alternatives are options expanded to a level of structure and detail that enable evaluation, comparison, and selection. In this step, the necessary conditions for success are established: project conditions and assumptions that must be maintained throughout the project's useful life.
- **Step III: Improve the Robustness of the Plan:** Using the necessary conditions for success, engineers can determine vulnerabilities and opportunities: deviations from those conditions that respectively reduce or improve the chances for project success. For each alternative, strategies for handling vulnerabilities and opportunities are then developed based on the level of certainty. Resiliency has been added to the basic DAP process, taking into consideration the possibility that deviations may exceed the ability of the preplanned strategies and actions to handle them. Resiliency strategies and actions enable the project to recover from failures relatively quickly and at a relatively low cost. These strategies and actions are limited by cost and the availability of options.
- **Step IV: Evaluate and Select the Best Alternative:** Using the information developed in Step III, the project team evaluates the alternatives. Alternative selection is based primarily on which alternative is the most robust and cost effective in achieving the project objectives and meeting the owner's requirements and specifications throughout the project life cycle.

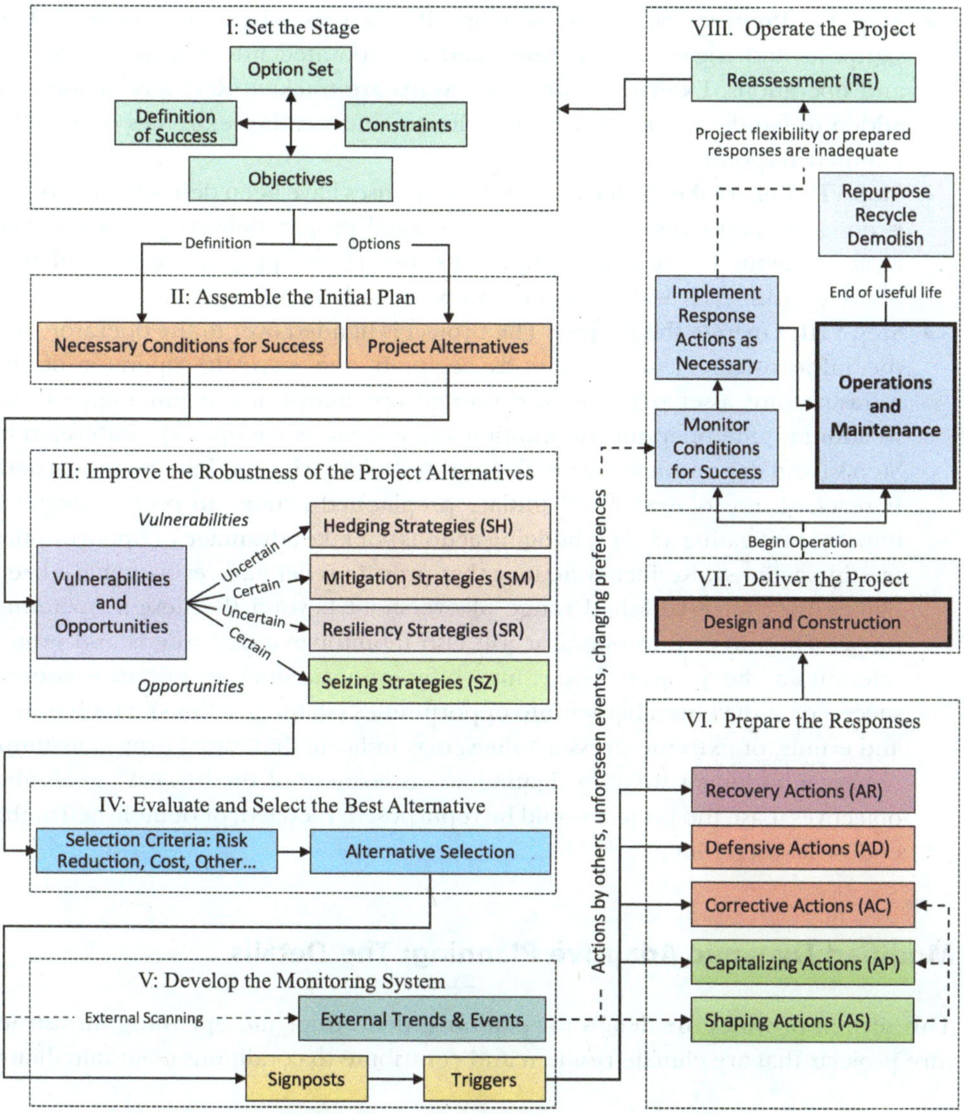

Figure 7-1. Modified dynamic adaptive planning process.

- **Step V: Develop the Monitoring System:** The actions associated with strategies determined in Step III for the selected alternative are now defined in terms of the variables or events to be monitored (signposts) and actions taken at specific threshold values or incidents (triggers). Monitoring of external events is also established. These are trends or events that may offer opportunities for devising shaping strategies (i.e., strategies to influence outcomes that improve performance by capitalizing on opportunities or reducing vulnerabilities).

- **Step VI: Prepare the Responses:** Specific actions based on pre-established signposts and triggers are defined and incorporated into the project design and operations. External trends and events are tracked. Recovery actions are added to handle instances in which climate and weather extremes overwhelm planned responses.
- **Step VII: Deliver the Project:** Once the responses have been defined, the project is designed and constructed using traditional project delivery processes. The work elements for designing and constructing the project are divided into discrete tasks, each with a defined scope, schedule, and cost.
- **Step VIII: Operate the Project:** The project is handed over to the operator, and the infrastructure asset operations begin. In this step, the operator of the infrastructure asset performs the normal operations and maintenance tasks. In addition, the operator will monitor the indicators (signposts) established in Step VI and determine if any indicator thresholds (triggers) have been reached. If reached, the operator will initiate preplanned actions to reduce negative impacts (mitigating actions, hedging actions) or take advantage of opportunities (seizing actions). Resiliency actions that provide relief and recovery have been planned for an established range of overshoot beyond the design operating range. The owner and operator will also monitor external trends and events relevant to the project. These may offer opportunities to enhance project success or turn vulnerabilities into opportunities (shaping actions). Some trends and events, or extreme stressor values, may indicate that initial project assumptions are no longer valid, instigating a reassessment of the project's goals and objectives. Last, the project would be repurposed, recycled, or demolished at the end of its useful life.

Modified Dynamic Adaptive Planning: The Details

This section provides the details for planning, delivering, and operating infrastructure projects that are climate resilient and contribute to conditions of sustainability.

Step I: Set the Stage

The components of this step are shown in Figure 7-2. This step consists of four tasks: (1) establish project goals and objectives, (2) determine project constraints, (3) define project success, and (4) develop an option set. These tasks are described in detail.

Completing these tasks is an iterative process involving interactions among the four parties: project owner, engineers and others on the project design team, public agencies with permitting and other project approval authorities, and key stakeholders. The outcomes of this step consist of (1) an initial set of project alternatives,

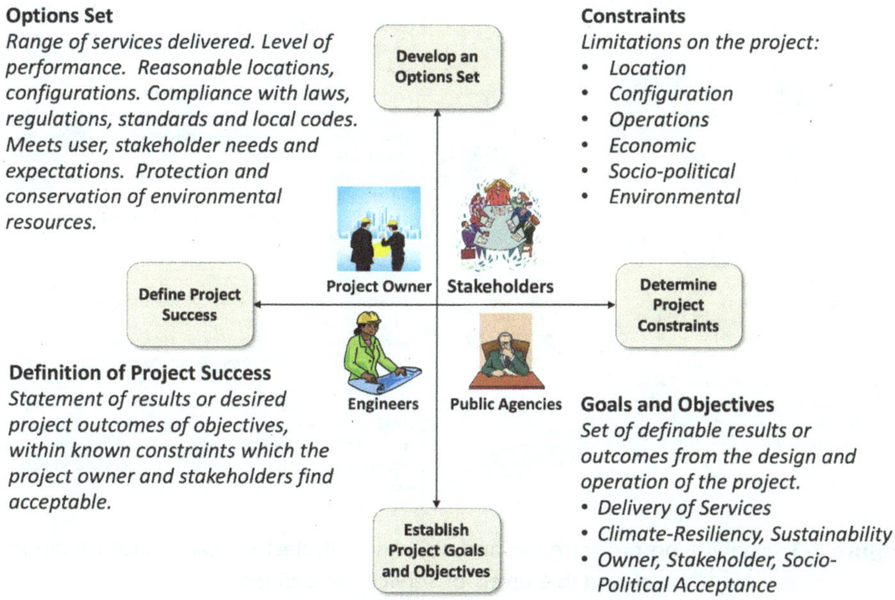

Figure 7-2. Set the stage components and outcomes.

project options that can achieve the goals and objectives within the defined project constraints, and (2) a definition of project success.

Step I A: Establish Project Goals and Objectives

Any project starts with a specific determination of the project's goals and objectives, a set of definable results and outcomes from the design and operation of the completed infrastructure asset. In this step, the results and outcomes are divided into three basic elements: (1) delivery of services, (2) climate resiliency and sustainability, and (3) owner, stakeholder, and sociopolitical acceptance. Depending on the specific project, additional elements or subelements could be added. The determination of project goals and objectives is depicted in Figure 7-3. These elements are outlined and described as follows.

Step I A.1: Delivery of Services. Provide the Required Functionality at Acceptable Performance Levels Consistent with Infrastructure Asset Criticality

The services delivered by the completed project must be safe, reliable, and meet the needs of the project owner, operator, users, and stakeholders. To make this determination, the owner and project team should do the following.

Define Project Functionality: What are the services the completed infrastructure project is expected to provide throughout its design life? When completed and operational, the project must meet the needs and expectations of the owner, operator, and affected communities. The affected communities include the host,

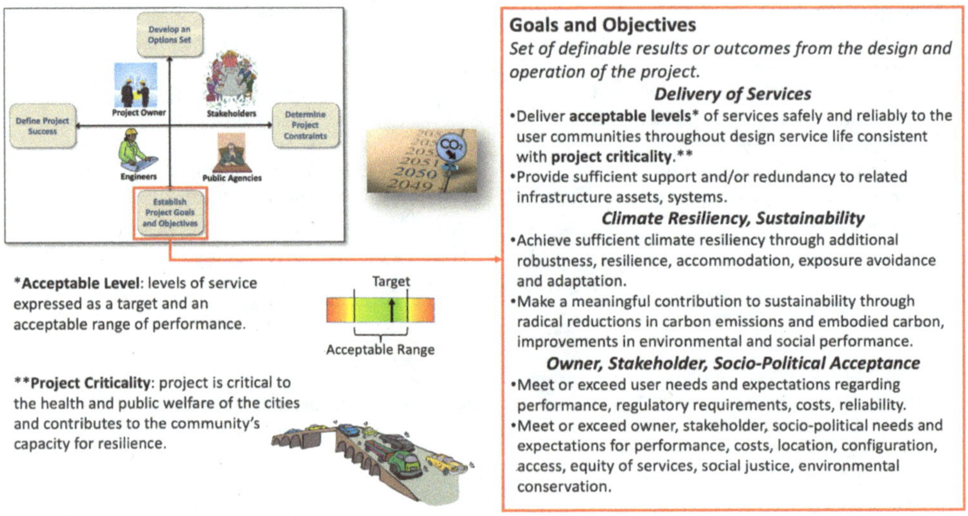

Figure 7-3. Establishing project goals and objectives. Project criticality and the range of acceptable levels of service are defined.

user, and others impacted by the project. In addition, as part of an infrastructure system, the completed project may support other infrastructure assets and provide service redundancy in case of failures.

Determine Project Criticality: How important is this project to the owner, stakeholders, users, and affected communities? In general terms, an infrastructure asset is seen as critical based on its importance to the health and welfare of the communities it serves, its contribution to their economies, its relationship to and effects on the environment, and its resilience in changing climate conditions.

Establishing the infrastructure asset criticality serves two purposes. First, it establishes acceptable levels of performance. The more critical the project to the communities it serves, the more reliable and durable project performance needs to be. And, in the case of failure, recovery needs to be fast and relatively low cost for a substantial range of stressor variable values beyond the design operating range. Service redundancy may be called for. Second, it determines climate scenario selection for establishing initial infrastructure project risk. The more critical the project, the more conservative the climate scenario selection should be, anticipating the possibility of and preparing for the worst.

In addition, the project may be part of a larger regional program for climate adaptation, one of a number of projects planned to be brought online in accordance with identified climate stressors and threshold values in a DAPP process. In these instances, the project delivery process presented here remains the same. However, the delivered project asset may be critical as an element of a DAPP regional plan designed to reduce the risk of exposure to specific climate hazards across a specified range of stressor values (see Chapter 8).

Determine Acceptable Levels of Project Performance: The performance of those services is further described in terms of output, reliability, durability, and other factors, all of which are to be delivered at acceptable levels of performance consistent with the asset criticality. Here, it is essential to define a range of acceptable performance as well as a specific performance target. This will allow engineers to make modifications in the design that reduce risk without making significant compromises in performance.

Determining Project Criticality

According to the Federal Highway Administration (FHWA 2014), project criticality can be determined in several ways. Criticality can be evaluated through

- **Numerical Analysis:** Desk reviews using quantitative information, numerical models, and decision support tools to determine asset importance. Although this approach is repeatable and transparent, it may miss nonquantifiable elements that have local importance.
- **Local Knowledge:** Project criticality can also be determined through stakeholders, people with local knowledge and expertise. This approach can provide a more detailed and informed understanding of matters important to the user communities. However, it can be subjective.
- **Hybrid Approach:** A better choice is a hybrid approach, combining local knowledge and understanding with quantitative information.

Criticality criteria for various infrastructure categories will vary because of the nature of the project. Water and wastewater projects are critical to community health and well-being, while transportation projects are essential for economic viability and other services. For example, in considering the importance of a highway project, criticality considerations could include

- Access to workplaces, healthcare, education, government, and residences;
- Transport of goods and services;
- Military, security access;
- Safety, fire and police, evacuation;
- Future community, regional goals; and
- Geographic and social diversity (CAMPO 2015).

Note that input for determining project criticality should be sought from key stakeholders and owners and operators of other related infrastructure assets. Relevant regulatory requirements and design standards can also provide guidance on project criticality.

Step I A.2: Climate Resiliency and Sustainability

In addition to the delivery of services, infrastructure projects must accomplish two things: (1) be sufficiently resilient to a changing climate and (2) make a meaningful contribution to conditions of sustainability over the project's design life. Achieving a sufficient degree of climate resiliency means anticipating and adapting to changing climate conditions in a way that results in reliable delivery of services at an acceptable level of performance and quality. Making a meaningful contribution to conditions of sustainability means reducing the project's ecological footprint while maintaining or enhancing (not degrading) the environmental, economic, and social performance of the user, host, and other affected communities.

- **Achieve Sufficient Climate Resiliency:** The completed project is expected to maintain functionality through additional robustness, resilience, accommodation, exposure avoidance, and adaptation during its entire design life. A sufficient level of climate resiliency is defined as the ability of the completed project to resist or absorb climate stressor hazards while maintaining its design functionality. This is accomplished through a combination of climate stressor hazard accommodation, avoidance, and protection for each project component, fully integrated into the infrastructure asset at the system level.

  The degree to which the project is designed to handle climate stressors is based on its determined criticality and exposure to climate stressor hazards at the project location. Designs should include some accommodation for resilient failure (i.e., failure for which recovery is quick and incurs a relatively low cost). Design for resilient failure (i.e., determining the transition point between resilient failure and catastrophic failure) should be commensurate with the cost of recovery from extreme events. Without enabling some degree of resilient failure, these events can result in costly repairs, long downtimes, and costs associated with the inoperability of the completed project. A frequently used example of a resilient failure design is a stormwater management system designed so that excessive floodwaters are channeled into a holding pond and away from populations or infrastructure assets. The design capacity of the holding pond is based on the risk and cost of damage to the populations and assets exposed to catastrophic flooding.

  In planning and designing for resiliency, the owner and project team should consider the possibility of new climate stressor extremes well beyond previous levels. Recalling the discussion in Chapter 4, climate and weather variable values are moving into uncharted territory, creating the possibility of knock-on effects: climate stressors that are compounding, reinforcing, or propagating. The rate at which conditions are changing should also be taken into account.
- **Make a Meaningful Contribution to Sustainability:** The completed project should be designed to significantly reduce its ecological footprint, primarily carbon, by minimizing carbon emissions and embodied carbon. Ideally, the project should have net-zero carbon emissions. Concurrently, the project should maintain or enhance other environmental, economic, and social aspects of the

affected communities. The quality of life of the affected communities should not be diminished.

To plan, design, construct, and operate an infrastructure project that contributes to conditions of sustainability, project owners should employ one or more of the sustainability benchmarking tools available. A helpful tool currently available is the Envision sustainable infrastructure rating system.[1]

Using a series of sustainable and resilience credits, Envision enables users to rank civil infrastructure performance achievements throughout the project life cycle across five categories: Quality of Life, Leadership, Resource Allocation, Natural World, and Climate and Resilience. The system offers incentives for performance improvement using qualitative levels of performance improvement: improved (lowest), enhanced, superior, conserving, and restorative (highest). Points are awarded based on performance achieved. The system also encourages innovation by offering bonus points for extending sustainable performance beyond expectations. A description of the Envision credits used to rate sustainable performance on infrastructure projects, including their intent and metrics, is provided in Appendix B. The Institute for Sustainable Infrastructure trains and qualifies Envision Sustainability Professionals (ENV SPs) who understand the Envision framework and can evaluate infrastructure projects on their performance achievements.

As of this writing, Envision is used by more than 200 cities, counties, towns, public agencies, and academic institutions across the United States. It also has global partners in Italy, Mexico, Canada, Spain, and East Africa. Envision has been applied to more than 300 infrastructure projects with a total net constructed value of more than $237 billion (ISI 2024). Envision also complements ASCE/COS 73-23: *Standard Practice for Sustainable Infrastructure*, a nonmandatory standard that provides guidance to project owners for developing and implementing civil infrastructure projects that contribute to conditions of sustainability (ASCE 2023).

Step I A.3: Owner, Stakeholder, Sociopolitical Acceptance

Over its life and in the face of changing conditions the project should be acceptable as defined by the project owner, engineers, regulators, and key stakeholders. Project acceptability means

- **Meeting User Needs and Expectations:** These needs and expectations include performance, costs, and reliability. The services provided by the infrastructure asset must meet predetermined user needs and expectations throughout the project life

[1]In the interest of full disclosure, the author developed and delivered Envision version 1.0 in 2011 under a contract to ASCE, one of the founding members of Institute for Sustainable Infrastructure. The author then worked with Anthony Kane from the Harvard Graduate School of Design in the Zofnass Program for Sustainable Infrastructure incorporating their work with Envision version 1.0 to produce Envision 2.0, released in 2012. Envision 3.0 released in 2018 is the latest version. Anthony Kane is currently the CEO of the Institute for Sustainable Infrastructure.

cycle. Basic user needs and expectations include service quality, reliability, and affordability. Meeting these needs and expectations will require a program for monitoring and responding to changed conditions, including climate.

- **Meeting Stakeholder Requirements:** These requirements include performance, costs, location, configuration, access, equity of services, social justice, and environmental conservation.
- **Meeting Required Project Permit and Approval Requirements:** Every substantial civil infrastructure project must obtain a broad range of permits and approvals from various federal, state, and municipal authorities. Such permits and approvals are an official form of sociopolitical acceptance. The project team will determine what agencies and authorities have project permitting and approval responsibilities. The team will also determine the processes and estimated timelines for obtaining these project approvals. In addition to the standard approvals and signoffs, the responsible federal, state, and municipal authorities may have additional requirements or recommendations for handling sustainability-related issues or climate-related risks. Project permitting and approval requirements will set many project constraints.

Step I B: Determine Project Constraints

In concert with establishing goals and objectives, the project participants must also determine project constraints: restrictions in design and operational choices posed by the project's location, setting, function, regulation, and configuration coupled with economic and sociopolitical conditions. These constraints will affect what project goals and objectives can be achieved and control the set of viable options. They will also control definitions of project success, acceptable outcomes, and results (Figure 7-4).

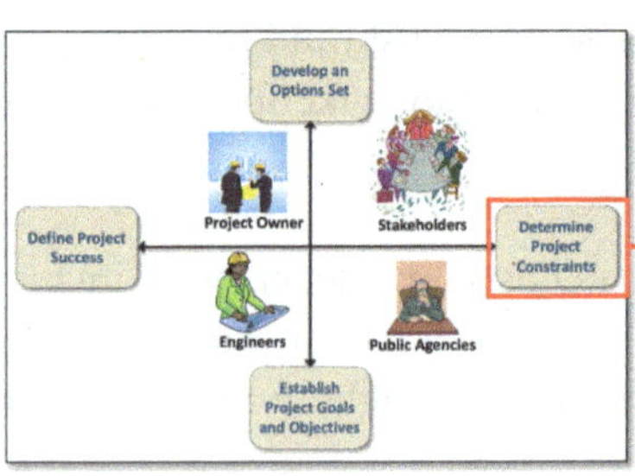

Imagery supplied by Clipart.com

Constraints
Limitations on the infrastructure project.
Location
Location, environmental setting, access to utilities, supplies, proximity to hazards.
Configuration
Layout requirements, asset footprint, confining features both natural and constructed.
Laws and Regulations
Laws and regulatory requirements, local codes and standards. Permit requirements. Zoning requirements. Operational limitations.
Operations
Operations and maintenance requirements. Reliance on support to and from other infrastructure systems. Worker and user safety.
Economic
Affordability, budgets, business and financial risk.
Socio-Political
Political risks, acceptance. Public, stakeholder acceptance.
Environmental
Potential impacts on natural resources and ecological systems.

Figure 7-4. Determine project constraints: limitations on the infrastructure project.

Project constraints can be organized into the following general categories. The project team should customize these categories, adding, subtracting, or modifying the elements to fit their project.

- **Location:** Location, environmental setting, access to utilities, supplies, and proximity to hazards.
- **Configuration:** Layout requirements, asset footprint, and confining features, natural or constructed.
- **Laws and Regulations:** Laws and regulatory requirements, local codes and standards. Permit requirements. Zoning requirements. Operational limitations.
- **Operations:** Operations and maintenance requirements, reliance on support to and from other infrastructure systems, worker, user, and public safety.
- **Economic:** Affordability, budgets, business, and financial risk.
- **Sociopolitical:** Political risks, acceptance. Public, stakeholder acceptance.
- **Environmental:** Potential impacts on natural resources and ecological systems.

Step I C: Definition of Success

Once project goals, objectives, and constraints are determined, the project participants can produce a clear picture of project success. Project success is defined as a specification of results or desired outcomes achieved within identified constraints that the project owner and stakeholders find acceptable over the project's life cycle.

Figure 7-5 depicts elements of the definition of success. These will be used in Step 2, Assemble the Initial Plan, to determine the conditions for success:

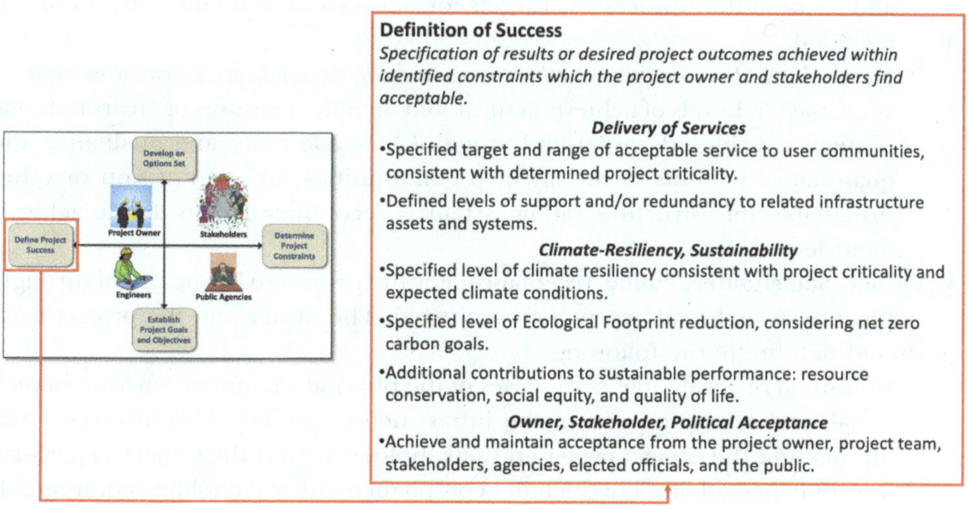

Figure 7-5. Definition of success.

assumptions and conditions that must remain valid to achieve what has been defined as project success.

Like the project goals and objectives, project success can be defined along three themes:

1. **Delivery of Services:** What services and level of service is the infrastructure project expected to deliver? Specifically,
   - Specified targets and ranges of acceptable service levels to affected communities, consistent with determined project criticality. Ideally, these should be quantitative specifications of the target and range of acceptable levels of services to the affected communities.
   - Defined levels of support and redundancy to related infrastructure assets and systems. These should be quantitative specifications of services provided to other infrastructure assets. They also may be backup services in situations of failure and extended downtime of other infrastructure assets or systems.
2. **Climate Resiliency and Sustainability:** What degree of climate resiliency and sustainability is expected in the infrastructure project? This should include the following.
   - *Specified level of climate resiliency consistent with project criticality and expected climate conditions.* This should include numerical ranges of operation for significant climate stressors and ranges of resiliency for infrastructure project components and systems (i.e., resilient failure to catastrophic failure transition points). The target and range of potential downtimes in cases of failure should also be stated.
   - *Specified level of ecological footprint reduction, considering net-zero carbon goals and objectives.* This should include using low- or no-carbon emission energy sources, using materials with low or no embodied carbon, and applying carbon capture and sequestration processes. Targets for net-zero carbon emissions should be included.
   - *Specific levels of achievement in other environmental, economic, and social dimensions of sustainability.* Levels of achievement should include measures of environmental protection achieved, target and range of life-cycle costs, and qualitative and quantitative benefits to the affected communities. Use of Envision or other sustainable infrastructure rating system is recommended to define achievement levels.
3. **Owner, Stakeholder, Public Acceptance:** For the project to be successful throughout its life cycle, broad acceptance is needed. The owner and the project team should determine the following.
   - Definition of acceptable boundaries of the physical, economic, environmental, social, and service aspects of the infrastructure project. This involves determining how the project owner and stakeholders expect the project to perform over its expected life. This includes compliance with stakeholder requirements

and limitations. Other aspects include location, performance, configuration, environmental impacts, access, equity of service delivery, and costs.

- Definition of acceptable boundaries regarding the risk of failure to perform or the risk of substandard performance. Expected levels of performance reliability over the project life cycle.

Step I D: Options Set

Figure 7-6 depicts the options set. Based on the project goals and objectives, constraints, and how the parties define project success, the owner and the designers can now develop a reasonable set of options that appear capable of producing an acceptable project outcome. The option set should be broad but able to meet the required performance within the predetermined acceptable range. The options should be described in enough detail to produce a workable set of alternatives for refinement and evaluation.

Although identified, project constraints are not necessarily applied at this juncture unless they create valid and insurmountable limitations. Unrealistic options (e.g., untried technologies, costs well in excess of resources) should be screened out.

Each option should address these elements, a combination of services delivered at a specific location, configuration, and operation within known constraints.

- **Service:** The services provided by the completed infrastructure project can meet owner requirements and specifications.
- **Location:** The location from where the services are to be provided is realistic.
- **Configuration:** Project configuration (e.g., size, layout, and footprint to deliver services) is within constraints.
- **Performance:** Performance levels for the services delivered are acceptable to the owner, permitting and approval agencies, and stakeholders.

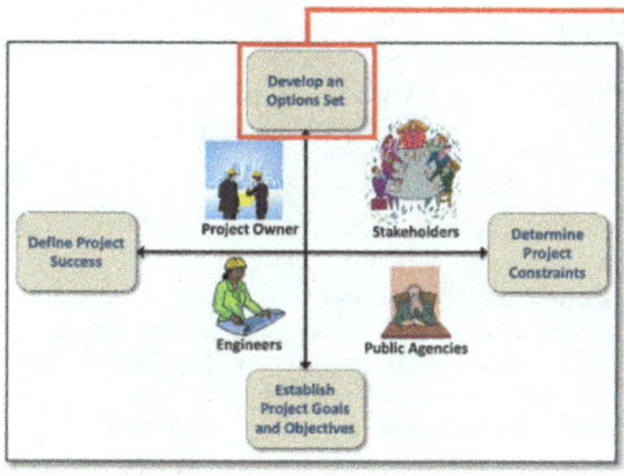

Options Set
A set of reasonable project options that could produce an acceptable project outcome.

Service. The services provided by the completed project can meet owner requirements and specifications.

Location. The project location is realistic.

Configuration. Project configuration is within known constraints.

Performance. Levels of performance are acceptable to the owner, permitting/approval agencies and stakeholders.

Laws and Regulations. Capable of meeting laws and regulations, permitting and zoning requirements.

Operations. Operates across a broad range of conditions taking climate change into consideration.

Economic. Designed, built and operated for a reasonable cost.

Socio-Political. Meets or exceed stakeholder, public expectations.

Environmental. Meets or exceeds expected environmental performance, public expectations.

Figure 7-6. Options set.

- **Laws and Regulations:** The option can meet applicable laws, regulations, codes, and zoning requirements, among other things.
- **Operations:** The option can operate in a broad range of conditions considering climate change.
- **Economic:** The option can be designed, built, and operated for a reasonable cost.
- **Sociopolitical:** The option can meet or exceed stakeholders' and public expectations.
- **Environmental:** The option can meet or exceed public expectations of environmental performance.

Results and Input for Step II

As input to the next step, the owner and the project team will have assembled a set of options based on the project goals and objectives modified by known project constraints. They will also have defined project success. These elements will be used to assemble an initial plan, a set of project alternatives that can be compared and evaluated against their reliability and ability to incorporate additional robustness cost-effectively.

Using the definition of project success, the owner and team will devise conditions for success. These are the conditions and assumptions that must remain valid throughout the project's life cycle for the project to be deemed successful.

Step II: Assemble the Initial Plan

This step is depicted in Figure 7-7. Assembling the initial plan consists of two tasks: (1) defining the necessary conditions for success and (2) devising project alternatives. The first task involves the development of the conditions for project success, conditions derived from the definition of project success. These conditions will

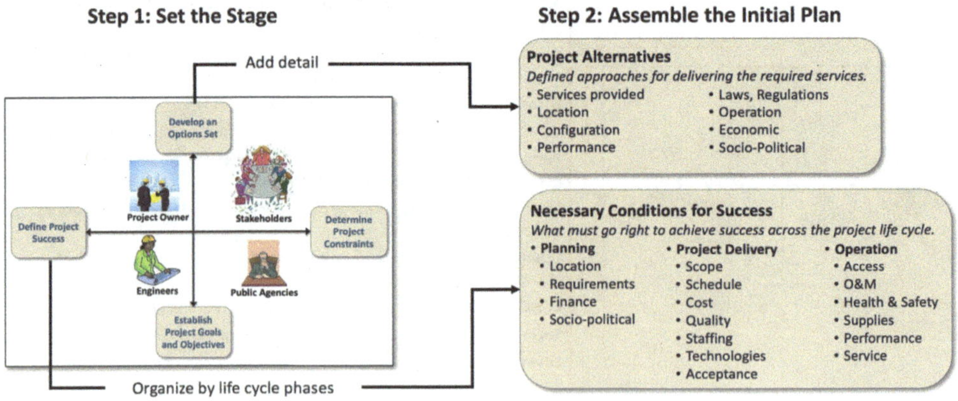

Figure 7-7. Step 2: Assemble the initial plan.

provide a testing ground for the alternatives: How will alternatives perform over their life cycle if actual conditions deviate significantly from the initial design and operating assumptions? Upcoming analyses will determine which alternative can best address those potential deviations, positive and negative.

The second task involves converting the options set produced in Step I into relatively detailed project alternatives from which a preferred alternative can be selected.

II.A: Defining the Necessary Conditions for Success

A basic set of conditions for success are outlined in Table 7-1. The conditions are divided into three phases, corresponding to the project life cycle. The *planning phase*

Table 7-1. Conditions for Success: What Has to Go Right with the Project Throughout Its Life Cycle?

Conditions for success	
Planning	
Devising the best project alternative based on initial available information prior to project initiation (time < 0)	
Aspect	**Assumptions, condition for success**
	Location
Property ownership	The property is free of any incumbrances that will affect the project or completed asset
Site conditions	The property is compatible with the infrastructure project. Site is free of significant quantities of regulated contaminants
Surroundings	The environment in the vicinity is suitable for the project and does not pose any significant hazards or threats
Adjacent properties, structures, facilities	Adjacent properties and facilities pose no threats to the project during and after completion
Climate, weather hazards, exposures	Significant climate and weather hazards, exposures at the project site location are successfully addressed
	Requirements
Laws, codes, standards, regulations, permits	No significant changes in applicable codes and regulations permit requirements. Net-zero emissions standards may be forthcoming
	Finance
Project budget	The authorized project budget is sufficient to deliver the project
Operations and management	Funding for infrastructure asset operations and maintenance (O&M) is sufficient
Insurance	Insurance will be available to cover the project throughout the life cycle

(Continued)

Table 7-1. Conditions for Success: What Has to Go Right with the Project Throughout Its Life Cycle? *(Continued)*

	Sociopolitical
Stakeholder acceptance	Project is acceptable to stakeholders
Administrative continuity	No changes in administration or elected officials resulting in loss of project support
Project delivery	
Completing the project as specified on time and within budget (time < 0)	
Scope	Scope of work is sufficient for the owner's specifications and requirements
Schedule	Project is delivered in accordance with agreed-on schedule
Cost	Cost of labor, materials, and equipment are the same as initially estimated
Quality	Project quality is maintained in accordance with the project quality management plan
Staffing	Project staff levels, key staff are maintained throughout the project delivery process
Technologies	Applied technologies work as expected
Acceptance	Acceptable to the affected communities during construction
Operations	
Completed project delivers the expected services, levels of services while handling change (time > 0)	
Access	Efficient user and operator access to the completed works are maintained
Operations and maintenance	Staff operate the completed works efficiently and cost effectively. Maintenance is performed as required to maintain efficient operation and functionality
Health and safety	User, operator health, and safety are protected
Utilities	Water, electricity, and sanitation are supplied as needed
Supplies	Supplies are available and delivered as needed
Performance	Project performs as specified upon completion and throughout the project life cycle
Service	Demand for services develops as initially forecast

Note: Time (t) = 0 is the project startup, the point at which the project is completed and begins operation, providing its intended services.

establishes that all the pieces are in place for the project to be successful. The *project delivery phase* consists of the elements the project team needs to accomplish in project design and construction. The *operations phase* consists of the elements necessary for delivering the intended services at the expected level of performance while managing changed conditions. In deriving the conditions for success, the project team should organize the conditions in accordance with the project timeline throughout its life cycle: planning, delivery, and operations. Users are encouraged to modify, expand, or add new conditions specific to their project and to the project alternatives

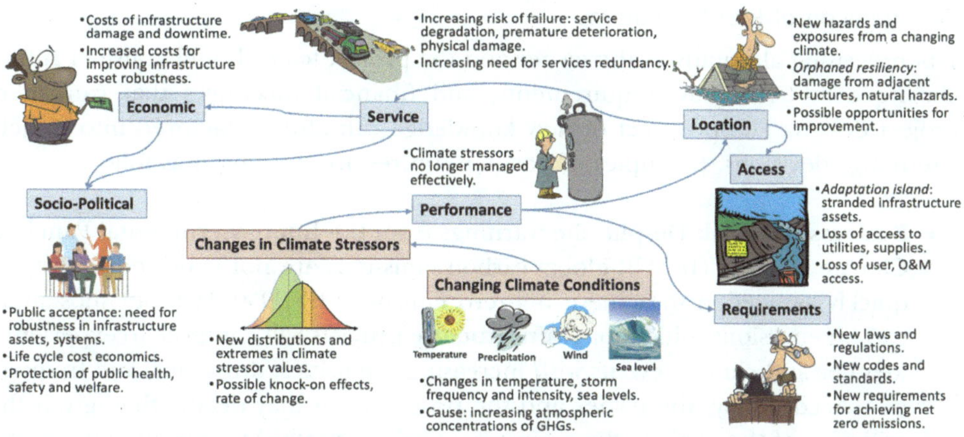

Figure 7-8. How a changing climate can cause deviations in conditions for success.

they are considering. As depicted in Figure 7-8, climate change is a significant source of deviations from conditions for success. In developing these project-specific conditions, the project team should consider how changes in climate conditions can influence project performance over the entire project life cycle.

II.B: Devising Project Alternatives

Working with the options developed in Step I, the project team works to refine the options, resulting in one or more project alternatives. Alternatives must be distinct and sufficiently detailed to make evaluations and comparisons.

Contribution to Net-Zero Carbon Emissions

The project alternatives should incorporate elements in the design to reduce carbon emissions, contributing to the goal of net-zero emissions by 2050. These elements should cover both embodied carbon and carbon emissions during operation. As noted in the Consideration of Possible Futures section of this chapter, future laws and regulations may require carbon emissions reduction.

Project Criticality

Project criticality determined in Step I is considered. Projects crucial to safety, access, and economic viability must operate reliably and effectively. The degree of importance or criticality of the completed project will determine its level of robustness (strength, durability, resilience, etc.). Redundancy should also be incorporated when practical. In the event of failure, other infrastructure assets could provide similar functions or services.

Consideration of Possible Futures

When devising alternatives, the owner and the project team should consider future climate conditions, new requirements, and financial incentives that can affect project design. The prospect of new knowledge can also be factored into project alternative decisions. Examples of possible futures are presented as follows.

- **Business as Usual:** Despite the warnings from the Intergovernmental Panel on Climate Change (IPCC), global carbon emissions are not expected to drop as quickly as needed to achieve net-zero emissions by 2050. Without substantial carbon emissions mitigation, infrastructure projects will continue to experience weather and climate events with increasing intensities, durations, and frequencies. Accordingly, the owner and the project team may decide that, given the criticality of the project, the appropriate design strategy is to assume the worst: business-as-usual climate scenario conditions. The IPCC has developed representative concentration pathways (RCP), a range of climate scenarios covering possible futures. These scenarios are presented in Table 7-2. RCP 8.5 is the "business-as-usual" scenario. Estimated climate conditions for several scenarios have been developed and can be downscaled to approximate conditions at the project locale.

Table 7-2. Representative Concentration Pathways (RCPs) and Expected Global Surface Temperature Changes.

Scenario name	Description	Concentrations (ppm CO_2 equivalent) by 2100	Global surface temperature change by 2100
RCP 2.6	Emissions reduced substantially from current pathway. Considered too optimistic for planning purposes	430–480	0.5–3.0 °F (0.3–1.7 °C)
RCP 4.5	Emissions reduced sufficiently so that the total radiative forcing is stabilized by 2100	580–720	2.0–4.7 °F (1.1–2.6 °C)
RCP 6.0	Emissions reduced sufficiently so that the total radiative forcing is stabilized by 2100	720–1,000	2.5–5.6 °F (1.4–3.1 °C)
RCP 8.5	High emissions continue through 2100. Most representative RCP of current track (i.e., business as usual), although significant efforts are underway to reduce emissions	>1,000	4.7–8.6 °F (2.6–4.8 °C)

Source: FHWA (2017), p. 29.

RCP and SSP Scenarios

The RCP scenarios complement the new shared socioeconomic pathways scenarios presented in a recent IPCC report (IPCC 2023). Shared socioeconomic pathways (SSPs) consist of five narratives of a broad range of socioeconomic trends showing how the world might evolve in the absence of climate policies. When combined with RCPs, shared socioeconomic pathways allow researchers to examine barriers and opportunities for carbon emissions mitigation, considering variations in population, economic growth, education, urbanization, and rate of technological development.

- **Less Carbon Emissions/Not Critical:** If the project is judged not critical or the IPCC warnings are expected to be heeded, the owner and the project team could also opt for a different emissions scenario (e.g., RCP 6 or RCP 4.5). Current policies and actions by the world's nations are expected to increase warming to 2.7 °C (3.7 °F) (CAT 2023). Current federal agency guidance states that the low-emissions scenario, RCP 2.6, should not be used for planning purposes (FHWA 2017).
- **New Rules:** New rules, regulations, or guidelines may be established to protect infrastructure assets, existing and planned. Mandated reductions in carbon emissions will likely affect infrastructure assets that require some form of heat or electrical power to operate. New rules for construction materials may include reductions in embodied carbon. For transportation assets, alternative forms of transportation and changes in demand may require new infrastructure designs. Based on these possibilities, the owner could incorporate some of these elements into the project. In addition, the owner could incorporate flexibility, the ability to modify the completed project to accommodate new requirements. Anticipating a push for lowering carbon emissions, the project could be designed to use renewable energy for heat and power. Some states have carbon emissions trading programs designed to limit future emissions.
- **New Incentives:** In the latter part of 2022, Congress passed and the president signed two pieces of legislation that provided US$1.25 trillion in investments for carbon emissions reduction and resilient, adaptive infrastructure. Funds in the form of grants and tax credits are being made available for federal agencies, states, and communities for projects in transportation, energy, buildings, natural resources, agriculture, water and wastewater, coastal protection, preparedness, and emergency response. Infrastructure projects will likely be required to meet various levels of carbon emissions reduction, resiliency, and adaptation to qualify for these incentives.
- **New Knowledge:** Given the concern over climate change coupled with the pace of change and technological development, the project team could design for flexibility. New knowledge gained after the project becomes operational could be used to improve project performance. Designing for flexibility in the

operations phase, such as the ability to raise flood barriers incrementally, may be a better strategy than fixed barrier designs against maximum future flood levels that may not come about.

Another approach for accommodating new knowledge is to design for a shorter useful life. Building a less durable project can lower the project's cost, enabling the owner to reserve funds for future replacement. This approach assumes that the completed project will be replaced relatively soon with a new design based on a better understanding of future conditions. Such an approach was used for the boardwalk at Percival Landing by the city of Olympia, Washington. The boardwalk was designed for a shorter useful life, saving money for a future boardwalk replacement based on new knowledge acquired during the initial boardwalk's design life. To handle sea level rise and occasional sea water inundation, the elevation of the initial boardwalk was made adjustable and used saltwater-tolerant materials in its construction. The city's recent sea level rise response plan has options for raising the elevation of the boardwalk and constructing a new boardwalk (City of Olympia 2019).

Step III: Improve the Robustness of the Project Alternatives

Step III takes the project alternatives developed in Step II and devises ways to improve the robustness of each alternative (i.e., their strength, durability, resiliency, and flexibility in the face of changing conditions). This process begins with identifying potential deviations from the conditions for success: changes in expected conditions or incorrect assumptions that produce vulnerabilities and opportunities that could affect project success. The approach for improving the robustness of the project alternatives is presented in Figure 7-9.

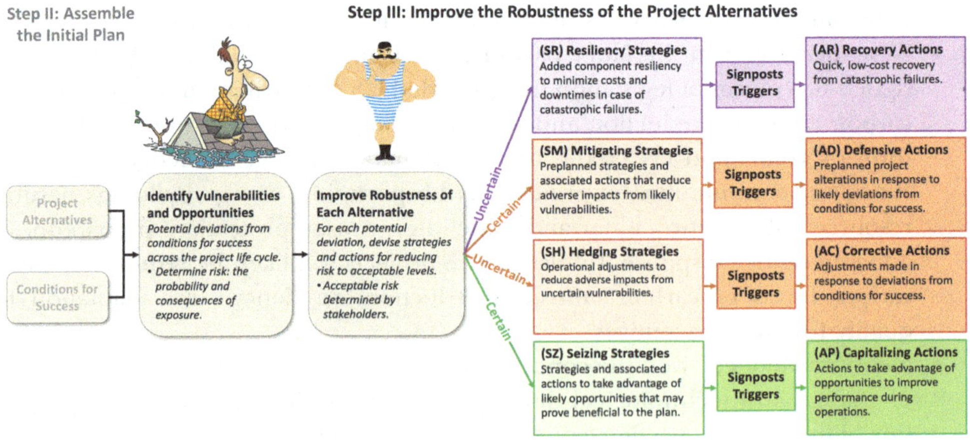

Figure 7-9. Step III: Improve the robustness of the project alternatives.

Some changes are more likely than others, suggesting that some adaptation actions can be formulated beforehand. Other changes are less certain. In those cases, operators must rely on flexibilities designed into the project to allow for future operational adjustments. These adjustments could include physical as well as operational changes. Finally, regardless of the efforts to strengthen and adapt to changed conditions, there is a lingering risk that future extremes can still overwhelm project defenses, resulting in failures. For these situations, some level of resilience—fast and low-cost recovery from extreme climate stressor values or events—should be incorporated. In addition, owners and operators should make plans for immediate relief and long-term recovery if stressor values exceed designed resiliency levels and catastrophic failures occur. The degree to which additional resiliency is incorporated into the project is a function of cost, the likelihood of extreme events, and project criticality. In cases of high project criticality, redundancy of the services provided by the project should be considered.

The selected project alternative should be the one that best handles deviations from conditions for success. Making this choice will involve stress testing the alternatives, pushing the deviations out to reasonable extremes, determining what can be done to manage the associated risks, and judging how the project alternative performs.

Step III A. Identify Vulnerabilities and Opportunities

Using the conditions for success developed in Step II, the owner and the project team will determine potential deviations from those conditions, deviations that could create significant vulnerabilities and opportunities throughout the project life cycle. Vulnerabilities are conditions that reduce the chances of project success. For example, scope creep because of project staff-initiated changes could increase project costs. In terms of climate, a severe rainstorm could increase water flow in a river and cause excessive bridge foundation scouring. In contrast, opportunities are conditions that improve the chances of success. For example, the cost of disposing of excess excavation materials could be reduced if a nearby project needs fill material. Unusually favorable weather conditions could reduce project construction time.

For each deviation, the project team, in collaboration with the owner, estimates the probability of each deviation. For simplification purposes, the estimates in this work are shown as qualitative. However, in actual practice, the owner and the project team could incorporate more detailed and quantitative estimates.

For deviations that are relatively certain, strategies and actions should be developed in the planning stage, ready to be executed as certain conditions arise. For example, sea level rise is happening and is expected to continue at some rate depending on a number of factors. Based on this level of certainty, the project should be designed to adapt to rising seas by adding a flood wall built to some prescribed height and incorporating the ability to increase wall height as sea levels rise. For relatively uncertain deviations, the project should be designed to allow for operational adjustments. For example, unexpected extreme heat waves can cause heat stress on the operating staff and limit services. To handle these extreme heat events, maintenance during operation could be rescheduled to evening hours when temperatures are cooler. Air conditioning equipment could also be added.

Step III B. Improve the Robustness of Each Alternative

For any project alternatives to succeed, they must be able to handle vulnerabilities by reducing the associated risk to acceptable levels at an acceptable cost. They should also take advantage of opportunities to reduce or avoid vulnerabilities or convert them into opportunities. Thus, the best project alternative will be the one that is most robust and cost effective in reducing vulnerabilities and taking advantage of opportunities.

Improving alternative robustness requires an understanding of the risks the project is expected to face throughout the project life cycle, as well as the level of risk deemed acceptable to the project stakeholders. In the past, these risks were generally well understood and taken into account in accepted codes and standards for infrastructure design. But, with changing climate conditions, many of the accepted codes and standards are no longer reliable for protecting infrastructure projects from changing climate conditions. However, they do provide guidance on what constitutes acceptable levels of risk. For example, a 1% risk of flooding in any given year has been a benchmark level of acceptable risk for flood insurance purposes. Unfortunately, with a changing climate, storm intensity, duration, and frequency have increased. Consequently, flood prevention designs based on past climate conditions can no longer protect to that level of risk. However, the potentially affected communities may still agree that a 1% chance of flooding in any given year is still an acceptable level of risk.

Strategies and Actions to Increase Robustness

Strategies and their associated actions to increase robustness are described in Table 7-3. For each strategy, there can be one or more actions based on the magnitude of the deviation and the available action choices.

- **Mitigating Strategies (SM):** Strategies for taking preplanned **Defensive Actions (AD)** that reduce adverse impacts from likely vulnerabilities. These are designed into the project, enabling the project owner or operator to construct and put in place additional features to mitigate risk. An often-used example of a mitigating action is to increase the height of a sea wall incrementally based on some predetermined increment of sea level rise.

  These preplanned actions align with the observational method used by geotechnical engineers to handle the uncertainties of subsurface soil conditions. In this method, a most likely model of subsurface conditions is developed

Table 7-3. Strategy Matrix for Handling Vulnerabilities and Opportunities Based on the Level of Certainty.

	Vulnerability	Opportunity
Certain	*Mitigating Strategies* (SM). Preplanned strategies and associated *Defensive Actions* (AD) that reduce adverse impacts from likely vulnerabilities. These can be physical or operational changes	*Seizing Strategies* (SZ). Strategies and associated *Capitalizing Actions* (AP) to take advantage of likely opportunities that may prove beneficial to the plan. These can be physical or operational changes
Uncertain	*Hedging Strategies* (SH). Strategies for making *Corrective Actions* (AC), physical and operational adjustments to reduce adverse impacts from uncertain vulnerabilities	*Exploiting Strategies* (SE): Strategies and associated *Capitalizing Actions* (AP) to take advantage of uncertain but possible developments that, if realized, could improve chances of success
	Resiliency Strategies (SR). Strategies for taking *Recovery Actions* (AR) to minimize costs and downtimes in case of failures that exceed the capacity or capability of mitigating or hedging strategies	*Not applicable*
	Shaping Actions (AS). Actions taken in response to external trends or events. These may be vulnerabilities or opportunities that are unforeseen and cannot be planned for as part of robustness improvement. However, the project owners and operators can monitor external trends and events and be prepared to take actions to influence the outcome. Actions may be defensive, corrective, or capitalizing	

along with other reasonable models. During construction, subsurface conditions are monitored for changed conditions, and preplanned actions are taken in accordance with the conditions encountered (Wallace et al. 2017, Ayyub 2018).

- **Hedging Strategies (SH):** Strategies for taking **Corrective Actions (AC)**. These are operational adjustments taken during the operational phase to reduce adverse impacts from uncertain vulnerabilities. An example of a hedging action is the early identification of alternative project funding sources for monitoring or management activities if operating funds are cut (Wall et al. 2015).
- **Seizing Strategies (SZ):** Strategies for taking **Capitalizing Actions (AP)** to take advantage of opportunities that may benefit the plan. An example of a seizing action would be an operator of an infrastructure asset deciding to include climate adaptation retrofit projects as an element of a county's overall asset management program (Wall et al. 2015).
- **Exploiting Strategies (SE):** To complete the matrix, the DAP model includes exploiting strategies to handle uncertain opportunities by implementing

Capitalizing Actions (AP). *This author chose not to incorporate this choice in this methodology as there are few examples in the literature in which practitioners have identified and devised strategies in the planning stage for handling uncertain opportunities.*

- **Resiliency Strategies (SR):** This proposed methodology incorporates resiliency into the mix of strategy choices. These strategies are associated with **Recovery Actions (AR)** developed to minimize costs and downtimes in instances of failures. These are done at the project planning and design phases to address the possibility that mitigating and hedging actions to handle a changing climate are insufficient. For example, the project team may put in place barriers to prevent flooding. In this case, designing for resiliency could mean waterproofing certain infrastructure asset components in the event those barriers are breached during an extreme storm event. Designing for resiliency can also mean increasing storage of operating supplies in case of extreme weather shipping interruptions or including backup electricity generation capacity in case of a power outage.
- **Shaping Actions (AS):** These are actions taken in response to external events. Owners and operators may be able to influence external trends or events to improve their chances of success.

Signposts and Triggers Associated with Strategies and Actions

Each action has one or more associated variables or events (signposts) that are measured at some predetermined frequency to detect deviations. Each variable has one or more values or incidents (triggers) that, if reached, will start an action in a predetermined sequence.

For example, to protect a coastal infrastructure asset from flooding, one signpost could be sea level rise using information from federal or state sources. If the sea level exceeds a specified trigger value, flood walls would be built to some predetermined height, a defensive action designed to protect against flooding from high tides and storm surges. If sea levels continue to increase, another higher trigger value would instigate a second defensive action, raising the height of the flood walls to another predetermined height. In the right situation, the flood wall height could be made adjustable (Figure 4-16 in Chapter 4).

Robustness Matrix

To compare alternatives, the project team should create a robustness matrix, a table covering the vulnerabilities and opportunities, including deviations from conditions for success over the entire project life cycle. An example of a robustness matrix is presented in Appendix C. Think of the robustness matrix as an expanded project risk register, one that not only spells out vulnerabilities but also project opportunities. The matrix outlines reasonable project risks and opportunities and shows how they will be managed.

For each identified condition for success, the project team should (1) determine potential deviations from conditions for success and their associated vulnerabilities or opportunities, (2) state the impacts along with their respective level of uncertainty, and (3) lay out the associated strategies and actions to be taken.

Creating a matrix for each alternative will provide the project team with a tool for Step IV, assessing and comparing the requirements for each alternative and selecting the best alternative. Alternative selection will be made based on the costs and perceived effectiveness of strategies and actions for handling deviations to these conditions.

Given the size of the matrix, entering vulnerabilities, strategies, and actions for each alternative may seem labor-intensive. However, unless the alternatives have major location, structural, or technology differences, their matrices will be similar. In evaluating alternatives, differences that arise will likely be confined to a limited number of conditions for success.

In preparing to move to the next step, the project team would assemble robustness matrices for each alternative, describing the strategies and actions to be taken for reasonable deviations from conditions for success.

Assessing the Risk

For each strategy and associated action, a risk assessment should be made to determine whether or not the risks associated with the deviations from conditions for success have been reduced to acceptable levels. Risk is defined as the likelihood of exposure to a hazard, in this case, a deviation from project success and the consequences of that exposure or deviation. Risks can be rated and presented qualitatively (CAMPO 2015), as depicted in Figure 7-10.

Estimates of initial risk and target risk are the starting points. Residual risk is the risk remaining after risk reduction strategies and actions have been applied. An

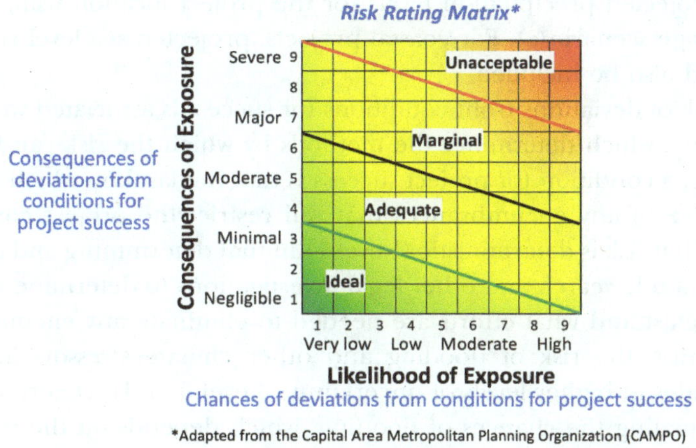

Figure 7-10. A qualitative risk rating matrix used by the Capital Area Metropolitan Planning Organization (CAMPO).
Source: Adapted from CAMPO (2015), Figure 29, pp. 6–8.

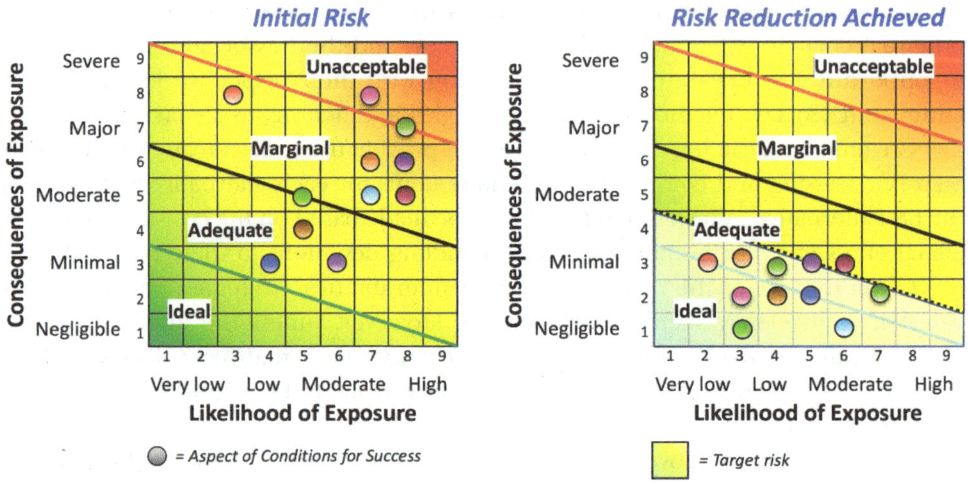

Figure 7-11. Initial and risk reduction achieved compared to target risk.

example plot of initial risk and residual risk (risk reduction achieved) compared to target risk is presented in Figure 7-11.

Initial Risk

The initial risk is the chances and consequences of deviation from project success if no actions are taken. For example, in the case of hazardous wastes being present in significant quantities and concentrations on the project site, the initial risk could be based on the results of site investigations. For flooding, the initial risk would be the likelihood of flooding over the design life of the project and the consequences of flooding. The likelihood would be determined by calculating the chances of flooding using projected precipitation levels for the project location using the selected climate change scenario(s). For coastal projects, projected sea level rise and storm surges would also be included.

Each risk of deviations from conditions for success is associated with some level of uncertainty, which determines the methods by which the risk can be mitigated. For example, a condition for project success is that the land on which the project is located is free of any encumbrances that will restrict the project construction or operation. That risk is deterministic (Level 1) in that determining and reducing that risk involves a title search and other legal investigations to determine whether such restrictions exist and what efforts are needed to eliminate any encumbrances.

In contrast, the risk of flooding and other climate stressors at the project location resides in higher levels of uncertainty (Level 3 or 4). Determining the risk involves estimating the chances of flooding, which depends on the frequency and intensity of extreme precipitation at the project location. Statistically and dynamically downscaled US climate data are available.

Global Climate Change Projected at the Local Level Designers are encouraged to use the National Climate Assessment Scenarios information on projected US temperatures and precipitation levels and RCP 4.5 and RCP 8.5 emission scenarios for three periods: early, mid, and late 21st century. The information is downscaled to 1/16th degree spatial resolution (Scenarios for the Local National Climate Assessment 2024). The choice of emission scenario should be based on the criticality of the infrastructure project. If the project is deemed critical, the owner and the project team can decide to be cautious and opt for a higher emissions scenario.

Data sets called localized constructed analogs (LOCA) are provided for multiple temperature and precipitation climate stressor variables. Examples include daily maximum temperature, annual number of days of icing, and annual days with precipitation greater than 3 in. An example of the graphic output available from LOCA is provided in Figure 7-12. Downscaled climate stressor information is also available in table form.

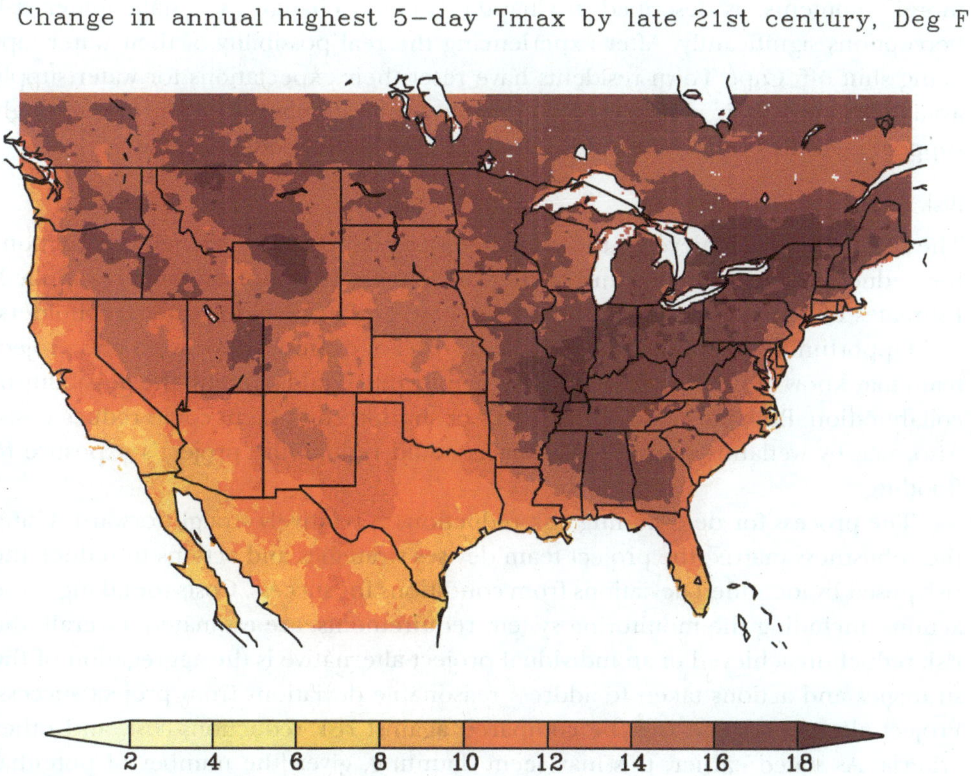

Figure 7-12. Example of localized constructed analogs (LOCA) for annual highest five-day maximum temperature, high emissions scenario, late 21st century United States compared to 1976–2005 average. Graphics and localized data sets are available for many temperature and precipitation scenarios.

Source: Scenarios for the Local National Climate Assessment (2024).

Target Risk

The target risk is the limit of acceptable risk as defined by the users, affected communities, and other stakeholders. In many, if not most, cases, target risks have already been determined and exist in current codes and standards. As noted earlier in this chapter, a frequently used example of an acceptable target risk is the concept of a 100-year flood plain. If the user/stakeholder perception of what constitutes acceptable risk has stayed the same, the target level of risk should remain constant throughout the project life cycle. For example, as climate changes produce more frequent and intense storms, the chances of flooding outside the current 100-year floodplain also increase. Stakeholder perception of acceptable risk could still remain as a 1% chance of flooding annually, but the 100-year floodplain map (i.e., which areas reside inside and outside the 100-year floodplain) would change based on new climate information and flooding experience.

That said, relying on existing standards and risk perceptions may not be sufficient. Projects in locations where the communities have experienced "No more!" moments, as described in Chapter 1, could change user/stakeholder risk perceptions significantly. After experiencing the real possibility of their water taps being shut off, Cape Town residents have reset their expectations for water supply availability, and their measure of acceptable water supply risk. They also strongly support measures to reduce water demands and bolster water supplies.

Risk Reduction Achieved

The risk reduction achieved is the estimated residual risk after strategies and actions for reducing vulnerabilities and taking advantage of opportunities are applied. Estimates of residual risk need to be presented and accepted by the stakeholders.

Opportunities can also reduce risk and cost. For example, the owner and project team may know of similar projects in the vicinity that could open up the possibility of collaboration. Bulk purchases of materials or sharing equipment could reduce costs. Also, nearby wetlands restoration projects could reduce the project's exposure to flooding.

The process for determining risk reductions is relatively straightforward. Using the robustness matrix, the project team devises strategies and actions to reduce the risk posed by identified deviations from conditions for success. Costs for taking these actions, including the monitoring system requirements, are estimated. Overall, the risk reduction achieved in an individual project alternative is the aggregation of the strategies and actions taken to address reasonable deviations from project success. Project alternatives can now be compared against risk reduction, cost, and other criteria. As stated earlier, this may seem daunting, given the number of potential deviations. However, many of the conditions for success and the strategies and actions proposed for handling them are likely common to many, if not most, of the alternatives. Therefore, comparisons among alternatives will likely be limited to a few distinctions.

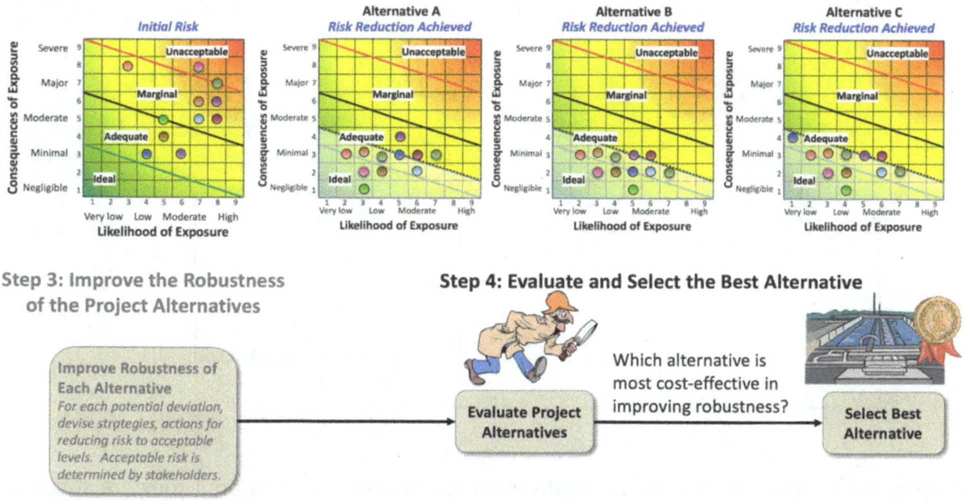

Figure 7-13. Step IV: Evaluate and select the best alternative. Comparison of risk reduction achieved for three example alternatives.

Step IV: Evaluate and Select the Best Alternative

Once the information is assembled, the alternatives can be compared against predetermined selection criteria, and a selection can be made.

Step IV A: Selection Criteria: Risk, Cost, Other

In this task, the project team, in coordination with and with the owner's approval, develops criteria for objectively comparing the project alternatives. The alternative selected will be the one that can most cost effectively reach the target risk levels (Figure 7-13).

Additional selection criteria are offered as follows.

- Flexibility;
- Performance efficiency and effectiveness;
- Compatibility with current engineering practices;
- Residual technical and business risks;
- Residual economic, social, and operational risks;
- Extent of sustainable performance improvement; and
- Ease in constructability.

Step IV B: Alternative Selection

Applying the selection criteria, the project owner selects the best alternative. Weights can be applied to the additional selection criteria.

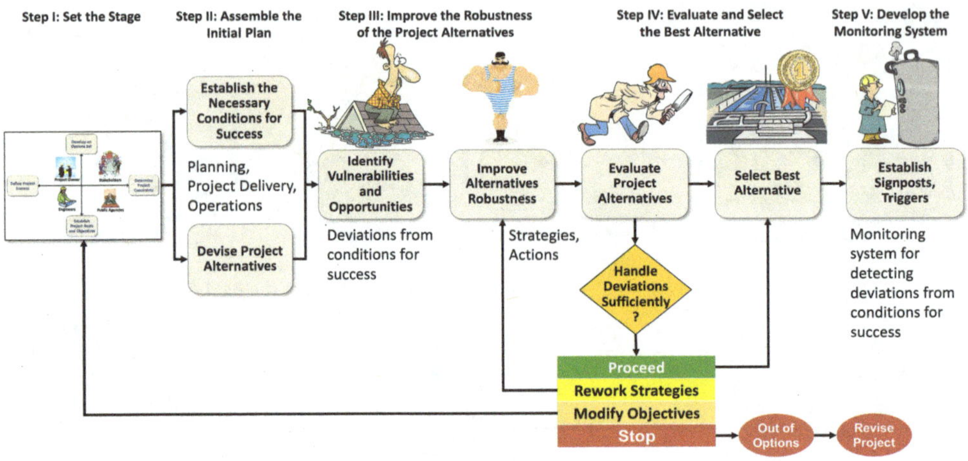

Figure 7-14. What to do if the alternatives are insufficient. What to do if they are sufficient.

What if the Alternatives are Not Sufficient?

If the alternatives cannot reduce risk to acceptable levels, the project team must revisit the strategies and actions used to add robustness to the project. If changes to those strategies and actions do not produce an acceptable result, the project team and the owner must revisit the project's goals and objectives.

Here, the owner and the team must review the range of acceptable service levels established in Phase 1 to see if reducing performance requirements while staying within the acceptable range can provide sufficient technological leeway to achieve acceptable levels of risk. If performance requirement modifications fail to provide the necessary flexibility, the owner and the project team must revise the project (Figure 7-14).

Step V: Develop the Monitoring System

The selected alternative was chosen on the basis that deviations from project success could be handled effectively by monitoring established signposts and taking preplanned actions. Each signpost has one or more triggers which, if reached, would instigate an action. The specific actions are discussed in Step VI.

The project team now incorporates those signposts and triggers into the project design and prepares plans for monitoring them at a predetermined frequency. Examples of signposts and their associated triggers are presented in the robustness matrix shown in Appendix C. Defensive and corrective actions handle vulnerabilities. Capitalizing actions address opportunities. Recovery actions handle situations where the defensive and corrective actions were insufficient to handle deviations.

In addition to the established signposts and triggers, the project team should establish a plan for scanning external trends and events that could affect the project's success. Such trends and events—actions by others, changing preferences, loss of current funding, or identification of new funding sources—can create new vulnerabilities and opportunities. These are covered in Step VI.

As noted earlier in this chapter, external trends and events may provide the owner or operator opportunities to take *Shaping Actions*: actions taken to shape outcomes and improve the chances of project success. For example, by learning about the availability of federal funds for adaptation, a project owner can pursue additional funds to increase flood resilience (Wall et al. 2015).

Step VI: Prepare the Responses

Simplified flowcharts of Step VI, preparing the responses, and Step VII, project delivery, are presented in Figures 7-15 and 7-16, respectively. At this juncture, all important decisions have been made, and the project can proceed as a conventional project. The purpose of this step is to convert decisions into a comprehensive set of project requirements and specifications. From this, the project owner can devise a scope of work from which the project can be designed and constructed. The work can be broken down into discrete tasks, each with a defined scope, schedule, and budget, all sequenced and implemented logically and efficiently.

Step VII: Deliver the Project

Once the project alternative is selected and the monitoring system is defined, the project can move forward as a traditional infrastructure project, focusing on efficient and cost-effective completion. The first task is to design the project and produce the

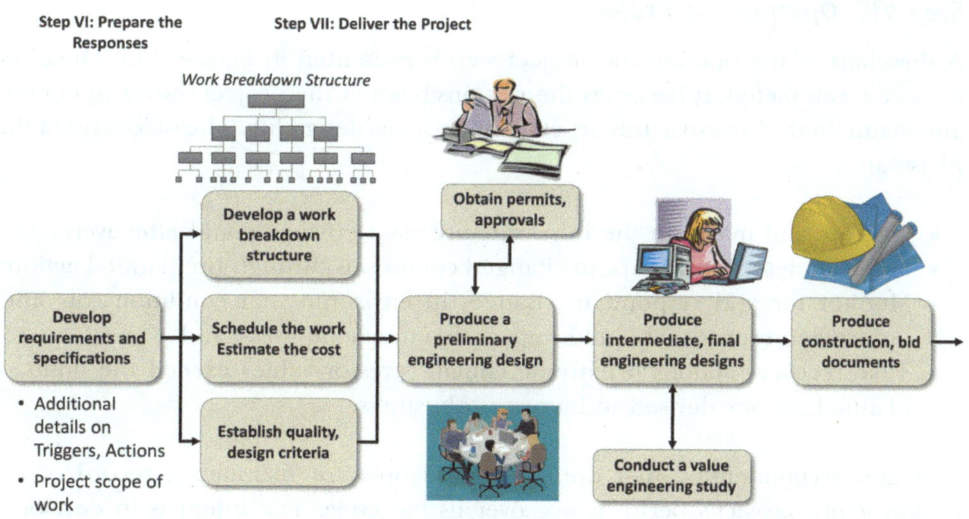

Figure 7-15. Step VI, preparing the responses, and Step VII, delivering the project, the initial tasks.

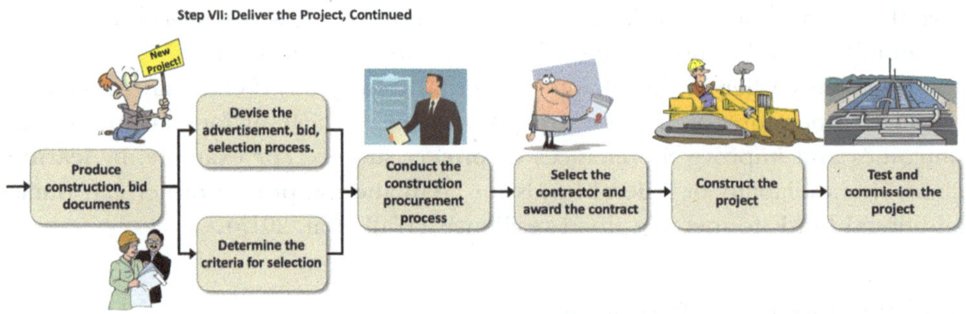

Figure 7-16. Step VII: deliver the project (continued from Figure 7-15).

necessary drawings and documentation to obtain the required permits and approvals. Once approved, final drawings and documentation will be produced to construct, test, and commission the project. Details for the monitoring requirements and the associated response actions are developed and incorporated into the scope of work.

These flowcharts highly simplify these processes, just showing the essential tasks. Delivering the project starts with organizing the work, producing preliminary designs for optimization, obtaining approvals, and designing for construction. Once the construction and bid documents are ready, the construction contract is advertised, and a construction contractor is selected. Then, the project is constructed, tested, and commissioned for operation. It is expected that public works departments and engineering companies will have more complex processes but essentially follow the same steps.

Step VIII: Operate the Project

A flowchart of the operate the project step is presented in Figure 7-17. Once the project is completed, it becomes the responsibility of the project owner to operate and maintain the infrastructure asset throughout its design life. The objectives of this phase are

- Operate and maintain the infrastructure asset efficiently and effectively;
- Monitor, detect and adapt to changed conditions through preplanned actions;
- Monitor for and respond to changes in environmental conditions or other unforeseen events that could impact design assumptions; and
- Take recovery actions if extreme climate stressor values exceed the limits of planned actions devised to increase robustness.

It is also recommended that the owner and operator maintain a record of the infrastructure project's performance over its life cycle. The intent is to develop a better understanding of how infrastructure projects perform under changing climate conditions and share this information throughout the industry. This information

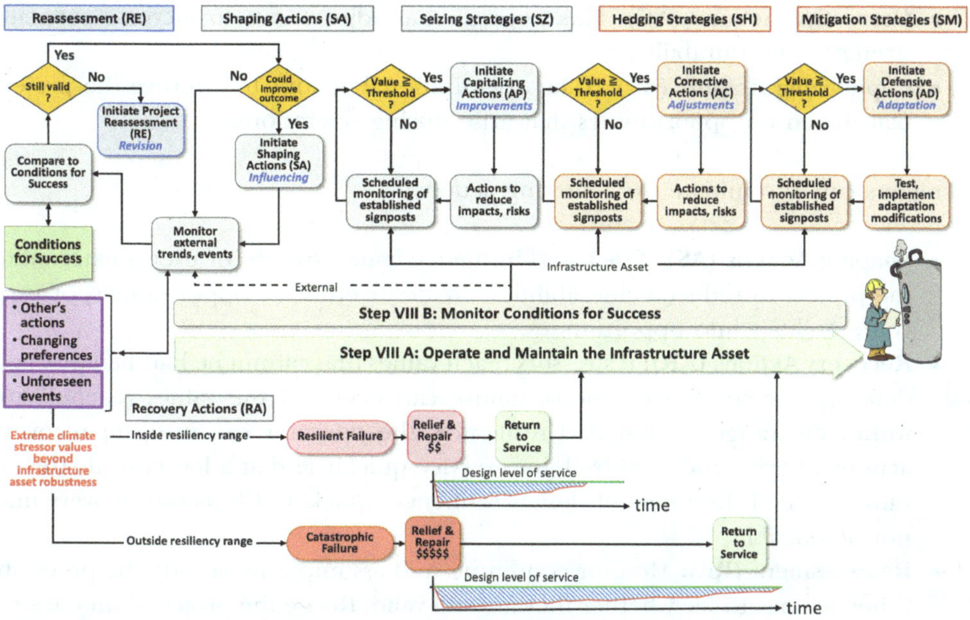

Figure 7-17. Step VIII: Operate the project.

should be collected and documented following established engineering quality standards and submitted to a well-maintained engineering knowledge base.

Step VIII A: Operations and Maintenance

The project owner will operate and maintain the infrastructure asset, as any other asset, as part of the community's infrastructure system. Based on the owner's specifications, contributions to sustainable performance may be included (i.e., contributions that significantly improve overall sustainable performance over conventional asset performance). These practices include energy and water conservation, carbon footprint reduction, emissions reduction, elimination of toxic or other hazardous materials, and solid waste reduction.

Step VIII B: Monitor Conditions for Success

Efforts to increase project robustness across the project life cycle are presented in Figure 7-17. The planned actions and responses are paired with the corresponding signposts and trigger values. Measuring systems are set up to monitor the signposts during operations and signal the operators when trigger values are reached. Specific actions corresponding to the trigger values are initiated:

- **Defensive Actions (AD):** Adapt to changing conditions by instigating pre-planned actions to increase capacity, strength, and durability.

- **Corrective Actions (AC):** Make operational adjustments to increase capacity, strength, and durability.
- **Capitalizing Actions (AP):** Take organizational actions to improve the project plan based on opportunities that arise during operations.

Triggers can also initiate the following actions:

- **Shaping Actions (AS):** Track and influence issues, trends, or events external to the project to reduce vulnerabilities, create or enhance opportunities, or turn vulnerabilities into opportunities.
- **Recovery Actions (AR):** If stressors reach values that cannot be handled through defensive or corrective actions, failure can occur. If the values reached are within the range of designed resilience, the operator can instigate recovery actions to bring the asset back into service quickly and at a low cost. If stressor values exceed the range of design resilience, quick and low-cost recovery may not be possible.
- **Reassessment (RE):** Monitor conditions and assumptions outside the preestablished actions to see whether they are still valid. Revise the project if they are no longer valid.

Step VIII C: Repurpose, Recycle, Demolish

At the project's end of life, it is desirable to make cost-effective use of the project materials and equipment. Also, the constructed works may be repurposed for other uses. The objectives for this phase are to minimize landfill disposal and make the best use of materials and equipment at the end of the project life. Figure 7-18 shows the reprocessing of concrete construction waste for reuse.

Step VIII D: Reassessment

If the project is insufficiently flexible to respond to changing conditions or the prepared responses are inadequate, it must be modified, as it can no longer meet the original objectives.

Conclusion

This modified version of DAP provides a methodology for planning, designing, constructing, and operating civil infrastructure projects in this era of climate uncertainty. Although the prospects of dealing with the deep uncertainty of climate conditions may seem intimidating, it turns out that engineers have always handled deep uncertainties in both their work and life. Indeed, multiple uncertainties exist

Figure 7-18. Excavator at landfill loading concrete waste in a mobile jaw crusher machine. Disposal of construction waste. Recycling concrete and asphalt from demolition. Re-use concrete after demolition.
Source: Maksim Safaniuk/Shutterstock.com.

throughout the life cycle of conventional projects. Key people might leave the project, resulting in a loss of critical skills. Buried hazardous wastes may be found during construction, causing project delays. Newly elected public officials with new priorities may jeopardize project funding.

Although these events are unpredictable, experienced project managers have learned to expect them and take action should they arise. A standard project management tool is a risk register, a way to identify potential risks and devise plans to deal with them should they become real. Similarly, the dangers posed by a changing climate can be identified along with strategies and actions to be taken when significant changes occur. The robustness matrix (Appendix C) is designed to identify and organize deviations from project success across the project life cycle. Project managers using this tool can develop strategies and actions for all life cycle phases: planning, project delivery, and operation. Effective monitoring systems are an essential element for achieving project success.

References

ASCE. 2023. "ASCE releases groundbreaking standard for sustainable infrastructure." October 19, 2023 Accessed April 19, 2024. https://www.asce.org/publications-and-news/civil-engineering-source/article/2023/10/19/asce-releases-groundbreaking-standard-for-sustainable-infrastructure

Ayyub, B. M., ed. 2018. "Climate resilient infrastructure: Adaptive design and management." Accessed February 6, 2024. https://sp360.asce.org/PersonifyEbusiness/Merchandise/Product-Details/productId/244232276.

CAMPO (Capital Area Metropolitan Planning Organization Office of Sustainability). 2015. "Central Texas extreme weather and climate change vulnerability assessment of regional transportation infrastructure: Final report." Accessed February 26, 2024. https://www.austintexas.gov/sites/default/files/files/CAMPO_Extreme_Weather_Vulnerability_Assessment_FINAL.pdf.

CAT (Climate Action Tracker). 2023. "The CAT thermometer." Accessed November 1, 2023. https://climateactiontracker.org/global/cat-thermometer/.

City of Olympia. 2019. "Sea level rise." Accessed February 6, 2024. https://www.olympiawagov/services/water_resources/storm___surface_water/sea_level_rise.php.

FHWA (Federal Highway Administration). 2014. "Assessing criticality in transportation adaptation planning." Accessed February 6, 2024. https://www.fhwa.dot.gov/environment/sustainability/resilience/tools/criticality_guidance/criticality_guidance.pdf.

FHWA. 2017. *Synthesis of approaches for addressing resilience in project development.* FHWA-HEP-17-082. Washington, DC: FHWA.

IPCC (Intergovernmental Panel on Climate Change). 2023. "Summary for policymakers." In *Climate change 2023: Synthesis report,* edited by H. Lee and J. Romero, 1–34. Geneva: IPCC. https://doi.org/10.59327/IPCC/AR6-9789291691647.001.

ISI (Institute for Sustainable Infrastructure). 2024. "Envision sustainable infrastructure framework." Accessed March 9, 2024. https://sustainableinfrastructure.org/.

Scenarios for the Local National Climate Assessment. 2024. "LOCA viewer." Accessed March 9, 2024. https://scenarios.globalchange.gov/loca-viewer/.

Wall, T. A., W. E. Walker, V. A. W. J. Marchau, and L. Bertolini. 2015. "Dynamic adaptive approach to transportation-infrastructure planning for climate change: San-Francisco-bay-area case study." *J. Infrastruct. Syst.* **21** (4): 05015004. https://doi.org/10.1061/(ASCE)IS.1943-555X.0000257.

Wallace, B., D. Ellison, and R. Daugherty. 2017. "Delivering safe, cost-effective, sustainable civil infrastructure projects under conditions of non-stationarity." In *Proc., Int. Conf. on Sustainable Infrastructure 2017: Technology.* New York. https://doi.org/10.1061/9780784481219.

Additional References (Not Cited)

FHWA. 2018. *Vulnerability assessment and adaptation framework.* FHWA-HEP-18-020. Washington, DC: FHWA.

Global Change.gov. 2023. "Fourth national climate assessment Vol. II: Impacts, risks, and adaptation in the United States." Accessed February 6, 2024. https://nca2018.globalchange.gov/.

CHAPTER 8

A Proposed Dynamic Adaptive Policy Pathways Methodology

"The measure of intelligence is the ability to change."

Albert Einstein

Introduction

This chapter presents a modified version of dynamic adaptive policy pathways (DAPP), summarized in Figure 8-1. Dynamic adaptive policy pathways methodology handles risk by reducing exposure to changing conditions in a defined area (e.g., city, region, or watershed). This is accomplished through a series of actions called *pathways*, each designed to handle a known range of conditions. The pathways are sequenced such that as conditions exceed the capability or capacity of the current pathway, a new pathway is taken. For example, to handle flooding from rising sea levels, a city might start by using sandbags to control occasional localized flooding. As the sea rises, sandbagging will eventually become ineffective. However, before reaching that point, the city could build flood walls to prevent flood damage. As the sea continues to rise, the city could incrementally raise the height of the flood walls and install additional pumps to move floodwaters away from infrastructure assets. Each of these actions would be implemented in the order dictated by their range of effectiveness. Implementation efforts would be started when monitoring determines that the current pathway is reaching its limit of effectiveness and that the next pathway will soon be required. These efforts must begin early enough to enable the next pathway to become operational before effectiveness limits are reached.

Haasnoot et al. (2013) noted that the DAP and DAPP have different approaches to decision support but are complementary. In this proposed DAPP methodology, DAP is an integral part of the overall DAPP methodology. Each DAPP pathway can be considered a project or a group of projects designed to contribute to the overall objective for a known range of conditions. And like a project, each pathway has its own conditions for success, which must be met if the pathway is to contribute to achieving the overall objective. As an individual pathway is taken, deviations from its conditions for success must be detected and trigger actions to keep the pathway on track. As applied in DAPP, pathways may be physical alterations, operational adjustments, or guiding policies controlling the system. Potential pathways can be

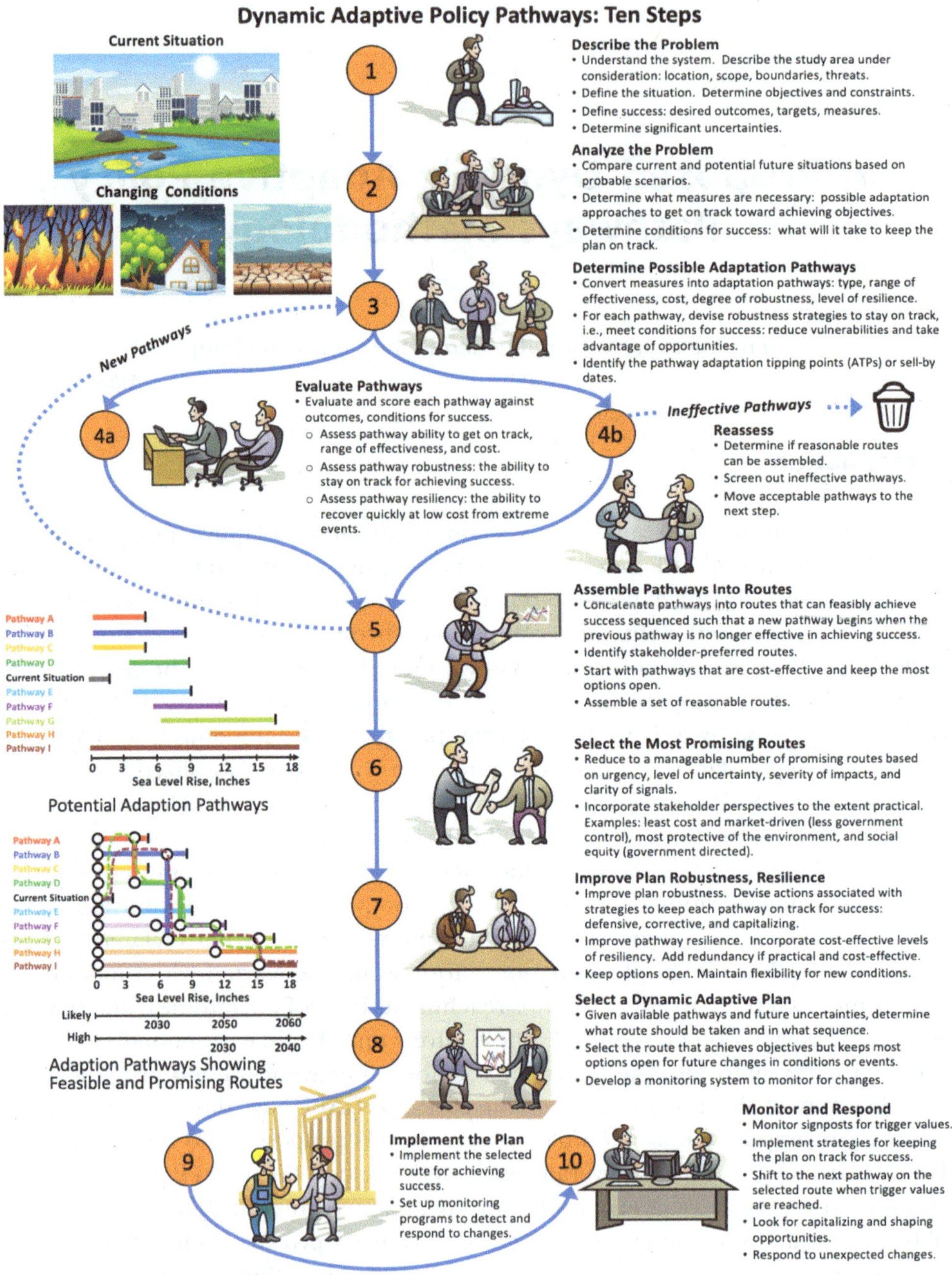

Figure 8-1. Dynamic adaptive policy pathways (DAPP) process steps.
Source: Adapted from Haasnoot et al. (2013).

derived based on various assumptions, objectives, and perspectives, applied as warranted by current conditions or as new knowledge is acquired. Pathways are selected based on their cost, robustness, and resiliency across their respective range of effectiveness. Pathway sequencing should be robust in its ability to retain the most options as new knowledge is acquired. Additional pathway selection criteria can be applied.

Case studies and applications of DAPP are showing up for many problems at many locations. As discussed in Chapter 6, stakeholders in the Algarve region of southern Portugal used DAPP to devise plans for managing water scarcity (Dias 2020). In the United States, DAPP was used to assess alternative flood mitigation strategies for the C-7 basin in Miami, Florida (Bouwer et al. 2017) and to explore long-term water system risks and alternatives for the San Francisco Public Utilities Commission (The Water Research Foundation 2021). The city of Olympia used a version of DAPP to address rising sea levels (City of Olympia/LOTT 2019). DAPP was cited in the *Post Hurricane Sandy Transportation Resilience Study in New York, New Jersey, and Connecticut* (FHWA 2017) as beneficial to guide investment decisions.

Others have developed case studies for how DAPP could be applied to various climate hazards. The Asian Development Bank (ADB 2022) presents a case study for improving flood resilience in key road segments in Cambodia. Another case study using DAPP (Vrinds 2021) involved balancing protection from flooding because of sea level rise and storm surges, freshwater demands, and shipping efficiency in the Netherlands waterways. Toimil et al. (2021) proposed DAPP to reduce the coastal erosion risk in northern Spain. Lawrence et al. (2019) used a DAPP and real options analysis hybrid to develop a 100-year coastal adaptation strategy for Hawke's Bay, New Zealand. Zeff et al. (2016) expanded on the work of Haasnoot et al. (2013) to handle water supply issues in the Research Triangle region of North Carolina. Manocha and Babovic (2018) applied DAPP to stormwater management for a tropical urban catchment in Singapore. Ramm et al. (2018) combined robust decision making and DAPP in a study on coastal flood risk for Lakes Entrance, a town in southeast Victoria, Australia.

Dynamic Adaptive Policy Pathways Terminology, Types, and Characteristics

Before describing the process, it is necessary to clarify some of the terminologies. The current technical writings describing the DAP and DAPP methodologies use *actions* as a blanket term covering multiple approaches, strategies, and actions taken to get and keep the guiding policies, operational adjustments, and physical alterations on track. To avoid confusion, this terminology is applied.

- **Measures** are potential adaptation approaches that have yet to be fully defined and verified in terms of the range of effectiveness, cost, robustness in achieving success, or the ability to recover from extreme events.

- **Pathways** are measures that are well defined in terms of their range of effectiveness in achieving the overall objectives. It can be a guiding policy (voluntary or enforced), an operational adjustment, or a physical alteration. Within their range of effectiveness, pathways should have a significant degree of robustness (i.e., the capability and capacity to stay on track to achieve the overall objectives). A pathway must successfully handle contingencies (deviations from conditions for success) that may occur within its range of effectiveness. A pathway also may have some level of resilience, the ability to get back on track if unexpected extreme conditions diminish its functionality. Pathway robustness and resilience will be discussed in more detail later in this chapter.
- **Strategies** are approaches for dealing with contingencies devised to get and keep a pathway (policy, physical, operational) on track toward success over its range of effectiveness. Following the DAP methodology, strategies come in multiple forms (mitigating, hedging, seizing, shaping) and are implemented through actions taken as contingencies arise.
- **Actions** are specific efforts made to get and keep the pathway on track. They carry out the strategies devised to deal with contingencies affecting a chosen pathway. Actions are executed based on predetermined stressor values or events. As in DAP, actions can be defensive, corrective, capitalizing, or recovery.
- The **Adaptation Signal** is the value or event detected and validated for one or more **Signposts.** When validated and the values reach predetermined levels (thresholds), the signal acts as a **Trigger**, initiating a preplanned response. Adaptation signals keep a pathway on track or initiate a transfer to a new pathway.
- **Routes** are sets of pathways that, if followed, can achieve success. Routes to achieving the same objectives and outcomes can vary based on the range of choices, resources, and stakeholder preferences.
  - **Feasible Routes** are routes composed of pathways that have passed in the initial evaluation.
  - **Preferred Routes** are those routes derived from differing stakeholder perspectives. Haasnoot et al. (2013) identified three perspectives for controlling flooding and water supply for a project in the Netherlands. The Hierarchist desires a large role in government, controlling water levels and enforcing water conservation policies. The Egalitarian is concerned with the environment and equity and prefers actions that reduce water demand through improved efficiency. The Individualist sees a smaller role for government, relying instead on markets and technological innovation to control water levels and protect water resources.
  - **Promising Routes** are routes composed of pathways that rank high in terms of cost, robustness, and resilience.
- **Plan** refers to one or more promising routes selected for achieving success. The routes contain pathways organized in a logical sequence. Conditions may change in unexpected ways. Plans must be robust in their ability to deviate from the initially selected routes, having other pathways available to respond to changed conditions.

Modified Dynamic Adaptive Policy Pathways: A Summary

To handle changing conditions in a specified geographic area, multiple approaches (pathways) are generally available. The Dynamic Adaptive Policy Pathways methodology organizes the pathways into a sensible order, starting with the most cost-effective approach for the given condition and leaving the most flexibility for change based on unforeseen events or new knowledge acquired over time. For example, drought episodes could be addressed early on by calls for water conservation. However, if the drought continues, voluntary water conservation may prove insufficient. In that case, voluntary water conservation could be replaced by water pricing based on consumer use. Efforts to locate new water supplies could be instigated. Additional reservoir capacity could be built. The determination of pathway order is based on their range of effectiveness and the flexibility retained for changing pathways as new knowledge is acquired. Individual pathway selection is based on cost, robustness, and resilience.

Setting the Scope, Objectives, and Limitations

The DAPP process starts with a determination of the area to be covered. Stakeholders and decision makers representing the defined area set the scope, objectives, and limitations for handling changing conditions. For example, the objective for the city of Olympia and its partners is to "protect downtown from sea level rise" through 2100. For the Netherlands, the objectives of their Delta Programme are "to protect the Netherlands from high water and flooding, to ensure a sufficient supply of fresh water, and to contribute to rendering the Netherlands climate-proof and water-resilient." The objectives of the Algarve region in southern Portugal are to maintain the region's Water Exploitation Index (WEI) at 43% while not decreasing the agricultural community's irrigation demands.

At the start, objectives and constraints are translated into targets and definitions of success. In the case of Olympia, success was defined as no flooding through the end of the century. For the Netherlands (see Figure 8-8), success was defined as "no floods occur, and there is enough fresh water during the next 100 years." The frequency of water shortage will be at least similar to the present situation: water shortage once in 10 years. Flooding avoidance is established in Dutch law as the probability per year that a flood would occur. Probabilities vary by dike ring area, defined land areas protected from flooding by multiple defenses: dikes, dunes, and hydraulic structures. Acceptable flood probabilities range from once every 1,250 years to once every 10,000 years (Jonkman et al. 2008, Haasnoot et al. 2013).

The selected route to achieving the objectives will be a set of pathways arranged by their range of effectiveness that will achieve success. As conditions change and approach the limit of pathway effectiveness, the plan transitions to the next pathway. Because of multiple external factors and knock-on effects, changes in conditions are uncertain. Therefore, the selected plan of pathway sequencing must be robust, i.e., leave options open (able to take other pathways) for unexpected changes. Moreover, the pathways themselves must be robust: able to handle contingencies (deviations from conditions for pathway success) across their specific range of effectiveness.

Types of Pathways

Dynamic Adaptive Policy Pathways methodologies consider three types of pathways to meet established objectives and achieve desired outcomes across a defined range of conditions. Each pathway has a unique cost and impact on objectives and outcomes. Each has an individual capacity and capability to stay on track once initiated and in effect.

- ***Physical pathways*** are generally constructed works such as barriers to flood waters, reservoirs to meet freshwater demands, or the relocation of communities in the face of rising seas;
- ***Operational pathways*** are adjustments made to existing operations. These adjustments could be changing maintenance schedules to avoid worker heat stress. They can also involve physical alterations that can be made during operations, such as waterproofing or additional cooling; and
- ***Policy pathways*** range from voluntary guidelines to enforceable rules imposed by the owner, operator, or governing organizations. For example, in cases of severe drought, a city may restrict certain water uses to maintain water supplies for critical functions.

Pathway Elements

An example of pathways and their elements is depicted in Figure 8-2. Each pathway has a range of effectiveness within which it is capable of handling changing conditions (e.g., flood levels, water stress, storm frequency, and intensity). As conditions change, the current pathway will eventually reach its adaptation tipping point (ATP), the limit of its effectiveness. Before that happens, the activity of handling that stressor must be transferred to another pathway that is effective in the new range of conditions. The trigger point for making this transfer is based on a predetermined adaptation signal from one or more signposts.

As shown in Figure 8-3, shifting to new pathways can be complex. First, the adaptation signal for making the transfer decision must be interpreted and verified as relevant, credible, and legitimate. The interpretation process must leave sufficient time for making the transfer before reaching the ATP. Some pathways, particularly those involving engineering design and construction, require long lead times and could be subject to delays.

In its basic form, the DAPP plan consists of a sequence of pathways initiated to address an identified changing condition, such as sea rise and water supply, for a relatively small area with well-defined elements (e.g., buildings, roads, critical

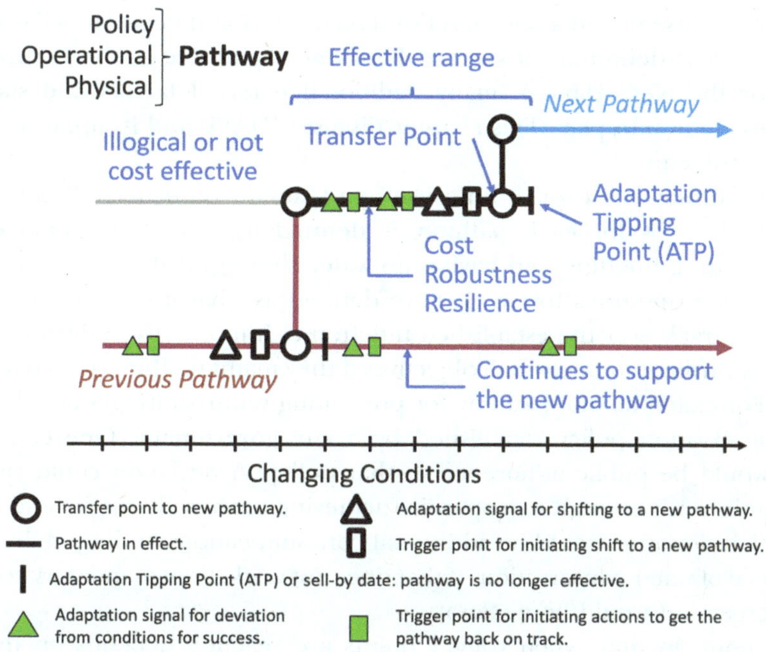

Figure 8-2. Example of sequence of pathways for addressing one or more objectives over a defined range of effectiveness.

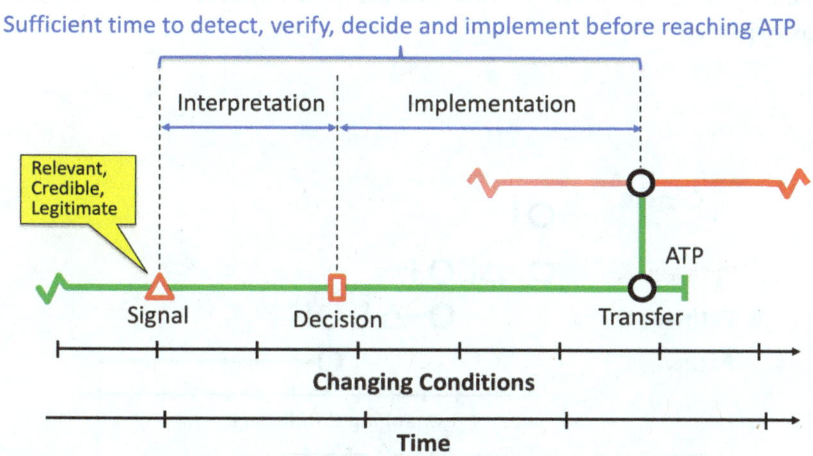

Figure 8-3. Signal interpretation and lead times required between signal detection, verification, decision, and implementation (transfer to new pathway before sell-by date). To compensate for a noisy environment, signals must be relevant, credible, and legitimate.

infrastructure assets) and a known set of stakeholders and decision makers who can arrive at a clear definition of success. For large regions (e.g., the Colorado River District or the Netherlands), many additional external factors and stakeholder influences come into play. This chapter discusses DAPP and its application to less complex problems.

Each pathway is a project unto itself and should be managed based on DAP principles. Success for each pathway is defined as achieving overall objectives (e.g., preventing flooding and having no water shortages) throughout its range of effectiveness. Conditions for success are defined as what it will take to keep each pathway on track over the established timeframe. Importantly, individual pathways can only contribute to the overall objectives if they maintain their own conditions for success. For example, one pathway for preventing water shortages could be a strict water conservation policy established by a city government. One condition for success would be public adherence to that policy. A deviation could violate that policy significantly enough to jeopardize achieving success. Strategies and actions to address that deviation could be additional communications to the public about the importance of water conservation. If that does not work, then mandatory compliance and enforcement would be necessary.

The point in time when trigger points are reached depends on the rate of change of the conditions being observed. In planning for changing conditions, it is useful to consider multiple scenarios and select the one most applicable to the objectives and pathways. Figure 8-4 shows a simple set of pathways compared against two scenarios, high and low. Transfer points in the high-end scenario will be reached sooner than in the low-end scenario. For critical assets, a high-end scenario might be selected, assuming for safety reasons that the pathway ATP will occur sooner than perhaps expected.

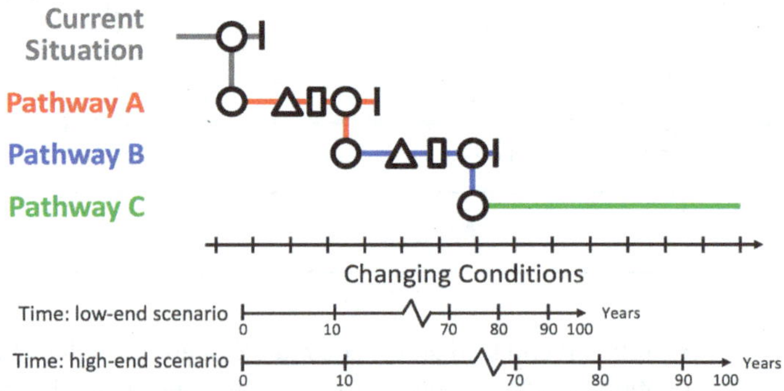

Figure 8-4. Concatenated pathways for handling changing conditions. The timing of the change to the next pathway (the sell-by date) is based on actual conditions, bracketed by the high- and low-end scenarios.

Pathway Objectives: Getting on Track and Staying on Track

Figure 8-5 is a snapshot of transitions among several pathways. As conditions change, the range of effectiveness of one pathway is reached. Signals received by the operator indicate that a transition to the next pathway is required. For a city or community, the operator could be a manager in a public works department. For areas with multiple jurisdictions, the operator could be a commission, board, or council of some representative body. When the signal is determined to be valid, the operator initiates the transition to the next pathway, putting in place a physical alteration, an operational adjustment, or a policy directive.

When the transition is made, the challenge for the new pathway is to stay on track throughout its range of effectiveness. It must be able to handle its own unique deviations from success, vulnerabilities, and opportunities that could undermine or improve chances for success for the pathway and, accordingly, the entire plan.

Figure 8-5 also shows a list of the conditions for success taken from the modified DAP process described in Chapter 7. However, their applicability varies based on the type of pathway. For example, a desalination project built and operated to alleviate water stress would have multiple conditions for success and be exposed to many more possible deviations than an enforceable water conservation policy. Policies, too, have important and relevant conditions for success. Their application must have adequate financial support, be operationally enforceable, and have sufficient public acceptance.

Conditions for Success		
Planning	**Project Delivery**	**Operations**
Location	Scope	Access
Requirements	Schedule	O&M
Finance	Cost	Health & Safety
Socio-Political	Quality	Utilities
	Staffing	Supplies
	Technologies	Performance
	Acceptance	Service

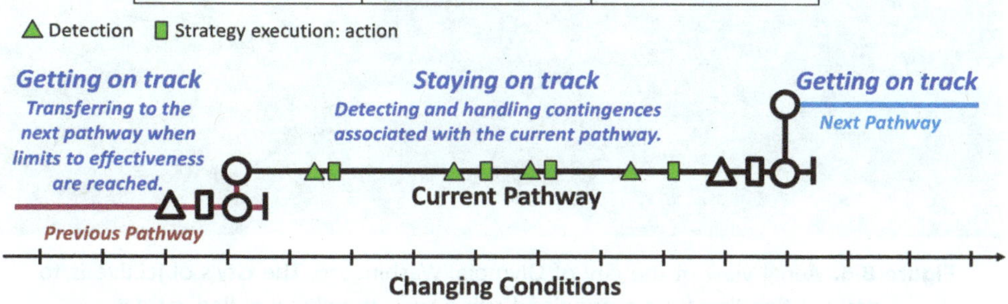

Figure 8-5. Getting on track and staying on track. Detection and handling of contingencies to stay on track, transferring to the next pathway when conditions warrant.

Practitioners and decision makers should use the DAP methodology for each pathway, determining which conditions for pathway success apply and devising strategies and actions to keep the pathway on track. An example of conditions for success, contingencies, strategies, and actions is summarized in Table 7-1 in Chapter 7. The robustness matrix, Appendix C, can be used as a tool for devising strategies and actions.

Pathway Characteristics

Each pathway has a range of effectiveness, the range of conditions within which the pathway is effective in achieving overall success. Beyond that range, an ATP is reached, and the subject pathway is no longer effective. Before the ATP, a transfer is made to another pathway with a range of effectiveness aligned with the new conditions (see Figure 8-2). For example, to protect the city of Olympia (Figure 8-6) from flooding from rising sea levels and storm surges, low-level flooding could be managed with sandbags placed at critical locations. Beyond a 3 in. sea level rise, sandbagging was judged to be no longer effective, to be replaced by constructed flood barriers. Sea rise above 6 in. would require additional flood barriers, consolidated outfalls, tide gates, and pump stations. An additional tide gate and pump station would be needed if the sea level rise reached 24 in. (City of Olympia 2011).

The plan then continues on the new pathway until its limits of effectiveness are reached. As with the previous pathway, the plan switches to another pathway with a higher range of effectiveness (Figure 8-5). Some pathways may continue to operate

Figure 8-6. Aerial view of the city of Olympia, Washington. The city's objective is to prevent flooding from storms and rising seas, thereby avoiding retreat (i.e., moving the downtown area to other locations).
Source: Always Wanderlust/Shutterstock.

after their ATP is reached. At that point, their contribution to success could continue as a supplement to the next pathway. For example, in the case of Olympia flood protection, a 6-inch rise in sea level would not result in dismantling flood barriers. Instead, tide gates, pumps, and additional flood barriers would be put in place to augment the flood protection provided by the existing barriers.

In addition to a range of effectiveness, pathways have the following characteristics:

- **Cost.** The cost of planning, designing, constructing, and operating the pathway;
- **Robustness.** This is the ability of the pathway to stay on track, i.e., maintain the pathway's specific conditions for success over its range of effectiveness; and
- **Resilience.** The ability of the pathway to recover relatively from extreme events quickly and at a relatively low cost.

These characteristics are further described as follows.

Cost

This is the cost of planning, designing, constructing, and operating the pathway throughout its life cycle. The cost includes maintaining pathway robustness (effective handling of contingencies and staying on track) and resilience (cost of recovery from failures and getting back on track).

Robustness

Within a pathway's range of effectiveness, there must be a degree of robustness, the ability to maintain conditions for success. This is defined as the extent to which contingencies, significant deviations from defined conditions for success, can be detected and managed successfully and cost effectively throughout its range of effectiveness, keeping the subject pathway on track and contributing to achieving the plan objectives.

Following the principles of DAP, robustness is achieved through defined strategies and contingency actions based on deviations from conditions for success. The greater the availability and range of cost-effective contingency actions that keep the plan on track, the greater the robustness.

Using DAP, several forms of strategy are applied.

Mitigating Strategies

Mitigating Strategies are strategies devised for a pathway that has known vulnerabilities. For these strategies, preplanned **Defensive Actions** against vulnerabilities are initiated when predetermined indicators (signposts) reach one or more predetermined thresholds (triggers). Signposts can be a stressor value or event. Furthermore, there can be multiple Defensive Actions within a Mitigating Strategy. For example, during a severe drought, a pathway could be a policy for maintaining water supplies by limiting water use. To maintain the effectiveness of that policy,

Mitigating Strategies would consist of pre-set water use rules (defensive actions) of increasing stringency based on the extent of deviation from the policy. Concurrently, there could be an action to restart groundwater pumping in dormant wells;

Hedging Strategies

For uncertain changes in stressor values or less probable stressor events, a series of **Corrective Actions**—multiple operational adjustments can be made to reduce or eliminate vulnerabilities. For example, if budget cuts threaten the operations of a facility designed to treat raw water, facility operators may adjust their operating costs or cut back on low-priority expenses;

Seizing Strategies

For anticipated (likely) opportunities, a series of **Capitalizing Actions** can be devised to improve chances for success. For example, broad public acceptance of policies to conserve freshwater resources can allow policymakers to strengthen other water conservation policies and diversify water supplies. Overall, this will increase the chances of success in reducing the effects of drought; and

Shaping Strategies

External trends or events may create new vulnerabilities that jeopardize success. Alternatively, they may create opportunities to improve chances for success. Shaping strategies are designed to mitigate vulnerabilities or take advantage of opportunities. For example, federal efforts to restore nearby wetlands could be leveraged to reduce the chances of flooding in the plan's defined area.

Resilience

If conditions extend beyond the expected deviations from success, the pathway may fail (i.e., not stay on track to achieve success). If resilience is limited, extreme conditions may cause the strategy to fail catastrophically, making recovery lengthy and costly. In addition, catastrophic failures could jeopardize the plan's overall success if subsequent strategies cannot be implemented fast enough or if alternative routes are unavailable. For example, planners may design flood walls with adjustable heights to prevent a predetermined level of flooding. However, an intense rainstorm may overwhelm flood walls and inundate subways and underground concourses, causing massive damage to the facilities and equipment.

As in DAP, planners should devise **Resiliency Strategies** for implementing **Recovery Actions**—fast and low-cost recovery from extreme stressor values or events resulting in pathway failure. To increase resilience, planners could waterproof the equipment and provide additional emergency flood barriers. Resilience may also be enhanced through system redundancies (i.e., having backup systems or parallel systems with extra capacity to handle the demands of changing conditions until the failed system is back in service).

Determining and Evaluating Routes for Achieving Objectives

Having identified a number of pathways that cover the range of expected conditions, the planners then determine and evaluate possible routes for achieving the overall objectives. The initial route determination is a matter of stringing pathways together into routes that cover the expected range of changing conditions throughout the planning period. While all are feasible, some routes may not be logical or cost effective.

Evaluation and Selection of Promising Routes

Each pathway in these feasible routes will have a cost to implement, plus a set of advantages and disadvantages. The promising routes should be a series of pathways, each possessing a high degree of robustness and resiliency. Overall, the cost of implementation of any promising routes should be reasonable and affordable.

The selected plan should be one or more promising routes that provide the most robustness and resiliency at an affordable cost. Furthermore, as plan implementation progresses, the plan should offer the most options should conditions change unexpectedly or new knowledge is acquired.

Modified Dynamic Adaptive Policy Pathways: The Details

This proposed process produces a plan that achieves the desired objectives and outcomes while maintaining conditions for success. It consists of 10 steps that produce multiple pathways that can be sequenced into one or more routes for achieving objectives. The pathways are selected based on their ability to achieve plan objectives and stay on track throughout their established range of effectiveness. Their sequencing is determined by their range of effectiveness and flexibility to change based upon changed conditions or new knowledge acquired. To stay on track, each pathway must contend with its own contingencies: potential deviations on the pathway that can undermine its individual conditions for success.

Step 1. Describe the Problem

Understand the system

The first step in the process is to define and describe the study area under consideration: its location, area boundaries, and the characteristics relevant to the problem to be solved. The boundaries of the study area are determined by the effectiveness of adaptive measures and the ability of the governing entity (or multiple entities) to implement those measures. For example, the study area could be a city concerned with flooding from rising sea levels and storm surges. The study area could also be a river basin covering multiple states and natural resource management jurisdictions concerned with future water resource availability. For this chapter, the scope of the systems under consideration will be confined to problems of limited complexity (i.e., relatively small areas with few objectives and outcomes).

Regardless of size, the study area faces threats common to communities within the area. The overall objective is to resolve the threats in a way that is satisfactory to the stakeholders, primarily the affected communities. For example, the Netherlands is concerned about flood damage from rising seas, freshwater supplies, and the availability of waterways for navigation.

The characteristics of the study area include a description of the physical, economic, and social makeup, i.e., the environment, infrastructure assets, commercial and industrial activities, and the social and cultural aspects of the communities. These will be used to determine plan objectives and constraints.

Define the situation

Once the system and its characteristics are understood, planners can define the situation (i.e., what problems, current and future, the system will face that could affect the stakeholders and communities within the system). A system, whether a city, a region, or a watershed, will likely have multiple problems created by a changing climate. Determine what problems need to be addressed and when actions need to happen.

Determine objectives and constraints

Once the situation and problems are defined, the planners should determine the objectives: what problems must be successfully addressed throughout the planning period and within what constraints. For example, the city of Olympia wanted to protect the city from rising sea levels and stormwater surges throughout the 21st century while retaining its current culture and configuration (City of Olympia/LOTT 2019).

Define success

Success can be defined by desired outcomes, specific targets associated with the objectives and constraints, and the measures against which success is determined. For example, protection from flooding can be specified as some low incidence of flood frequency. The Thames Barrier (Figure 8-7) that protects the City of London from flooding because of storm surges was designed with an annual exceedance probability (AEP) of 0.001 [i.e., a 0.1% chance of flooding per year (Halla et al. 2019)]. Also, protection from drought can be specified in terms of the level of water stress deemed acceptable by the stakeholders. The region of Algarve in southern Portugal determined that success in managing water scarcity would be to maintain a WEI of 43% through the remainder of this century, assuming the worst case, a "business-as-usual" climate scenario (Representative Concentration Pathway, RCP 8.5).

Determine the significant uncertainties

Knowing the objectives and constraints and having a definition of success or desired outcomes, planners now identify the uncertainties that could affect success. First and foremost, there are the uncertainties associated with overall objectives and

Figure 8-7. Thames Barrier, which protects the city of London from flooding because of storm surges, was designed with an annual exceedance probability of 0.001 (i.e., a 0.1% chance of flooding in any year).
Source: Alexey Fedodrenko/Shutterstock.

constraints. Considering the objective of protecting a community from flooding and its definition of success (a selected AEP), the significant uncertainties are sea level rise, storm intensity, and storm surge. Second, there are uncertainties associated with the types of measures that can be taken to achieve success. For example, financial support to fund reservoir construction could be reduced. Public support for wastewater reuse may weaken. Populations and their demand for services (e.g., freshwater) could increase. As measures are identified, and pathways are assessed, other uncertainties will likely emerge.

Step 2. Analyze the Problem

Compare current and possible future situations to objectives and desired outcomes

Without knowing the future degree of climate change effects and mitigation, planners seek to bracket the possibilities using scenarios of future conditions and their estimated impacts. For example, at the high-end scenario (RCP 8.5), little or no reduction in carbon emissions will result in a rise in sea levels and increased storm intensities in a shorter period. At the low end (zero net carbon by 2050 scenario),

Figure 8-8. Ijsselmeer area in the Netherlands. Flood protection- and water supply-related uncertainties include changes in temperature and precipitation, population growth, and rising sea levels.
Source: Mauvries/Shutterstock.

nations engaging in extensive carbon emissions mitigation may slow sea level rise and moderate future storm intensities. For example, the stakeholders in Algarve used the RCP 8.5 (business-as-usual) scenario to establish the timeline of future conditions to the end of the 21st century. They could have considered a more aggressive carbon emission reduction scenario (e.g., RCP 6.0) but decided to plan for the worst case. Table 7-2 in Chapter 7 lists RCP scenarios.

Determine what measures are necessary to achieve objectives

Assuming that both high and low scenarios (and some in between) are possibilities, planners determine what and when measures are needed to meet the objectives. These measures can be policies, operational adjustments, or constructed works implemented within the defined area. These measures must be effective across some range of changing conditions in reducing vulnerabilities (threats to achieving objectives) or taking advantage of opportunities (helping achieve objectives). For example, constructing a desalination plant can reduce the threat of water scarcity for a community. Also, collaboration with new federal agency programs to protect wetlands can be an opportunity to reduce flood risk within the system.

If the measures cannot address the expected range of climate stressor values with a reasonable chance of success, planners should consider retreat as an adaptation option. For the City of Olympia (City of Olympia/LOTT 2019), wholesale retreat (i.e., moving buildings and infrastructure out of Olympia's 450-acre downtown area to other locations) was considered. However, based on available options for the planning period out to the year 2100, retreat was deemed unnecessary.

Determine conditions for success

Drawing from DAP, for success to be achieved, certain conditions and assumptions must remain valid throughout the planning period. In implementing the plan, potential deviations from conditions for success must be identified and addressed. Two types of conditions are considered: conditions for getting the plan on track and conditions for keeping the plan on track.

To get the plan on track, the planners should determine the conditions that affect the pathway selection. For instance, in the Algarve example discussed in a previous chapter, achieving the target level of water stress was conditioned on stakeholder acceptance. One option for getting on track was to reduce agricultural water needs by cutting back on agricultural irrigation and switching to drought-resistant crops. This option was not economically or socially acceptable to the agricultural community.

To keep the plan on track, it must be recognized that each pathway has its own conditions for success, which must be addressed. Significant deviations must be detected, and actions must be taken to keep each pathway and the overall plan on track. In the Algarve example, for increasing landscape water retention, all water retention projects need to be working effectively throughout the planning period.

Step 3. Determine Possible Adaptation Pathways

Convert measures into adaptation pathways

Planners now convert the measures determined in Step 2 into adaptive pathways, as depicted in Figure 8-9. Each pathway should be identified by type (policy, operational, or physical) and characterized by a range of effectiveness, degree of robustness, and level of resilience. Each pathway must be able to contribute to achieving the overall objective across some applicable range of changing conditions (get on track) and must possess an adequate level of robustness and resiliency to maintain its effectiveness (stay on track). Its degree of robustness would be based on the pathway's individual conditions for success. These conditions would include the ability to meet variable demands, stakeholder acceptance, availability of financial support, ability to handle exposures and damage from climate stressors, and more. Staying on track for overall success means being able to handle contingencies: possible deviations from its own conditions for success. Therefore, each pathway must develop its own strategies and actions for handling contingencies. Its level of resilience would be its ability to return quickly to full functionality at relatively low recovery costs if any stressors were to exceed the design levels of robustness.

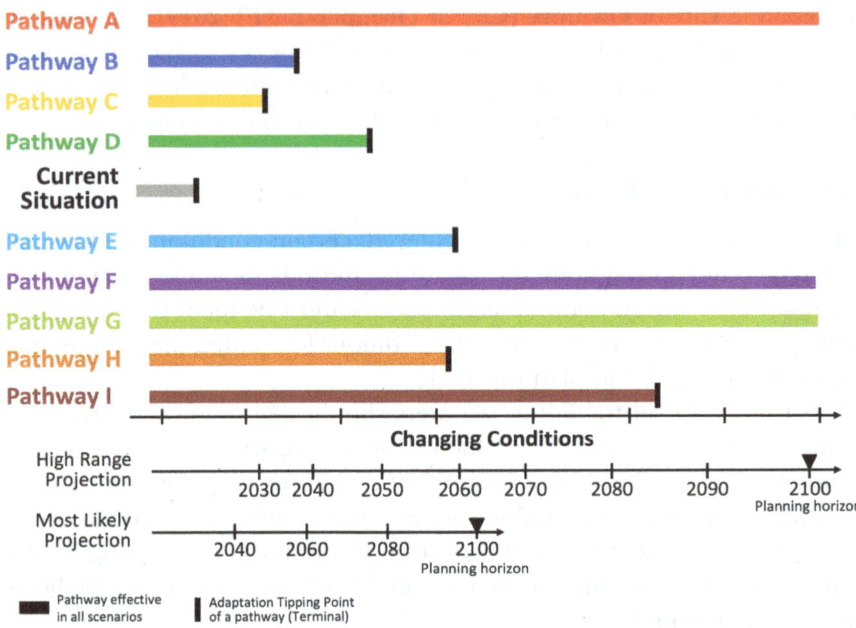

Figure 8-9. Multiple pathways for achieving success across high and most likely projections for changed conditions.

A reasonable pathway might be constructing a desalination plant (physical) of a specific water delivery capacity. That capacity would be based on expected water demands, land availability, proximity to the coast, cost, and stakeholder acceptance. As such, the desalination plant would be effective in reducing water stress based on the extent of the drought and the community's demand for water.

Identify pathway adaptation tipping points or sell-by dates

Using the chosen scenarios, planners must identify each pathway's ATPs: the points in time at which the pathways are no longer effective for a given scenario. Two example scenarios, high and most likely, are shown in Figure 8-10. These dates will be used to determine the transfer points: the points in the plan when a transfer to a new pathway is required. Timing is important because moving to the next effective pathway will require preparation. As the selected plan progresses, planners will better understand the actual rate of change and the likely sell-by dates.

Step 4a. Evaluate Pathways

Evaluate and score each pathway against outcomes and conditions for success

Once the possible pathways are determined and arranged in accordance with their range of effectiveness, the pathways should be evaluated and scored based on their

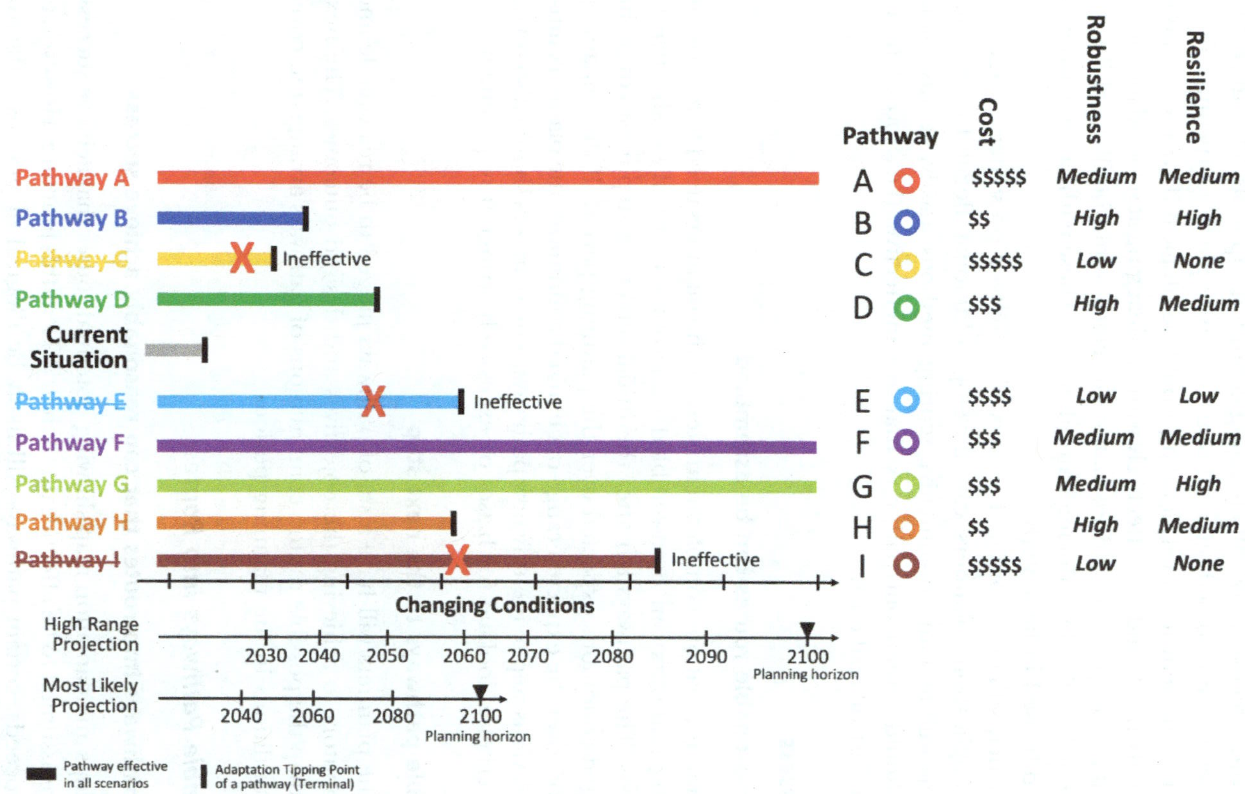

Figure 8-10. Example of evaluating pathways qualitatively for cost, robustness, and resilience. In this example, three pathways were judged as ineffective.

ability to achieve overall success, that is, for achieving objectives and outcomes. The evaluation should cover cost, robustness, and resilience.

At this point, the assumption embedded in this evaluation is that each pathway is equally logical and effective across all changing conditions up to its ATP. In reality, pathway effectiveness varies based on conditions. For example, constructing a flood wall to handle intermittent low-level flooding would be illogical if such conditions could be managed quickly and cost effectively with sandbagging at selected locations. Building a desalination plant to address water shortages would be illogical if such shortages could be handled by reducing water demand. These pathway effectiveness variations are addressed in later steps.

An example of qualitative scoring of a series of pathways is presented in Figure 8-10. In actual application, planners can develop additional detail for pathway characteristics as well as more quantitative scoring methods. Lastly, at this point, the planners should screen out pathways that are ineffective (i.e., too costly or insufficiently robust or resilient).

Step 4b. Reassess

Determine if reasonable routes can be assembled

Given the pathway evaluations completed in Step 4a, the planners need to determine if the remaining pathways can be assembled into routes that will be effective in achieving success. The pathways left after the initial screening step must cover the range of changed conditions expected over the planning period. If the remaining pathways cannot cover the expected range of changed conditions, the planners must return to the previous step to identify new pathways. The planners should also revisit the evaluation criteria, looking for biases or excessively stringent requirements.

Move acceptable pathways to the next step

At this point, the planners will have a set of pathways judged to be reasonable and effective contributors to achieving the objectives and desired outcomes. The next step will be to map the possible routes: combinations of pathways arranged to cover the range of conditions in the planning period.

Step 5. Assemble Pathways into Routes

Concatenate pathways into routes that can reasonably achieve success

This step develops an adaptation map showing potential routes that achieve success. All pathways are assumed to be effective up to their ATP at this juncture. However, as the process proceeds, certain pathways will likely be illogical or not cost effective under certain conditions. An example of a map of all feasible pathways that emerged following the screening in Steps 4a and 4b is presented in Figure 8-11. Here, it is assumed that all pathways can be employed effectively at the start.

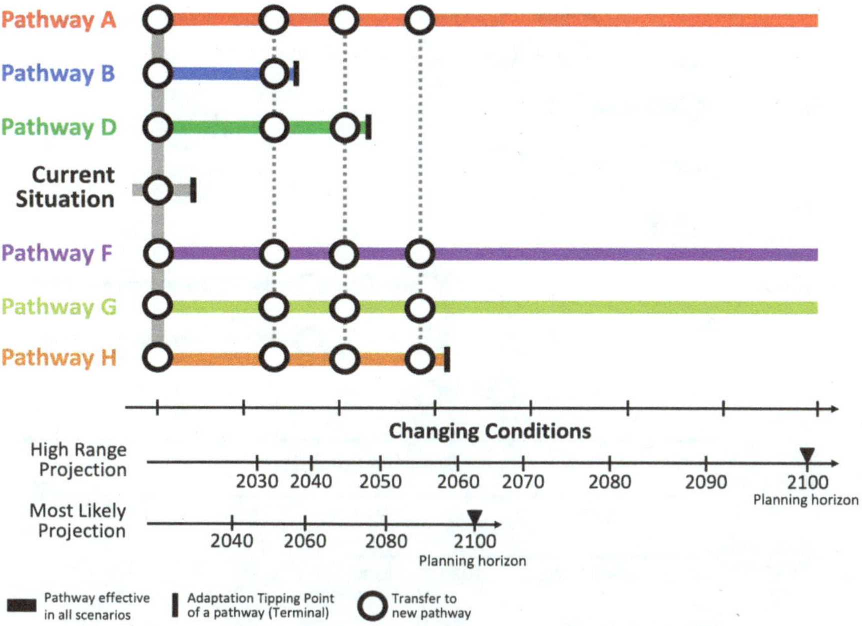

Figure 8-11. Map showing all feasible routes for handling defined changing conditions out to a planning horizon of 2100.

Assemble a set of reasonable routes

Using the acceptable pathways established in Steps 4a and 4b, planners can assemble one or more routes to achieve the desired objectives and outcomes. Routes are deemed reasonable if (1) the pathways can be sequenced in a way that achieves success across the range of changing conditions, (2) the selected pathways received sufficiently high scores in the pathway evaluation criteria presented in Step 4a, (3) the pathways are logical and effective in the range of conditions for which they will be used, and (4) the routes keep options open. Figure 8-12 depicts three example routes. An example of placements of adaptation signals and trigger points for shifting to the next pathway is also shown. In devising the routes, pathways are sequenced such that new pathways begin at a point when the previous pathway is closing in on its ATP. Care should be taken that the adaptation signal and trigger point are timed so that the next pathway can be made operational and ready for transfer.

Incorporate preferred routes as needed

For more complex problems with multiple stakeholders, Haasnoot et al. (2013) suggest that a manageable number of preferred routes be developed based on stakeholder perspectives and that these routes form the basis of the plan. This approach incorporates and weighs stakeholder acceptance as part of the pathway robustness evaluation. However, it seems helpful to devise both reasonable and

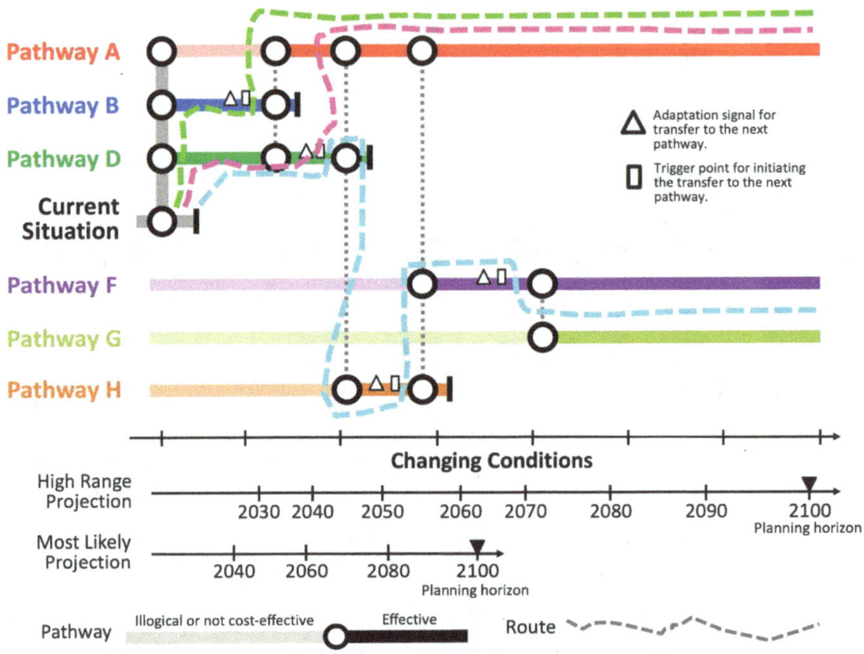

Figure 8-12. Examples of three potential routes developed by evaluating individual pathway cost, robustness, and resilience.

preferred routes as a way for planners and stakeholders to visualize and examine the differences.

Step 6. Select the Most Promising Routes

Reduce to a manageable number of promising routes

Drawing from the routes assembled in Step 5, planners should now select the most promising routes. These routes should be chosen based not only on the overall scores received in Step 4a but also on additional factors.

- *Urgency of action.* If changing conditions warrant immediate action, the initial pathways of the selected routes should be easy to initiate.
- *Degree of uncertainty of changing conditions.* If there is a high degree of uncertainty regarding changing conditions (e.g., the rate of change, impacts, and knock-on effects), routes that keep more future options open are a better choice.
- *The severity of impacts if unable to keep on track.* If the impacts of changing conditions are severe, then routes composed of pathways that are highly effective and robust are a better choice.

- *Clarity of signals.* Signals for change should have little ambiguity, both for handling contingencies within the pathway and transferring to a new pathway before reaching ATP.

Incorporate stakeholder preferences to the extent practical

Given the importance of stakeholder acceptance, planners should incorporate pathways that meet stakeholder preferences if those pathways are reasonable. Those pathways should have high evaluation scores and be effective in the region of changing conditions in which they are applied.

Step 7. Improve Plan Robustness, Resilience

Improve pathway robustness within the promising routes

Define the strategies and associated actions for keeping the pathways on track. To stay on track, each pathway will have conditions for success, conditions or assumptions that must remain valid throughout its range of effectiveness. To increase pathway robustness, planners should consider potential deviations from pathway success and define the strategies and actions that will be taken to keep the pathway on track. Drawing from DAP, the choice of strategies will be mitigating, hedging, seizing, or shaping. Signposts will be established for monitoring and instigating defensive, corrective, or capitalizing trigger actions.

Improve pathway resilience

Actual conditions may overwhelm the chosen pathway. As in DAP, resilience can be incorporated to get the pathway back on track quickly and at a relatively low cost. For example, extreme storms may defeat flood controls. However, flood damage may be reduced through additional waterproofing of equipment and temporary flood barricades protecting downstream facilities. Functionality can be retained through operational redundancy.

Keep options open

The routes under consideration should be flexible in anticipation of new conditions or knowledge that would alter the plan.

Step 8. Select a Dynamic Adaptive Plan

Decide what pathways should be taken immediately

Given the current conditions and in anticipation of changing conditions, planners will determine what route should be selected (Figure 8-13), drawing from one or more promising routes. The selected pathways will be those that can respond effectively while leaving the most options open for future conditions.

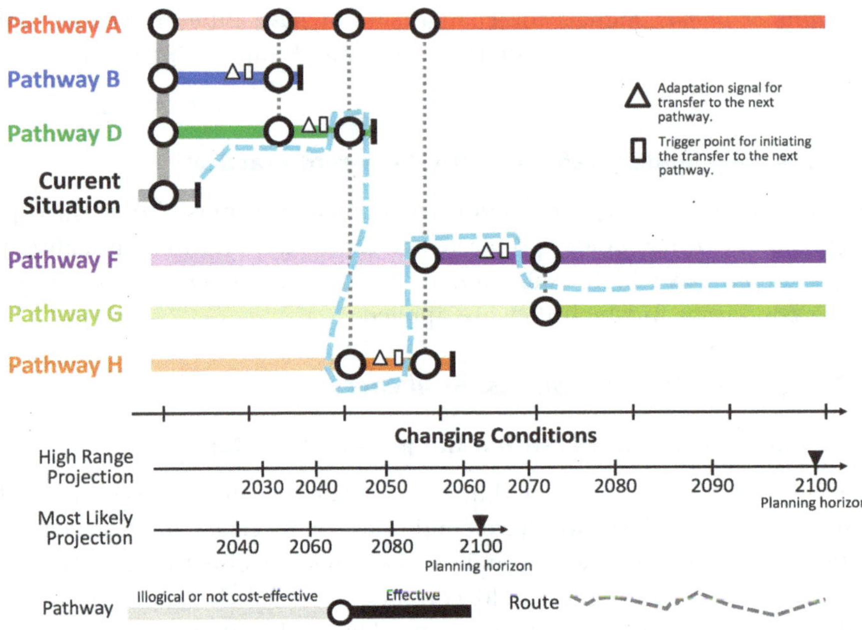

Figure 8-13. Selected plan.

Develop a monitoring system

Devise a system for monitoring conditions and events that would alter the plan. Include signposts, triggers, and actions.

Step 9. Implement the Plan

Implement the selected plan

Begin implementation of the chosen pathway or pathways.

Set up monitoring systems to detect changes

For the selected pathways, set up monitoring systems to detect signals and triggers for transfer to subsequent pathways. Signals and triggers should be designed to provide sufficient lead time to shift to a new pathway. For example, if the next pathway involves a significant physical alteration, such as a levee or a reservoir, the lead time required from detection to full operation could be years.

Step 10. Monitor and Respond

Monitor signposts for trigger values

For each active pathway, the operators will monitor signposts, looking for established values of changed conditions or deviations from conditions for success.

Implement strategies for keeping pathways on track

If deviations from conditions for pathway success reach preestablished values or events occur, either of which could harm chances for success, operators will initiate preplanned strategies.

Shift to the next pathway when pathway trigger values are reached

Operators will monitor signposts to determine when a shift to the next pathway or pathways is warranted. Signals received must be relevant, credible, and legitimate indications for transfer to the subsequent pathways. If valid, operators will initiate the transfer to the next pathway.

Look for capitalizing and shaping opportunities

Operators should not ignore the possibility of opportunities arising from new or changed circumstances or external trends or events that could be influenced in favor of one or more pathways. Budgets may increase, making funds available for improvements in robustness. As the plan proceeds, operators should look for capitalizing and shaping opportunities, opportunities to improve or influence chances for pathway or plan success.

Respond to unexpected changes

Given the huge uncertainties, climate, and otherwise, operators should monitor conditions, looking for unexpected changes that could disrupt the overall plan. They should keep options open to change course as needed.

Conclusion

The DAPP methodology is useful for addressing one or more climate stressors in a defined area, such as a city, region, or watershed. Alternative approaches (pathways) for adapting to stressors are presented in sequence based on their range of effectiveness. As the limit of effectiveness is reached, a transfer to the next pathway is made. Versions of DAPP are being applied at many locations around the world.

References

ADB (Asian Development Bank). 2022. "Disaster-resilient infrastructure: Unlocking opportunities for Asia and the Pacific." Accessed February 7, 2024. https://www.adb.org/publications/disaster-resilient-infrastructure-opportunities-asia-pacific

Bouwer, L., M. Haasnoot, D. Wagenaar, and K. Roscoe. 2017. "Assessment of alternative flood mitigation strategies for the C-7 Basin in Miami, Florida." Accessed February 7, 2024.

https://www.researchgate.net/publication/368756570_Assessment_of_alternative_flood_mitigation_strategies_for_the_C-7_Basin_in_Miami_Florida.

City of Olympia. 2011. "City of Olympia: Engineered response to sea level rise." Accessed February 7, 2024. https://cms7files.revize.com/olympia/Document_center/Community/Climate%20Change%20Response/SLR/Other/SLR-Report-2011.pdf.

City of Olympia/LOTT. 2019. "Olympia sea level rise response plan." Accessed February 7, 2024. https://cms7files.revize.com/olympia/Document_center/Community/Climate%20Change%20Response/SLR-Plan-Complete.pdf.

Dias, L. F., B. A. Aparício, J. P. Nunes, I. Morais, et al. 2020. "Integrating a hydrological model into regional water policies: Co-creation of climate change dynamic adaptive policy pathways for water resources in southern Portugal." *Env. Sci. Policy* **114** (2020): 519–532.

FHWA (Federal Highway Administration). 2017. *Post Hurricane Sandy transportation resilience study in NY, NJ, and CT.* FHWA-HEP-17-097. Washington, DC: FHWA.

Haasnoot, M., J. H. Kwakkel, W. E. Walker, and J. Ter Maat. 2013. "Dynamic adaptive policy pathways: A method for crafting robust decisions for a deeply uncertain world." *Global Environ. Change* **23** (2): 485–498. https://doi.org/10.1016/j.gloenvcha.2012.12.006.

Halla, J. W., H. Harvey, and L. J. Manning. 2019. "Adaptation thresholds and pathways for tidal flood risk management in London." *Clim. Risk Manage.* **24** (2019): 47–58. https://doi.org/10.1016/j.crm.2019.04.001.

Jonkman, S. N., M. Kok, and J. K. Vrijling. 2008. "Flood risk assessment in the Netherlands: A case study for dike ring South Holland." *Risk Anal.* **28** (5): 1357–1374. https://doi.org/10.1111/j.1539-6924.2008.01103.x.

Lawrence, J., R. Bell, and A. Stroombergen. 2019. "A hybrid process to address uncertainty and changing climate risk in coastal areas using dynamic adaptive pathways planning, multi-criteria decision analysis and real options analysis: A New Zealand application." *Sustainability* **11** (2): 406. https://doi.org/10.3390/su11020406.

Manocha, N., and V. Babovic. 2018. "Real options, multi-objective optimization and the development of dynamically robust adaptive pathways." *Environ. Sci. Policy* **90** (2018): 11–18. https://doi.org/10.1016/j.envsci.2018.09.012.

Ramm, T. D., C. S. Watson, and C. J. White. 2018. "Strategic adaptation pathway planning to manage sea-level rise and changing coastal flood risk." *Environ. Sci. Policy* **87** (2018): 92–101. https://doi.org/10.1016/j.envsci.2018.06.001.

The Water Research Foundation. 2021. "Long term vulnerability assessment and adaptation plan for the San Francisco public utilities commission water enterprise—Phase I." Accessed February 7, 2024. https://sfpuc.org/sites/default/files/about-us/policies-reports/LTVA_AdaptationPlanSFPUC_Phase1.pdf.

Toimil, A., I. J. Losada, J. Hinkel, and R. J. Nicholls. 2021. "Using quantitative dynamic adaptive policy pathways to manage climate change-induced coastal erosion." *Clim. Risk Manage.* **33** (2021): 100342. https://doi.org/10.1016/j.crm.2021.100342.

Vrinds, T. 2021. "Adaptive design of flood defence systems." Ph.D. thesis, TU Delft University of Technology, Civil Engineering and Geosciences. https://repository.tudelft.nl/islandora/object/uuid:4c24fee4-7457-41bc-a7de-5afea051ee60.

Zeff, H. B., J. D. Herman, P. M. Reed, and G. W. Characklis. 2016. "Cooperative drought adaptation: Integrating infrastructure development, conservation, and water transfers into adaptive policy pathways." *Water Resour. Res.* **52** (9): 7327–7346. https://doi.org/10.1002/2016WR018771.

Additional References (Not Cited)

Carstens, C., K. Mossberg Sonnek, R. Räty, P. Wikman-Svahn, A. Carlsson-Kanyama, and J. Metzger. 2019. "Insights from testing a modified dynamic adaptive policy pathways approach for spatial planning at the municipal level." *Sustainability* **11** (2): 433. https://doi.org/10.3390/su11020433.

National Delta Programme. 2024. "Now for the future." Accessed April 20, 2024. https://english.deltaprogramma.nl/documents/publications/2023/09/19/brochure-outlines-delta-programme-2024-english.

Nunes, J. P., L. Felipe Dias, B. A. Aparicio, I. Morais, A. Lúcia Fonseca, and A. Valérie Pastor. 2020. "Using dynamic adaptive policy pathways and hydrological modeling to co-create water resource adaptation policies for climate change: A practical example for southern Portugal." Accessed February 7, 2024. https://meetingorganizer.copernicus.org/EGU2020/EGU2020-11015.html.

CHAPTER 9

Preparing for an Uncertain Future

"Adults keep saying: 'We owe it to the young people to give them hope.' But I don't want your hope. I don't want you to be hopeful. I want you to panic. I want you to feel the fear I feel every day. And then I want you to act. I want you to act as you would in a crisis. I want you to act as if our house is on fire. Because it is."

Greta Thunberg, Swedish climate activist, speaking at the World Economic Forum, Davos, Switzerland, January 25, 2019

Climate Disasters Continue

Extreme weather events continue to rise. For the United States, NOAA predicted in early 2021, after what they termed a record-setting start, an above-average hurricane season from June 1 through November 30, 2021. NOAA's predictions came true. On August 29, Hurricane Ida landed in Louisiana, bringing sustained winds of 150 mph, causing flooding and structural damage. Power failures affected 1.2 million Louisiana residents (EIA 2021). After suffering through a severe cold wave in February that crippled its power grid (Henson 2021), Texas electricity customers lost power again, this time in September from Hurricane Nicholas (Texas Tribune 2021). Also, Ida, the fifth strongest hurricane on record, moved up the east coast to New Jersey and New York, causing devastating flooding (Washington Post 2021). At the end of September 2021, the number of billion-dollar weather disasters hit a new record high of 18, costing $104.8 billion, more than $4 billion higher than all of 2020 (NOAA 2021) (Figure 9-1).

Wildfires in the western United States also broke records. As of September 28, 2021, 11 western states reported 64 active large fires spread over 3.1 million ac. In 2021, more than 46,000 large and small fires covered almost 6 million ac. The Dixie Fire became the second-largest wildfire in California's history, just below the 2020 August Complex fire that burned more than 1 million ac (Smith 2021). An extended drought and heat wave provided the conditions for wildfires to develop and expand rapidly.

The United States was not unique. In the summer of 2021, the world experienced extreme climate and weather events, including storms, flooding, heat waves, wildfires, and drought (Figure 9-2). The northern hemisphere suffered 50 hurricanes, typhoons, and other storm events. In China, Typhoon In-fa caused record-breaking flooding, displacing more than 1 million people and killing 300. Other

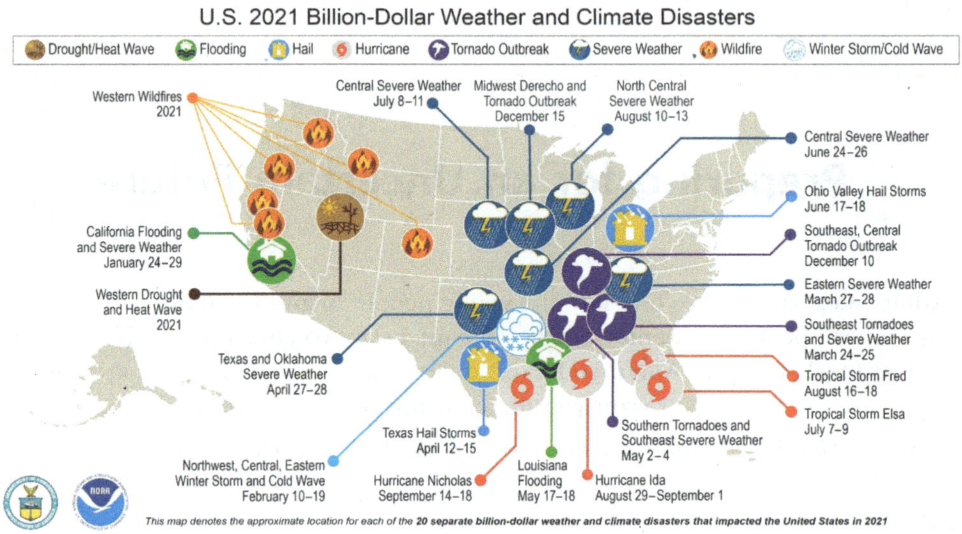

Figure 9-1. US 2021 billion-dollar weather and climate disasters.
Source: NOAA (2021).

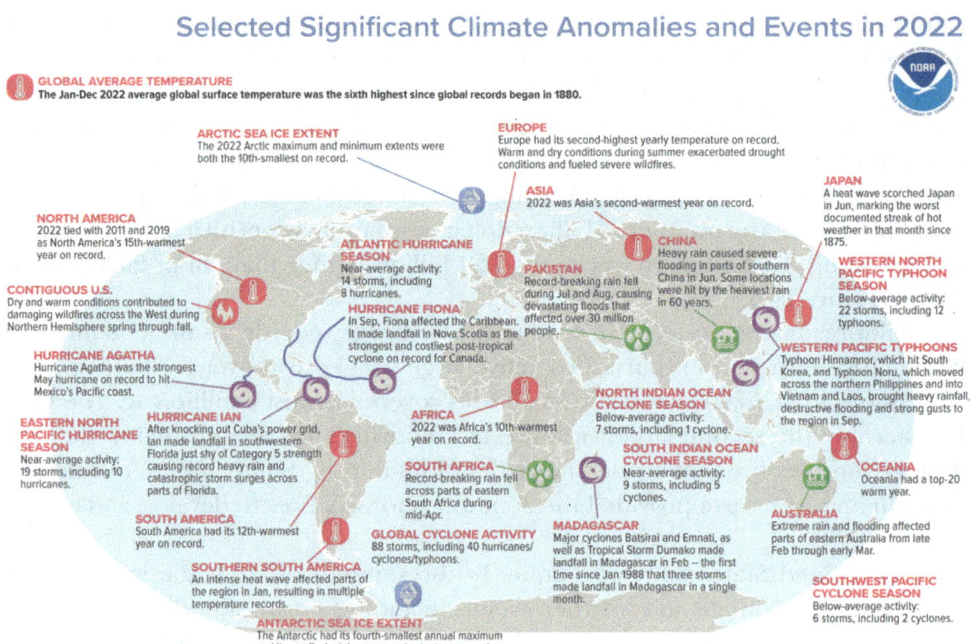

Figure 9-2. Selected significant climate anomalies and events in 2022.
Source: NOAA NCEI (2022).

flooding events in the northeastern states of Jigawa, Bauchi, and Adamawa, Nigeria, caused a bridge to collapse, killing 21 people. Heavy rains and flooding in India and Turkey caused landslides, while in Europe, record rainfall resulted in extreme river flooding in Germany, Belgium, and the Netherlands (Brimicombe et al. 2021). In total, at least 85% of the world's population has been affected by climate change. Four-fifths of the world's land areas have been impacted by global warming. The impacts included crop failures, floods, and heat waves (Timsit and Kaplan 2021).

While this was happening, the Amundsen–Scott South Pole Station in Antarctica reached record cold temperatures, averaging −61 °C (−78 °F), 4.5 °C lower than its 30-year average. Scientists are calling this a blip, an anomaly in the overall warming trend for Antarctica (Samenow and Patel 2021). To paraphrase author and environmentalist Paul Hawken, "It's not just global warming. It's also global weirding."

The Earth's average temperature is now 1.3 °C (2.3 °F) warmer than in the 1800s. That is also one-tenth of a degree warmer than what was reported when I began working on this book about 3 years ago. Since 1970, a half-century ago, global temperatures have risen faster than any other 50-year period in the last 2,000 years (European Environment Agency 2023).

As shown in Figure 9-2, disasters continued in 2022. Pakistan had the worst flood in the country's history, killing more than 1,700 people and causing more than US $30 billion in damage and economic losses (NDMA 2022). The floods displaced about 7.9 million people, leaving one-third of the country under water (Nabi 2023).

In Europe, summer temperatures in 2022 were the hottest on record (Copernicus 2022). In June of that year, Japan had the hottest heat wave ever recorded (BBC News 2022). In Africa, three nations—Nigeria, South Africa, and Somalia—had their most costly weather and climate disasters. Floods in Nigeria and South Africa caused US$7.7 billion in damages. Drought in Somalia from 2021 to late 2022 resulted in damages of US$1.1 billion, the most costly disaster in that nation's history (Masters 2023).

Global temperatures have broken records again. July 2023 was the Earth's hottest month ever, 0.24 °C (0.43 °F) warmer than any other July in NASA's records (O'Shea 2023). September 2023 became the warmest September on record, hitting 16.4 °C (61.5 °F), 0.9 °C (1.6 °F) above the 1991–2020 average for September. Not only was this a record high, but a record temperature anomaly: 1.8 °C (3.2 °F) higher than any September average for the preindustrial referenced period of 1850–1900. As of this writing, 2023 is the warmest year ever since temperature record-keeping began in 1850 (NOAA NCEI 2024).

Despite increasing temperatures and the scale and frequency of climate and weather disasters, relatively little climate change mitigation has been accomplished. Carbon emissions continue to rise, recovering from a short COVID-19-induced economic downturn and a corresponding dip in the demand for fossil energy.

Governments and industries went about their business, seeing the end of the pandemic and a return to business normalcy. Meanwhile, climate disasters continue, the Earth being unaware and indifferent to recent slippages in carbon emission rates.

The Tragedy of the Horizon

UN Special Envoy on Climate Finance Mark Carney identified another climate-related tragedy, self-interest diverting attention away from future financial instability. Speaking at Lloyd's of London in September 2015, Carney named this lack of attention "The Tragedy of the Horizon," a derivation of Garrett Harden's 1968 work noted in his article, "The Tragedy of the Commons." Carney stated:

> We don't need an army of actuaries to tell us that the catastrophic impacts of climate change will be felt beyond the traditional horizons of most actors – imposing a cost on future generations that the current generation has no direct incentive to fix.

He identified the traditional horizons. Business is focused on quarterly earnings. Terms of political office are 2 to 6 years. In finance, monetary policies cover 2 to 3 years; credit ratings 3 to 5 years. These are short-term relative to climate risk (Carney 2015).

The Tragedy of the Commons

The causes of climate change are an example of what Garrett Hardin called "The Tragedy of the Commons" (1968), a situation in which persons, acting in their self-interest, cause the depletion and, ultimately, the exhaustion of a shared resource. Harden used the example of sheep herders using a common grazing land. In this case, each herder finds adding more sheep to their herd to increase their income advantageous since all herders will share the damage caused by overgrazing. In the case of climate change, nations burning fossil fuels for heat, mobility, and electrical energy production are damaging the Earth's common ecosystem service, climate regulation. Their actions are driven by self-interest. Why switch to low- or no-carbon energy sources? Why pay that switching cost and suffer the inconvenience when nothing stops us from doing what we have always done? What if we switch and no one else does? That could make us uncompetitive.

Carney also identified three broad channels through which climate change could negatively affect financial stability. These are

- **Physical Risks:** Damage caused by storms, flooding, and wildfires;
- **Liability Risks:** Parties suffering damage will seek compensation; and
- **Ambition, Transition Risks:** Cost and lack of ambition to transition to a low-carbon economy.

Continuing on this theme, despite increasing climate catastrophes, national governments and businesses continue to work within horizons in which climate is not a priority. The developed nations tend to follow the business and political cycles, wanting to keep things just as they are. Similarly, the priority for developing nations is to improve their economies. To them, climate impacts are important but of less concern than economic growth. Finally, people in underdeveloped nations are worried about day-to-day survival. Although they are most affected by climate change, they have the least ability to do anything about it except to become climate refugees.

Revisiting the Brundtland Commission report's definition of sustainable development is instructive. The report stated that sustainable development is "development that meets the needs of the present without compromising the ability of future generations to meet their own needs." In this definition, there is a clear presumption that jeopardizing the fate of future generations would be sufficiently compelling for the current generation to make its economic development sustainable. Apparently, it is not.

Shifting from Intent to Action

The situation may be changing. Global indolence has shifted to guarded ambition, and not too soon. Starting with the 2015 Paris Agreement, most nations have committed to reducing greenhouse gas (GHG) emissions, seeking to achieve net-zero emissions by 2050. While similar promises have been made before, this time seems different. National actions will be verified, tracked, and reported against their stated commitments. A Conference of the Parties (COP) timeline showing recent and significant decisions and actions is presented in Figure 9-3.

COP 21 and the Paris Agreement

In late 2015, the nations of the world gathered in Paris for COP 21, the United Nations Climate Change Conference. After multiple decades of debate and weak resolutions, the international community signed the Paris Agreement, a legally binding treaty for dramatic reductions in GHG emissions. The goal of the agreement is to limit global warming well below 2 °C (3.6 °F) and preferably below 1.5 °C (2.7 °F) compared to preindustrial ambient temperature levels. The Paris Agreement and its requirements are part of the SDGs' climate actions in Goal No. 13, specifically Target 13.2: Integrate climate change measures into policy and planning.

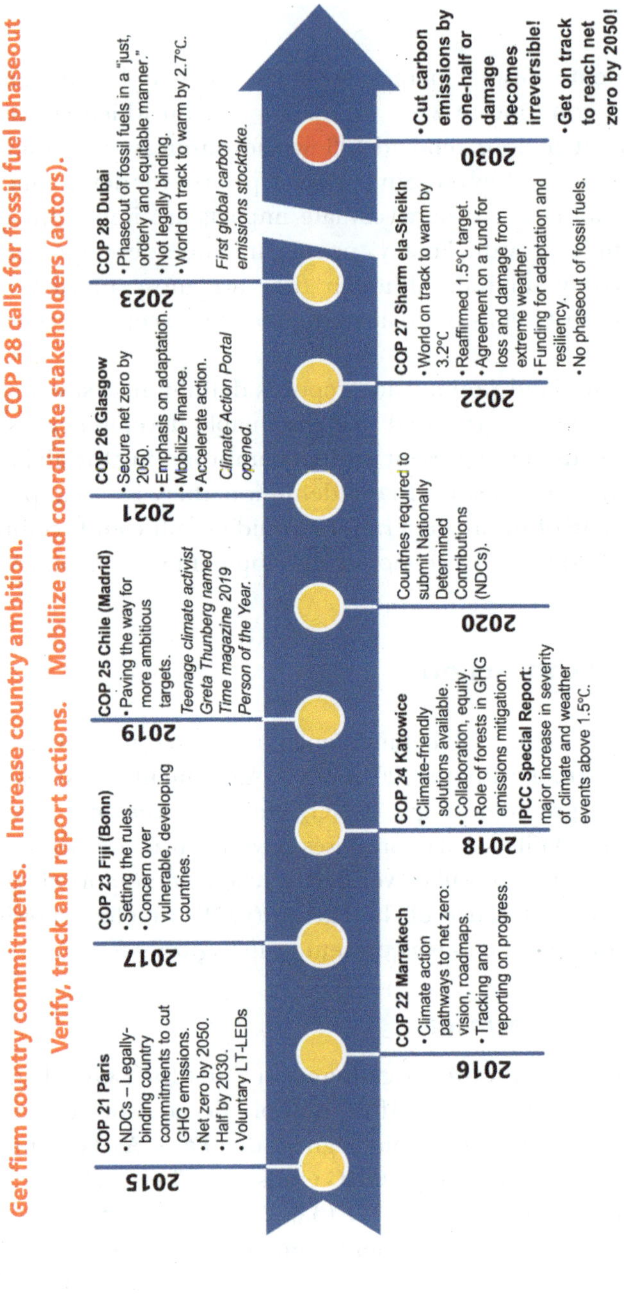

Figure 9-3. Timeline for climate action.

Limiting warming to these levels was essential. Climate scientists have warned that temperatures above 1.5 °C (2.7 °F) would bring about more devastation: storms of increasing severity, loss of species, extended droughts, rising sea levels, and more (Table 4-1). Keeping warming below 2 °C (3.6 °F), preferably below 1.5 °C (2.7 °F), would require enormous cuts in carbon emissions, forcing what amounts to a phase-out of fossil fuel-based energy sources.

Nationally Determined Contributions

The Paris Agreement was a turning point, the first forceful global attempt to prevent warming beyond 1.5 °C (2.7 °F) and to adapt to changing climate conditions. Nations are expected to develop bold climate actions: meaningful contributions to reaching net-zero carbon emissions. By 2020, each country is required to develop and submit plans detailing its goals and intended actions. Called Nationally Determined Contributions or NDCs, the Paris Agreement requires countries to show their intended actions: GHG emissions mitigation and adaptation to address the impacts of rising temperatures and extreme weather events. Guidance for NDC development is presented in Table 9-1. The initial NDC submissions were anticipated to be inadequate, producing offers based on what was economically and politically achievable domestically.

Table 9-1. Good Practice for 10 Key Elements of National Net-Zero Target Setting.

Good practice for ten key elements of national net zero target setting

	Element	Good practice
Scope	Target year	
Scope	Emissions coverage	All sectors and gases covered
Scope	International aviation and shipping	The net zero target fully covers emissions from international aviation and shipping
Scope	Reductions or removals outside of own borders	Reaching net zero within own borders
Architecture	Legal status	Legally binding target
Architecture	Separate reduction & removal targets	Separate targets for emission reductions and removals
Architecture	Review process	Legally binding review of target and progress against it at regular intervals
Transparency	Carbon dioxide removal	Transparent & scientifically robust assumptions on LULUCF and carbon removals & storage
Transparency	Comprehensive planning	Transparent and scientifically robust pathway / intermediate targets with clear measures for achieving net zero
Transparency	Clarity on fairness of target	Clear statement on why the target is fair

Source: ©2021 by Climate Analytics and NewClimate Institute. All rights reserved. Reproduced by permission.

Tracking and Assessing Progress toward a Country's Nationally Determined Contributions

The Paris Agreement, combined with the latest IPCC reports, its Special Report on Global Warming of 1.5 °C (IPCC 2018), and the Sixth Assessment Report (IPCC 2023) appear to have generated enough urgency for the world's nations to take climate change seriously. The challenge now is to set in motion sufficient actions to reduce global carbon emissions to net zero by 2050 while adapting to new climate stressor extremes. Achieving net zero can be accomplished if nations establish effective plans to achieve meaningful contributions to carbon emission reductions domestically and by providing their fair share of financial assistance to lesser-developed nations.

Progress in global carbon emissions reductions is being tracked and reported by Climate Action Tracker (CAT), a collaboration between two organizations to qualify and evaluate national climate change mitigation targets, policies, and actions. Climate Action Tracker covers all the big emitter nations plus a sample of smaller emitters. The actions tracked are (1) the effect of climate policies and action on emissions, (2) the impact of pledges, targets, and NDCs, and (3) whether the nation is doing its "fair share" compared with others to limit global warming. Policies and actions are analyzed, and climate consequences are projected using accepted climate models. Gaps between what is projected and what is needed are also identified (Climate Analytics 2023). The 2030 emissions gap, the difference between what policies and actions are required and current national policies and actions, is shown in Figure 9-4.

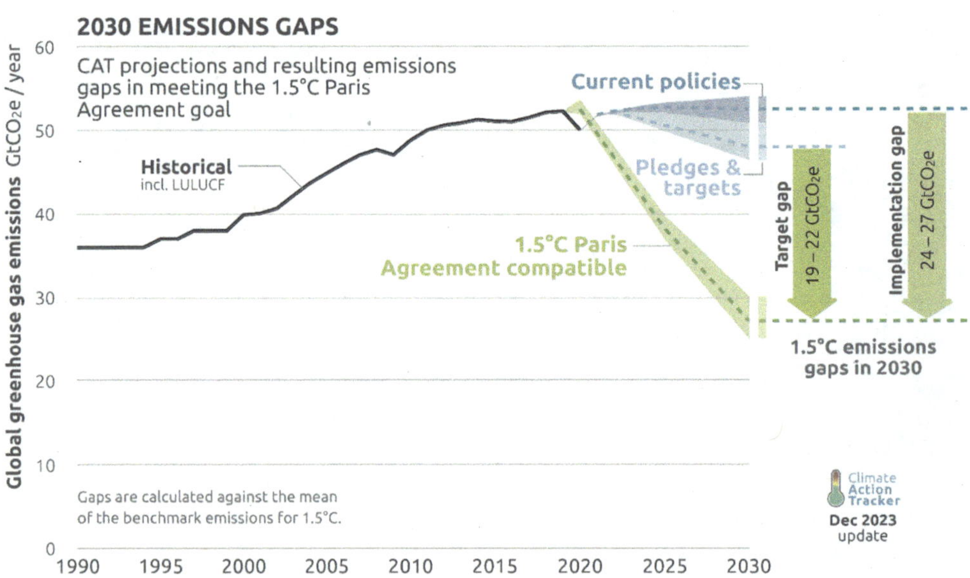

Figure 9-4. Climate Action Tracker. 2030 emissions gap: CAT projections and resulting emissions gap in meeting the 1.5 °C Paris Agreement goal.

Source: ©2023 by Climate Analytics and NewClimate Institute. All rights reserved. Reproduced by permission.

The overall country rating is a combination of the established government policies and actions covering domestic pathways to achieve net-zero emissions and the country's "fair share." The fair share is the range of emission targets based on considerations of (1) the country's historical responsibility for the current atmospheric concentrations of GHGs, (2) the capability of the country to reduce emissions, and (3) the country's right to develop.

Using its country rating system, CAT provides an overall rating of country climate action. None of the 40 countries rated as of November 2023 were compatible with the 1.5 °C Paris Agreement. These countries account for 81% of global emissions. Sufficiency ratings are based on the projected country temperature pathway. The CAT website provides access to individual country ratings, detailing actions and targets for 2030 and rating them relative to achieving the 1.5 °C limit. The country ratings are described in Table 9-2.

Table 9-2. Climate Action Tracker (CAT) Rating Definitions and Country Ratings Relative to the Paris Agreement.

Rating categories	Temperature limit	Description	Countries
1.5 °C compatible	≤1.5 °C	Consistent with the Paris Agreement goal of 1.5 °C or below	NONE
Almost sufficient	<2 °C	Not yet consistent with the Paris Agreement goal but could be with moderate improvements	Bhutan, Costa Rica, Ethiopia, Kenya, Morocco, Nepal, Nigeria, Norway, The Gambia
Insufficient	<3 °C	Policies and commitments need substantial improvements	Australia, Brazil, Chile, Colombia, European Union, Germany, Japan, Kazakhstan, Peru, Philippines, South Africa, Switzerland, United Arab Emirates, United Kingdom, United States
Highly insufficient	<4 °C	Policies and commitments inconsistent and could lead to increased emissions	Canada, China, Egypt, India, Indonesia, New Zealand, Saudi Arabia, South Korea
Critically insufficient	≥4 °C	Policies and commitments reflect minimal or no action	Argentina, Iran, Mexico, Russian Federation, Singapore, Thailand, Turkey, Vietnam

Source: Climate Action Tracker (2023).

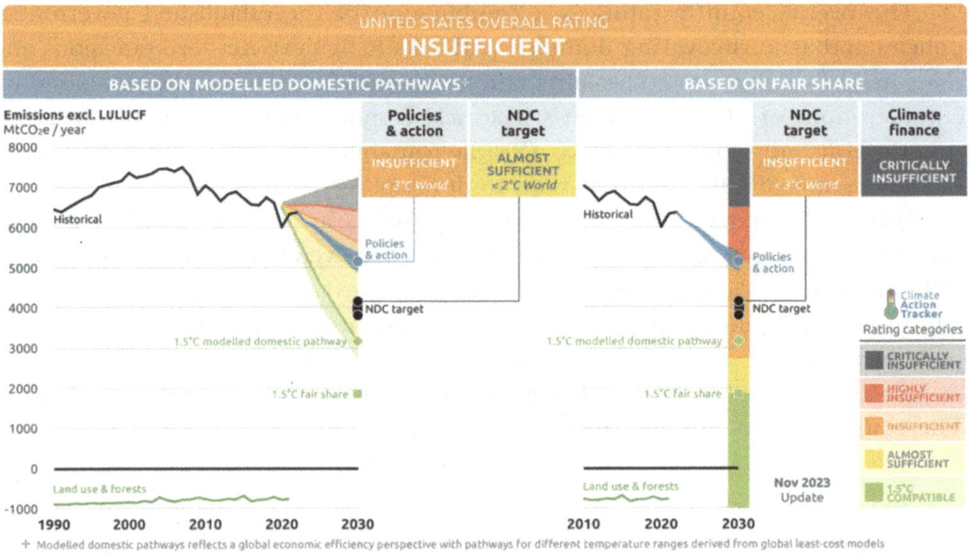

Figure 9-5. Overall US rating based on domestic pathways to 2030 and fair share. *Source*: Climate Action Tracker (2023);

The first two tranches of NDCs, submitted initially in 2016 and updated in 2020, yielded some emissions reductions but were well short of the level of reduction needed by 2030 to be on track to meet the Paris Agreement goal of 1.5 °C. According to CAT's analysis, pledges and targets as of December 2023 only narrow the 2030 emissions gap slightly (Figure 9-4).

Notably, the US rating has changed dramatically, moving up two grades from "Critically Insufficient" to "Insufficient." This change is because of a significant shift from the policies and actions of the previous Trump administration to those of the current Biden administration. The United States has rejoined the Paris Agreement and set more ambitious climate action targets (Figure 9-5). Legislation passed and signed into law in late 2022 will bring further improvements.

Long-Term Low Greenhouse Gas Emission Development Strategies

In addition to the required NDCs, countries are to submit voluntarily their long-term low greenhouse gas emission development strategies (LT-LEDS). Recognizing that politics and vested interests can disrupt or extinguish efforts to achieve a country's NDC, an LT-LEDS, also called long-term strategy (LTS), can add durability to NDC actions that outlast political shifts (Waisman et al. 2016), products of "The Tragedy of the Horizon". For example, the UK has set up an independent Climate Change Committee that advises the national government in GHG reduction target setting (Levin and Fransen 2019). As of September 2023, 75 parties have submitted long-term strategies (UN 2024).

The steps for the national development of an LTS include

- Securing a political decision to initiate the process and establish institutional arrangements;
- Defining the objectives of the strategy;
- Selecting the elements to be included in the strategy;
- Collecting inputs, including data, modeling, scenarios, and complementary qualitative approaches;
- Developing the strategy;
- Providing for transparent communication; and
- Establishing review and revision processes (Vener et al. 2019).

The development of long-term GHG strategies could be improved by using DAPP methodologies. Getting on track to achieve its NDC will require a country to put in place a combination of laws, policies, investments, and actions, all in a specific sequence and each at a sufficient level of robustness. Like the DAPP pathways, each element is subject to significant uncertainties, including the actual rate of GHG emissions reduction achieved, public acceptance of policies, policy effectiveness, degree of political polarization, effects of lobbying by vested fossil fuel interests, effectiveness of regulatory enforcement, national budget constraints, and new carbon mitigation technologies.

To achieve success, policy makers need to understand their country's conditions for success (i.e., what conditions and assumptions must be maintained if the NDC is to be reached). Strategies need to be developed not only to get on track to achieve their stated NDC but also to stay on track (i.e., handling possible deviations from conditions for success). Applying the DAPP methodology, voluntary policies or incentives (pathways) may be employed early on and relatively easily (low regret options) to begin GHG reduction. However, these may only be effective for a limited range of cuts. Policy makers would likely anticipate that these pathways will need to be replaced at some point (before their adaptation tipping point (ATP) or sell-by date) with stronger actions effective for an extended range of emissions reductions (e.g., laws and regulatory requirements).

Progress through Global Climate Action

While COP 21 in Paris established the objectives and the operating environment for success, COP 22 in Marrakech, Morocco, sought to harness the multiple voluntary efforts, initiatives, and coalitions, organizations taking multiple actions to reduce carbon emissions at the rates necessary to achieve net zero by 2050. To be effective, these organizations must be assembled, directed, and coordinated so that their collective actions produce the desired outcome. Moreover, individual organizations

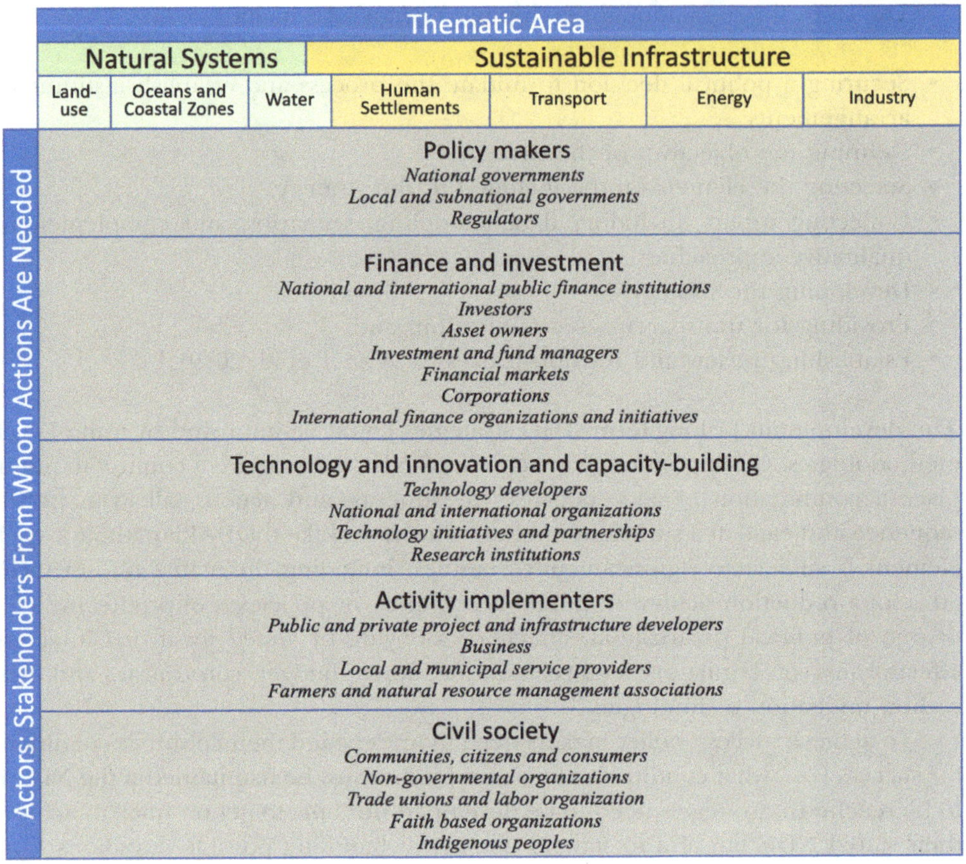

Figure 9-6. Multistakeholder engagement: Proposed thematic approach.
Source: Adapted from UNFCCC (2023b).

must be productive and effective in their efforts, and vehicles for emissions reduction must be worthy of investment.

The work of organizational harnessing falls on the Marrakech Partnership for Global Climate Action. Made personal by so-called high-level champions, the partnership provides a framework. Pathways, roadmaps toward achieving net-zero emissions, are organized by thematic areas in the natural and built environment. The actors, stakeholders from which actions are needed, are organized by their roles in government, business, finance, technology, and civil society (Figure 9-6). In the actions taken in all the thematic areas, the partnership states that consideration will also be given to gender, education, health, and decent work.

Each thematic area offers a vision and executive summary of how net zero will be achieved. An action table provides milestones for carbon emissions mitigation and climate change adaptation for each relevant actor or stakeholder (UNFCCC 2023b).

The work of civil engineers and other practitioners is identified primarily within the thematic area of Human Settlements.

Tracking and Validating Credible Climate Action Progress

In September 2021, the United Nations opened the Global Climate Action Portal, an online platform where all actors can present their efforts to reduce carbon emissions (UNFCCC 2024b). The portal opens with an active world map that provides drill-down features to examine individual countries and cooperative initiatives. The portal categorizes actors within and across countries into cooperative initiatives and individual actions. Cooperative initiatives are individual organizations taking coordinated action against climate change, grouped around a common theme. For example, the Carbon Neutral Cities Alliance consists of 20 international cities and seeks to reduce GHG emissions by at least 80% by 2050 or sooner.

Individual actions are organizations with a carbon footprint taking both internal and cooperative actions to reduce their GHG emissions. For example, the 3M Company has one cooperative and three individual climate action commitments. One of its individual commitments was to fulfill 50% of electricity consumption from renewable sources by 2025. That commitment was achieved in 2022 (UNFCCC 2024a).

Cooperative initiatives and individual actions are tagged to one or more thematic areas defined in Figure 9-6. These can be further expanded into discrete actions of cities, regions, companies, organizations, and investors, with additional details about what action is being taken and its relationship to one or more thematic areas.

Actions and achievements that have a direct impact on the engineering industry can be found under Human Settlements. This thematic area is divided into Built Environment and Waste and Consumption. These are further divided into impact areas (i.e., specific initiatives in which stakeholder efforts can produce needed results for carbon emissions mitigation and adaptation). The work of civil engineers is captured under Business and Service Providers. Developers, architects, engineers, contractors, infrastructure asset managers, and owners have specific roles in producing carbon mitigation and climate adaptation results (Figure 9-7).

Country Actions

Country maps are linked to a country profile, a detailed list of actions taken by the country and individuals organized by stakeholder type: cities, regions, companies, organizations, and investors. Additional details are available for each actor regarding cooperative initiatives and individual actions.

Cooperative initiatives are collected, profiled, and analyzed on the Global Climate Action portal. Links to individual initiatives are organized by the participant type, thematic areas, climate focus (mitigation, adaptation, resilience), function, and

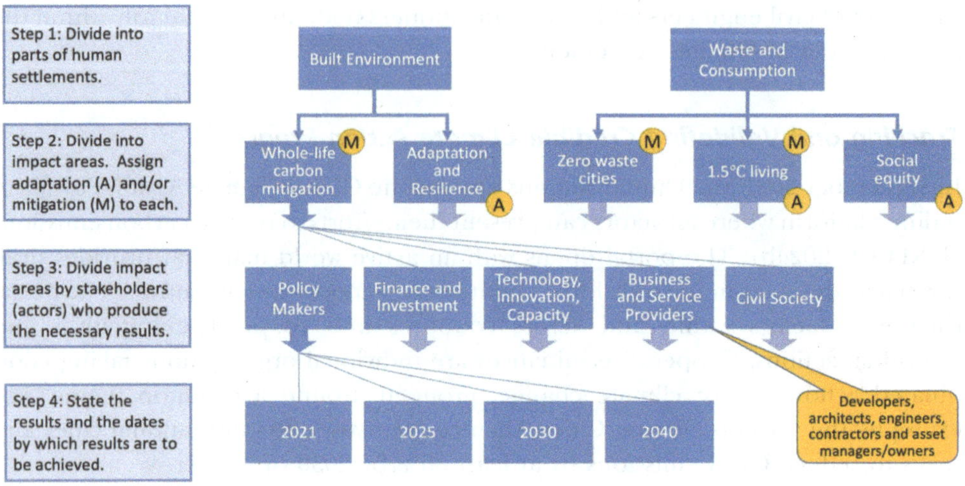

Figure 9-7. Construction of the Human Settlements climate action table. The mitigation (M) and adaptation (A) impact areas are marked. An example breakout for whole-life carbon management is shown.
Source: Adapted from UNFCCC (2023a).

relation to the Sustainable Development Goals. Progress is described in these terms as a way of judging improvements in effectiveness:

- **Ambition:** Goals and targets;
- **Robustness:** Dedicated staff, monitoring arrangements, participation criteria, budget; and
- **Delivery:** Actions undertaken, deliverables and output, challenges, and opportunities.

Individual Actions

Individual actions are also collected, profiled, and analyzed on the Global Climate Action portal. Actions can be accessed individually or by country. The number of participants is large, comprised of 34,636 actors in 194 countries as of early 2024 (UNFCCC 2024b).

Climate Action Pathways

To reach net-zero carbon emissions by 2050, the Marrakech Partnership identified the actions needed by certain dates to achieve specific adaptation and carbon mitigation impacts. These are organized by the sectorial themes shown in Figure 9-6 and by participating stakeholders. Additional cross-cutting themes, resilience and finance, are also included. Strategies for each thematic area are provided, supplemented with action tables specifying what achievements are needed by which actors

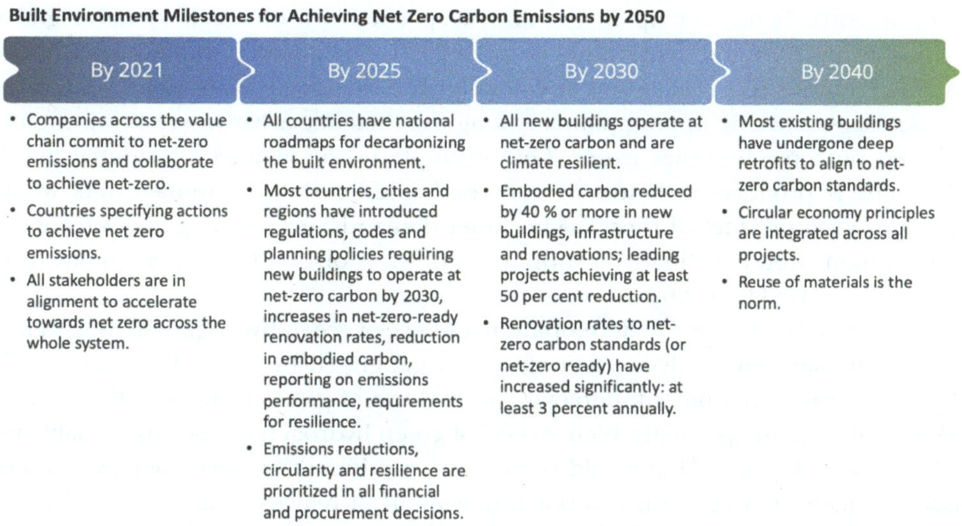

Figure 9-8. Built environment milestones toward 2050.
Source: Adapted from UNFCCC (2023a).

(stakeholders) before 2050, organized by date: 2021, 2025, 2030, and 2040 (Figure 9-8). The plan is to update these strategies and actions annually (UNFCCC 2023a).

Strategies and Actions for the Built Environment

Strategies and actions related to the built environment are also organized under the Human Settlements theme. Regarding global carbon mitigation, the buildings and construction sector produces nearly 40% of the energy and process-related carbon emissions. As to resilience, it is estimated that by 2050, more than 970 cities will experience extreme heat, 500 will have severe water shortages, and more than 570 could be impacted by rising sea levels. It is estimated that about 40% of the world's buildings and 75% of its infrastructure expected to exist in 2050 have yet to be built. New standards and methodologies will be required to make them safe and carbon-neutral. Those that remain will need to be retrofitted (UNFCCC 2023a).

The steps toward achieving the needed resilience and carbon emissions reductions in the built environment are shown in the timeline in Figure 9-8. Unfortunately for the United States, two achievements scheduled for 2021, company commitments and stakeholder alignment, were not reached. However, as of this writing, the current US administration intends to achieve net-zero emissions by 2050.

Rallying Support: Climate Ambition Alliance—Race to Zero

One of the cooperative initiatives is the Climate Ambition Alliance—Race to Zero, a United Nations-backed global campaign to mobilize actors outside of national governments. It was set up to get and keep countries on track to halve global emissions by 2030 and achieve net zero by 2050. The alliance studied the Climate Action Pathways sectorial visions and actions. It developed a set of what was termed *breakthroughs,* specific tipping points among the thematic areas where achieving the desired outcome becomes easier. One example of the alliance's potential breakthroughs is products with low or zero embodied carbon. Manufacturers would produce such products if there is consumer demand, supporting policies for using zero-carbon products, financial support from investors, and encouragement from civil society (UNFCCC 2020).

Another example is the development and use of green hydrogen: hydrogen fuel created by water electrolysis using electricity generated by low-carbon sources. To accelerate the development, deployment, and use of green hydrogen, the alliance identified a tipping point at which 25 GW of green hydrogen capacity is available for US$1.5 per kilogram. That would require a commitment by developers, end-users, policy makers, and financiers to reach that point. If realized, the net-zero goal for green hydrogen deployment (500 to 800 million tonnes) use would be achieved by 2050 (UNFCCC 2020).

The Built Environment Systems Map

The alliance also developed a Built Environment Systems Map, a digital platform for policy makers, businesses, investors, innovators, and the public to explore and visualize their individual and collective roles in transitioning to a net-zero built environment. The map is organized around the value chain in the built environment that delivers goods and services to society. It shows how governments, as owners, policy makers, rule setters, enforcers, stimulators, and conveners, can work to achieve net-zero emissions. It also shows the role of finance in decisions affecting net zero (Figure 9-9).

Need for Radical Collaboration

Although the system map and the associated climate action pathways identify necessary outcomes and the roles of the various actors, they do not describe how those outcomes will be accomplished nor the difficulties the actors will face in fulfilling their respective roles. Indeed, the Human Settlements action table calls for "radical collaboration" within engineering industry competitors, across industries, sectors, and across the value chain (UNFCCC 2023a) (Figure 9-10).

At this time, radical collaboration within the engineering industry is wishful thinking. Collaboration among engineering industry competitors happens only when teaming arrangements among firms are necessary to pursue and win projects. Furthermore, collaboration between the project owners and the architecture, engineering, and construction (AEC) firms selected to design and construct the owner's projects is limited by the controls typically applied to such projects. Legally

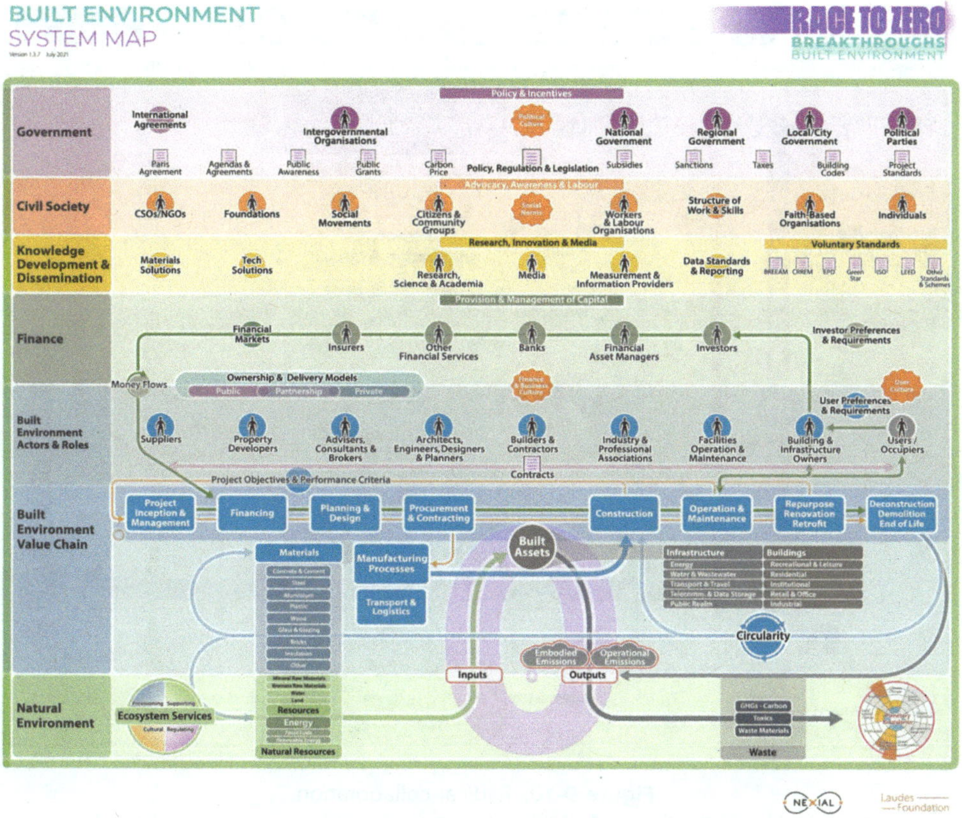

Figure 9-9. Climate Ambition Alliance's built environment systems map.
Source: Developed in collaboration with the UN Climate Champions team and Nexial, and commissioned by Laudes Foundation; Creative Commons BY-SA 4.0.

binding and enforceable contractual agreements specify what services AEC firms are expected to deliver and the responsibilities and obligations of each of the parties in the delivery process. Conflicts arise if any of the parties fail to meet the terms of the contract if the initial assumptions are found to be invalid or void by unexpected or unplanned events. And in this era of climate change, infrastructure project delivery is made more challenging because of changing climate conditions.

That said, radical collaboration among the parties is essential for handling changing climate conditions. At this time, achieving NDCs requires the application of new low- or no-carbon-emitting technologies and infrastructure design or retrofit for climate change adaptation. While the Climate Action Pathways spell out what needs to be accomplished, there is little guidance on how to make it happen. What needs to happen is to set aside the typical contractual relationships, replacing them with partnerships among the parties built on trust and dedicated to meeting the Climate Action Pathway milestones.

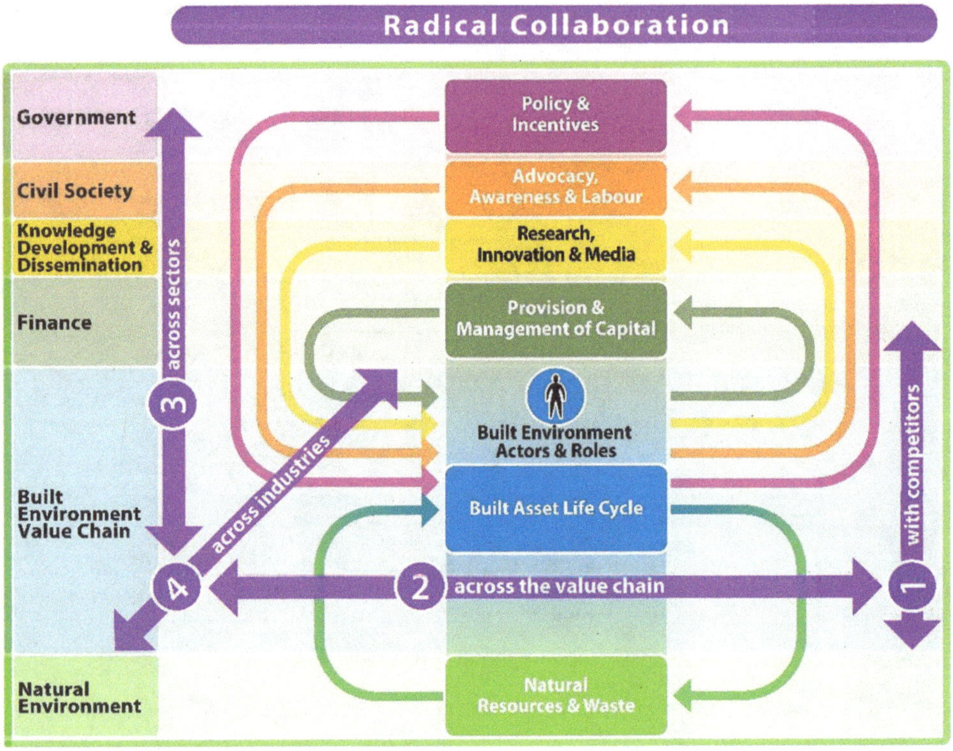

Figure 9-10. Radical collaboration.
Source: Human Settlements action table. Developed in collaboration with the UN Climate Champions team and Nexial, and commissioned by Laudes Foundation. Creative Commons BY-SA 4.0.

What the United States Is Doing

As noted earlier, the United States has made great strides in meeting its NDCs. In August of 2022, the 117th Congress passed two landmark pieces of legislation: the Infrastructure Investment and Jobs Act (IIJA), also known as the Bipartisan Infrastructure Law or BIL, and the Inflation Reduction Act (IRA). Combined, they represent an investment of US$1.25 trillion in carbon emissions reduction and resilient, adaptive infrastructure (Tomer 2022).

If fully applied, the IRA could substantially reduce US GHG emissions, achieving reductions of 31%–44% (King et al. 2022). This would change the "Policies and Actions" curve for the United States shown in Figure 9-5, bringing it closer to the 1.5 °C "fair share" target. The US rating for 2030 GHG emissions rating would also change from "Insufficient" to "Almost Sufficient." Combined with additional efforts from various federal agencies, cities, and states, the United States could meet its carbon emissions reduction commitments of 50%–52% emissions reductions below 2005 levels, bringing the "1.5 °C Compatible" rating within reach.

The IRA also provides US$119 billion in tax credits for electric vehicle production and US$267 for upgrading roads and bridges. This represents additional carbon emissions reductions and opportunities for transportation infrastructure adaptation (Tomer 2022).

The BIL provides a broader investment in infrastructure, US$550 billion in new investment for roads, bridges, water and wastewater, high-speed internet, clean energy transmission, power, airports, rail, ports, waterways, and public transit (The White House 2023). To implement the BIL, President Biden issued an Executive Order (EO 14052 2021) (Federal Register 2021) prioritizing investments in "building infrastructure that is resilient, and that helps combat the crisis of climate change."

In their assessment of BIL, the Georgetown Climate Center points out that while the BIL Guidebook (The White House 2022) identifies just $50 billion for resilience investments, funding for resilience and carbon emissions reduction is spread throughout the entire law. Grants and tax credits are available for federal agencies, states, and communities for transportation, energy, buildings, natural resources, agriculture, water and wastewater, coastal protection, preparedness, and emergency response projects. In aggregate, these projects represent a major US infrastructure rebuild. For the engineering industry, it follows that competitive selections for these projects will be based in large part on knowledge and experience in applying resilient, adaptive, and low-carbon technologies.

ASCE 73-23: Standard Practice for Sustainable Infrastructure

In late 2023, ASCE published ASCE 73-23, *Standard Practice for Sustainable Infrastructure*, a nonmandatory standard designed to guide infrastructure design, delivery, and operation through the entire life cycle. The Envision sustainable infrastructure rating system was used as the framework for developing this standard.

The standard covers the following elements of sustainable infrastructure.

- **Sustainability Leadership:** Provisions for leadership, commitment, stakeholder engagement, and collaboration throughout the project life cycle;
- **Quality of Life:** Contributions to the health, safety, equity, and well-being of the host and affected communities;
- **Resource Allocation:** Conservation of materials and energy resources, and pollution reduction;
- **Natural World:** Preservation of ecosystem resources and services;
- **Greenhouse Gas Emissions:** Reduction of operational and embodied carbon; and
- **Resilience:** Ability to withstand fast, cost-effective recovery from natural and human-constructed hazards.

The standard states that life-cycle cost analysis should be applied to compare and select infrastructure design alternatives.

How Civil Engineers Can Make a Difference

So, what to do? In preparation for COP 26 in Glasgow, the United Kingdom's Institution of Civil Engineers (ICE) took a bold step, stating how its members could seize the opportunity to use their knowledge and skills to bring about the needed changes. Its *State of the Nation 2021* report (ICE 2021) offered six ways for civil engineers to take action on climate change. These are presented and amplified as follows:

1. **This Is an Emergency—Ttreat It Like One**. Civil engineers work throughout the infrastructure project life cycle and throughout the infrastructure supply chain. They need to prepare themselves for those roles and responsibilities by doing the following.
   (a) *Understand the science and communicate the urgency to everyone you work with.* Learn the causes and the problems created by a changing climate. Make the case for taking action now.
   (b) *Understand carbon declarations or commitments made by your organization or your organization's clients.* What are these organizations doing about reducing GHG emissions?
   (c) *Understand and show how you are contributing to solving the crisis.* Demonstrate your commitments. Be able to show how you are contributing to solving the climate crisis. Understand its relevancy.
   (d) *Join a local collective working to create change.* Become part of the solution. Actively participate in groups working to make the necessary changes.
2. **Bring Carbon into Every Conversation**. As is done with safety, discuss the implications of carbon emissions in the planning, design, construction, and operation of all infrastructure assets. Specifically:
   (a) *Refuse "business as usual": Keep carbon reduction on every project agenda.* Communicate the options and benefits of a low-carbon approach.
   (b) *Be the expert: Keep yourself informed about how approaches to the climate crisis are evolving at the highest levels—internationally, nationally, environmentally, socially, and economically—and research the implications for your own sector.* Laxo (2023) noted that clients generally assume that the engineers they hire will apply the right information and make sound project decisions, trusting that the client's lack of knowledge will be filled by the professional they hired. Clients expect you to inform them about all reasonable risks and opportunities. Become the expert your clients expect you to be.
   (c) *Be aware of current standards and commitments. Look for examples of best practice, seek out gaps in knowledge and exchange your ideas with professionals.* Understand the

carbon emissions reduction state of play in your industry and in your clients' industries. Expand the engineers' body of knowledge on climate change.

(d) *Be willing to share your successes, big or small.* As noted earlier, engineer–client partnerships and radical collaboration among all parties are required. Knowledge sharing is also required among AEC firms. Build a knowledge base of sustainable and climate-resilient projects, processes, and technologies.

3. **Understand and Influence End Users.** How civil infrastructure is designed, constructed, and operated greatly impacts people's choices (i.e., how quickly and easily they will adapt to a low-carbon future).
   (a) *Acknowledge that your work has an impact on people's behavior.* Recognize the relationship and how behaviors can be improved.
   (b) *Know who your end users are and involve them by treating them as experts.* They know best how your projects work or do not work.
   (c) *Consider the needs of everyone affected by your infrastructure and engage them.* Get the user and affected communities involved in project decisions.
   (d) *Understand the EAST principles and apply them at your workplace.* People's habits can be changed if the new behaviors are Easy, Attractive, Social, and Timely (EAST). Civil engineers can design infrastructure systems that can move behaviors to low-carbon alternatives. For example, low-carbon mobility choices such as walking or bicycling could be made easy and attractive by providing safe and contiguous pathways, shelters, and places for secure bicycle storage.
4. **Design and Build for the "Right" Outcomes.** Create outcomes that meet the needs of all stakeholders. As I stated in a 2009 presentation to the ASCE Industry Leaders Council and incorporated into the Envision sustainable infrastructure rating system (Wallace 2011), the job of the civil engineer is not only to do the project right but also to do the right project.
   (a) *Understand and apply an outcome-focused approach, for example, by using project 13 principles.* Understand the purpose of the infrastructure project (i.e., what services the project is intended to provide). Look for other low-carbon approaches for delivering those services. Use Project 13 principles to improve project delivery. Project 13 is a new business model—based on an enterprise, not on traditional transactional arrangements—to boost certainty and productivity in delivery, improve whole-life outcomes in operation, and support a more sustainable, innovative, highly skilled industry.
   (b) *Define the outcomes you want and make low-carbon behavior one of the outcomes.* These outcomes must meet the needs of the project stakeholders and include low embodied carbon and low-carbon emissions.
   (c) *Where possible, implement, promote, or suggest a commercial model that incentivizes outcome-based solutions.* Reducing carbon emissions at the source is a better solution. For example, reducing vehicle miles traveled is more effective than reducing carbon emissions in road construction.

(d) *Make your infrastructure—whether creating a new asset or operating an existing asset—work harder to achieve multiple outcomes.* Design and operate infrastructure assets to meet a wide variety of outcomes. For example, a bicycle lane or walkway infrastructure can incorporate trees, vegetated swales, rain gardens, and other natural features to create a more resilient and attractive landscape.

5. **Strive for Creative Solutions**. The work of civil engineers is generally portrayed as large-scale assets: tall buildings, high-capacity highways, and signature bridges. While large-scale assets are impressive, they are not necessarily the most economical solutions for climate change adaptation or for reducing carbon emissions. Creative solutions are needed.

   (a) *Be prepared to be radical. Learn how to reuse and adapt existing infrastructure.* Adapting to a changing climate does not necessarily require replacing existing infrastructure. Civil engineers can better serve society by figuring out how to reuse and adapt infrastructure rather than tearing down and building new infrastructure.

   (b) *Understand how infrastructure is a complex system of systems.* Infrastructure assets are part of an integrated system, providing services to and supporting multiple assets and users. Infrastructure assets can be designed to provide redundancy and extra capacity in cases of failure or to manage extreme conditions.

   (c) *Understand, quantify, and seek to mitigate whole-life carbon impacts and consider climate change in every project.* Reduce carbon emissions throughout the entire life cycle of the project. Take both embodied and operational carbon into account. Assess the effects of a changing climate on the completed project across its entire design life.

   (d) *Define yourself not by what you build but by what you achieve.* Civil engineers must judge what they accomplish, defining themselves not on the projects' scale but on what those projects achieve.

6. **Be Responsible for Resilience**. A changing climate has exposed existing infrastructure assets to climate stressor extremes, often well beyond what the designers have anticipated. Importantly, these conditions are expected to intensify for decades to come. Thus, it is essential to build infrastructure resilience. Incorporating resilience into existing assets must become a priority.

   (a) *Consider the total impact and value of your project. For example, what effect will a project have on the resilience of other nearby infrastructure?* Redefine boundaries. Look beyond the scope of the infrastructure project. How could this project benefit or harm other infrastructure assets inside and outside the current system?

   (b) *Adopt a systems approach to infrastructure delivery.* Infrastructure assets are part of a system, a network of other assets and systems providing services to communities. To be resilient, these systems should be designed and operated not only to perform their function but also to provide support to other system assets.

(c) *Understand that it is in the early stages of the design process that we can have the most influence on resilience. For example, consider project location and conceptual design.* Consider resilience at the early stages of the project. The ability to withstand and recover from extreme weather and climate events must be considered at an infrastructure project's planning and conceptual design stages. Examples include project location, exposure to climate hazards, additional capacities, and system redundancies.

(d) *Stress test your plans with respect to a wide range of possible climate extremes that we could see in the future. Analyze what your system could resist.* Consider climate stressor extremes. Designers should consider a wide range of climate stressor variable values, especially values that could produce knock-on effects.

Can Engineers Respond Effectively?

Hopefully, with the aforementioned new and historic US legislation, work can begin in earnest. Barriers have been broken. Funds and tax credits are available for a civil infrastructure rebuild that reduces carbon emissions and adapts to changing climate conditions. Grants to states and communities are to be made based on justifiable improvements in hazard and vulnerability assessment, risk reduction, resilience, equity, resource and energy conservation, carbon emissions reduction, and ecosystem restoration.

Still in question is how the civil engineering community will respond. As of this writing, climate-related civil engineering standards have been identified, and work has begun to make the needed revisions, recognizing that the assumption of climate stationarity no longer applies. However, standards revision is typically a multiyear process, too slow for the challenges at hand.

What is needed is a new level of collaboration, radical collaboration, as pictured in Figure 9-10. Given the problem's complexities, new competencies are required among the stakeholders (Chester and Allenby 2019).

- **Shared Understanding:** Stakeholders need to know each other's goals and concerns and have developed a sense of trust and understanding. They have a shared commitment to the project's directions, goals, solutions, and needed actions;
- **Manage, Not Solve:** Solutions must constantly evolve in response to changing conditions, climate and otherwise;
- **Try, Learn, Adapt:** Future conditions are not predictable. Solutions can only evolve through experimentation, trying new adaptation approaches, and learning from successes and failures; and
- **Complexity Mindset:** A complexity mindset recognizes that relying on what worked before is not a viable solution for changing conditions. Complexity requires constant scanning to understand what changes may occur and identify the actions necessary to adapt.

An Extended Role for Engineers?

My purpose in writing this book is to help engineers and practitioners—anyone working in the built environment—transition from complicated to complex infrastructure projects. Following the recommendations of Chester and Allenby (2019), designing, constructing, and operating complex projects would change the traditional role of engineers and their relationships with owners, constructors, and operators. Unlike traditional infrastructure projects, the engineers who designed the project are needed at all stages of the project design life.

Traditional projects are structured as a series of arm's-length product handoffs, each structured so that the project owner's requirements are translated accurately and fulfilled each step of the way. Each product is expected to be complete, with some latitude given for clarifications, interpretations, or unexpected site conditions. The project owner delivers a detailed scope of work to the designer, a product with requirements and specifications to be completed within an agreed schedule and budget. The designer's products are contract documents, plans, and specifications for the constructor, which, if followed, produce a finished product that meets the owner's requirements. The constructor constructs the project in accordance with contract documents within an established budget and timeframe. Third-party construction inspections ensure the contractor complies with the requirements, standards, regulations, and quality. Upon completion, the constructor turns the project (the product) over to the project owner, who operates it or turns it over to an operator. Maintenance is performed as required to keep the project operational throughout its design life. Typically, the engineer who designed the project is only involved in the design phase, after which they move on to another design project.

Complex projects cannot be delivered and operated under the same construct. Engineers should be engaged throughout, starting in the early project stages and continuing through operation. Engineers are essential stakeholders, helping to create that shared understanding about the climate hazards the project will face, what can be done to reduce the hazards and exposures, and the uncertainties of future climate conditions. Engineers must collaborate with climate scientists to develop better predictions of future climate hazards and explore the uncharted territory of new climate and weather extremes. Adaptation to climate hazards is a series of engineered responses, some preplanned, others devised on the fly as new and unexpected conditions emerge. Successes and failures in robustness, resilience, and adaptation design must be shared throughout the engineering community as part of a lessons-learned knowledge base.

Closing Comments

As I stated upfront, my B.H.A.G. (Big, Hairy, Audacious Goal) in writing this book is to instigate a major overhaul in civil engineering standards and practices, an overhaul required by the consequences of a warming and increasingly harsh climate.

These consequences—higher ambient temperatures, increasing frequency and intensity of storms, extended droughts and heat waves, urban flooding, and rising sea levels—have stressed, damaged, or destroyed civil infrastructure. These are the structures and systems that communities depend on to support their economies and their quality of life.

This so-called B.H.A.G. may seem like an overreach. Regardless, something needs to be done. Today, civil engineers and others continue to design and deliver projects based on currently accepted standards and practices that do not account for changing climate conditions. They do not press their clients or bosses to do otherwise. Maybe it is because they themselves do not understand the issues and thus do not feel confident in making the arguments. It could also be about exposure to claims and liability for not following accepted standards and practices.

This book should be used as a resource, a basis for asserting that climate change is a real and existential threat to our economy and quality of life. Although future climate conditions are uncertain, what is certain is that following accepted standards and practices where climate-related hazards are involved can create risks to public health, safety, and welfare. Investments in US infrastructure funded by recent legislation set resilience as a basis for awards. Approaches for handling climate uncertainties in civil infrastructure design and operation are available and being applied. In fact, engineers have always dealt effectively with uncertainties both in their work and in their daily lives. Now, climate has been added to the mix.

References

BBC News. 2022. "Japan swelters in its worst heatwave ever recorded." Accessed February 28, 2023. https://www.bbc.com/news/world-asia-61976937.

Brimicombe, C., C. Di Napoli, H. Cloke, and M. Wanzala. 2021. "Guest post: Reviewing the summer of extreme weather in 2021." Accessed September 28, 2021. https://www.carbonbrief.org/guest-post-reviewing-the-summer-of-extreme-weather-in-2021.

Carney, M. 2015. "Breaking the tragedy of the horizon—Climate change and financial stability." Accessed February 26, 2023. https://www.bankofengland.co.uk/speech/2015/breaking-the-tragedy-of-the-horizon-climate-change-and-financial-stability.

Chester, M. V., and B. Allenby. 2019. "Infrastructure as a wicked complex process." *Elem. Sci. Anth.* **7** (2019): 21. https://doi.org/10.1525/elementa.360.

Climate Action Tracker. 2023. "Find your country." Accessed November 4, 2023. https://climateactiontracker.org/countries/.

Climate Analytics. 2023. "Carbon action tracker: The CAT thermometer." Accessed February 26, 2023. https://climateactiontracker.org/global/cat-emissions-gaps/.

Copernicus. 2022. "Summer 2022 Europe's hottest on record." Accessed February 28, 2023. https://climate.copernicus.eu/copernicus-summer-2022-europes-hottest-record.

EIA (US Energy Information Agency). 2021. "Hurricane Ida caused at least 1.2 million electricity customers to lose power." Accessed February 26, 2023. https://www.eia.gov/todayinenergy/detail.php?id=49556.

European Environment Agency. 2023. "Global and European temperatures." June 29, 2023. Accessed March 11, 2024. https://www.eea.europa.eu/en/analysis/indicators/global-and-european-temperatures?activeAccordion=ecdb3bcf-bbe9-4978-b5cf-0b136399d9f8.

Federal Register. 2021. "Implementation of the infrastructure investment and jobs act." Accessed February 7, 2024. https://www.federalregister.gov/documents/2021/11/18/2021-25286/implementation-of-the-infrastructure-investment-and-jobs-act.

Georgetown Climate Center. n.d. "Resilience in the infrastructure investment and jobs act (IIJA)." Accessed March 1, 2023. https://www.georgetownclimate.org/adaptation/toolkits/resilient-infrastructure-investments/what-funding-opportunities-does-iija-offer-for-building-resilience-across-sectors.html.

Hardin, G. 1968. "The tragedy of the commons: The population problem has no technical solution; it requires a fundamental extension in morality." *Science* **162** (3859): 1243–1248. https://doi.org/10.1126/science.162.3859.1243.

Henson, B. 2021. "Why the power is out in Texas . . . and why other states are vulnerable too." Accessed February 26, 2023. https://yaleclimateconnections.org/2021/02/why-the-power-is-out-in-texas-and-why-other-states-are-vulnerable-too/.

ICE (Institution of Civil Engineers). 2021. "State of the Nation 2021: Six ways for civil engineers to act on climate change." Accessed May 19, 2024. https://www.ice.org.uk/engineering-resources/briefing-sheets/six-ways-for-civil-engineers-to-act-on-climate-change.

IPCC. 2018. "Summary for policymakers". In *Global warming of 1.5°C. IPCC special report on the impacts of global warming of 1.5°C above pre-industrial levels and related global greenhouse gas emission pathways, in the context of strengthening the global response to the threat of climate change, sustainable development, and efforts to eradicate poverty.* Cambridge, MA: Cambridge University Press, 3–24.

IPCC. 2023. *Climate Change 2023: Synthesis Report.* Contribution of working groups I, II and III to the Sixth Assessment Report of the Intergovernmental Panel on Climate Change. Geneva: IPCC. https://doi.org/10.59327/IPCC/AR6-9789291691.

King, B., J. Larson, and H. Kolus. 2022. "A congressional climate breakthrough." Accessed February 7, 2024. https://rhg.com/research/inflation-reduction-act/.

Laxo, A. 2023. "Climate forward? How architects and engineers are(n't) using climate projections to inform design." Accessed February 7, 2024. https://hga.com/climate-forward.

Levin, K., and T. Fransen. 2019. "Climate action for today and tomorrow: The relationship between NDCs and LTSs." Accessed October 16, 2021. https://www.wri.org/insights/climate-action-today-and-tomorrow-relationship-between-ndcs-and-ltss.

Masters, J. 2023. "Dozens of billion-dollar weather disasters hit earth in 2022." Accessed November 3, 2023. https://yaleclimateconnections.org/2023/01/dozens-of-billion-dollar-weather-disasters-hit-earth-in-2022/.

Nabi, I. 2023. "Responding to Pakistan floods." Accessed March 16, 2023. https://www.brookings.edu/blog/future-development/2023/02/10/pakistan-floods/.

NDMA (National Disaster Management Authority). 2022. "Pakistan, NDMA floods SITREP 2022." Accessed February 28, 2023. https://cms.ndma.gov.pk/storage/app/public/situation-reports/November2022/N2n1eEarMt6q6Rb8ZYwn.pdf.

NOAA (National Oceanic and Atmospheric Administration). 2021. "US hit with 18 billion-dollar disasters so far this year." Accessed October 15, 2021. https://www.noaa.gov/news/us-hit-with-18-billion-dollar-disasters-so-far-year.

NOAA NCEI (National Centers for Environmental Information). 2022. "Annual 2022 global report." Accessed March 20, 2024. https://www.ncei.noaa.gov/access/monitoring/monthly-report/global/202213.

NOAA NCEI. 2024. "2023 was the warmest year in the modern temperature record." January 17, 2024. Accessed March 11, 2024. https://www.climate.gov/news-features/featured-images/2023-was-warmest-year-modern-temperature-record.

O'Shea, C. A. 2023. "NASA clocks July 2023 as hottest month on record ever since 1880." Accessed November 2, 2023. https://www.nasa.gov/news-release/nasa-clocks-july-2023-as-hottest-month-on-record-ever-since-1880/.

Samenow, J., and K. Patel. 2021. "South pole posts most severe cold season on record, a surprise in a warming world." Accessed February 26, 2023. https://www.washingtonpost.com/weather/2021/10/01/south-pole-coldest-winter-record/.

Smith, H. 2021. "The Dixie fire threatened to pass 1 million acres, then was stopped in its tracks. Here's how." Accessed October 29, 2021. https://www.latimes.com/california/story/2021-09-26/dixie-fire-stopped-just-shy-of-1-million-acres-heres-how.

Texas Tribune. 2021. "More than 100,000 Texas electricity customers without power after Nicholas. Restoring it to everyone could take days." Accessed February 26, 2023. https://www.texastribune.org/2021/09/13/texas-tropical-storm-nicholas-updates/.

The White House. 2022. "Building a better America: A guidebook to the bipartisan infrastructure law for state, local. Tribal, and territorial governments, and other partners." Accessed February 7, 2024. https://www.whitehouse.gov/wp-content/uploads/2022/05/BUILDING-A-BETTER-AMERICA-V2.pdf.

The White House. 2023. "Updates fact sheet: Bipartisan infrastructure investment and jobs act." Accessed February 7, 2024. https://www.whitehouse.gov/briefing-room/statements-releases/2021/08/02/updated-fact-sheet-bipartisan-infrastructure-investment-and-jobs-act/.

Timsit, A., and S. Kaplan. 2021. "At least 85 % of the world's population has been affected by human-induced climate change, new study shows." Accessed February 26, 2023. https://www.washingtonpost.com/climate-environment/2021/10/11/85-percent-population-climate-impacts/.

Tomer, A. 2022. "Will the infrastructure law and inflation reduction act transform American transportation? It's complicated." Accessed February 7, 2024. https://www.brookings.edu/research/will-the-infrastructure-law-and-inflation-reduction-act-.

UN (United Nations). 2023. "Climate action fast facts." Assessed February 26, 2023. https://www.un.org/en/climatechange/science/key-findings.

UN. 2024. "LT-LEDS Synthesis report." September 25, 2023. Accessed March 11, 2024. https://unfccc.int/lt-leds-synthesis-report.

UNFCCC (United Nations Framework Convention on Climate Change). 2020. Climate action pathway: Energy executive summary. 2020. Accessed March 11, 2024. https://unfccc.int/sites/default/files/resource/ExecSumm_Energy_1.pdf.

UNFCCC. 2023a. "Climate action pathways. Human settlements." Accessed March 11, 2024. https://unfccc.int/sites/default/files/resource/HS_ActionTable_2.1.pdf.

UNFCCC. 2023b. "Climate action pathways. Marrakech partnership for global climate action." Accessed March 11, 2024. https://unfccc.int/files/paris_agreement/application/pdf/marrakech_partnership_for_global_climate_action.pdf.

UNFCCC. 2024a. "Global climate action. NAZCA 3M Company" Accessed March 11, 2024. https://climateaction.unfccc.int/Actors/Companies/GCAP1.

UNFCCC. 2024b. "Global climate action." NAZCA. Accessed March 11, 2024. https://climateaction.unfccc.int/.

Vener, J., T. K. Fransen, J. Levin, C. Baumwoll, C. Elliott, and K. Ross. 2019. "Scaling up ambition: Leveraging nationally determined contributions and long-term strategies to achieve the Paris agreement goals." Accessed February 26, 2023. https://www.wri.org/publication/scaling-up-ambition.

Waisman, H., T. Spencer, and M. Colombier. 2016. "Long-term low emissions development strategies and the Paris Agreement—Why, what and how?" Accessed February 26, 2023. https://www.iddri.org/sites/default/files/import/publications/pb0616_ts-et-al_lt-leds.pdf.

Wallace, W. A. 2011. *Envision 1.0 systems manual.* Washington, DC: Institute for Sustainable Infrastructure.

Washington Post. 2021. "Ida's impact from the Gulf Coast to Northeast—By the numbers." Accessed September 28, 2021. https://www.washingtonpost.com/weather/2021/09/03/hurricane-ida-numbers-surge-wind-pressure-damage/.

Additional References (Not Cited)

Climate Analytics and NewClimate Institute. 2021. "Climate action tracker: Net zero targets." Accessed November 15, 2022. https://climateactiontracker.org/methodology/net-zero-targets/.

Climate Analytics and NewClimate Institute. 2022. "Climate action tracker: USA." Accessed November 4, 2023. https://climateactiontracker.org/countries/usa/.

Climate Analytics and NewClimate Institute. 2023. "Climate action tracker: USA." Accessed February 7, 2024. https://climateactiontracker.org/countries/usa/.

CNCA (Carbon Neutral Cities Alliance). n.d. "Mobilizing transformative climate action in cities." Accessed February 7, 2024. https://carbonneutralcities.org/.

National Interagency Fire Center. 2021. "National fire news." Accessed September 28, 2021. https://www.nifc.gov/fire-information/nfn.

Project 13. n.d. Accessed February 7, 2024. https://www.project13.info/.

UNFCCC (United Nations Framework Convention on Climate Change). 2021a. "Launch of the UN race to zero emissions breakthroughs." Accessed October 18, 2021. https://racetozero.unfccc.int/breakthroughs/.

UNFCCC. 2021b. "The Paris agreement. What is the Paris agreement?" Accessed October 10, 2021. https://unfccc.int/process-and-meetings/the-paris-agreement/the-paris-agreement.

USAID (United States Agency for International Development). 2022. "Fact sheet No. 8: Pakistan – Floods." Accessed February 28, 2023. https://www.usaid.gov/sites/default/files/2022-12/2022-09-30_USG_Pakistan_Floods_Fact_Sheet_8.pdf.

APPENDIX A

Climate Stressor Effects by Infrastructure Category

Aviation

Aviation consists of aircraft, airfield buildings and structures, runways, taxiways, aprons, and navigational aids. It also includes operations, maintenance, and safety. A total of 3,304 small and large airports carry more than 1 billion passengers per year.

	Effects	
Climate stressor	**Physical**	**Operations and maintenance**
Increased temperatures, extreme heat	• Pavement buckling on runways, taxiways, aprons. Reduced pavement lifespan • Softening of asphalt paving, making it unfit for use • More runway space needed for aircraft takeoffs	• Limits the types of aircraft that can use airport • Flight delays and cancellations • Increase in building cooling demands, cooling of cargo • Reduction in heating demands • As warming increases, reductions in cost for snow and ice removal • Improved outdoor worker productivity with increasing warmer days. Reduced productivity on extremely hot days • Increased incidents of outdoor worker heat stress • Increased water demand for outdoor workers and site operations • More frequent surface treatments on airport paving. Increased cost of materials and labor

(Continued)

(Continued)

	Effects	
Climate stressor	**Physical**	**Operations and maintenance**
Extreme cold	• Airports in normally warm climates are more affected • Waterlines must be emptied or heated • Equipment must be heated or left running • Additives to fluids to lower freezing point	• Decreased outdoor worker productivity on extremely cold days. Shortened work shifts • Increased energy costs for heating
Extreme precipitation, inland flooding	• Depth of flooding can damage aircraft • *Compounding*: Multiple flood events (increased frequency and duration) can damage pavement subbase • Additional requirements for flood prevention • Increase stringency of design standards for flooding, building on saturated soils	• *Adaptation Island*: Airport access roads flooded preventing customers, users, operations, and maintenance personnel from reaching airport • More intense rainstorms produce standing water on airport pavements causing flight delays, cancellations • Fewer days with snow and ice could mean lower removal costs • Less frequent but more intense winter storms require clearing of runways and aircraft deicing resulting in flight delays. Aircraft must maneuver around snow and ice • Damage to electrical equipment. Loss of electrical power
Sea level rise, extreme high tides, storm surge	• Airports located near the coast are vulnerable to flooding. Thirteen of the largest US airports have at least 1 runway within 12 ft of current sea levels • *Accelerating*: sea level rise may rise faster than anticipated because of large ice losses in Antarctica and Greenland. Airports located near the coast may have problems with erosion, shorter airport life spans	• *Adaptation Island*: Airport access roads flooded preventing customers, users, operations and maintenance personnel from reaching airport • Flight delays, cancellations

(Continued)

(Continued)

	Effects	
Climate stressor	**Physical**	**Operations and maintenance**
	• *Reinforcing*: Sea level rise combined with storm surge, high tide can magnify flooding	
Wind, dust storms	• Aircraft damage from debris • Increase in structural strength required for aircraft control towers, emergency hangers. These are considered essential facilities	• Flight delays, cancellations because of high winds. Loss of visibility because of dust storms • Damage to airport infrastructure • Worker exposure to dust
Drought	• May need to switch over to other water sources, use of recycled water	• Stress on water supplies • Increased cost of water
Wildfires	• Damage to facilities and aircraft	• Loss of visibility because of smoke. Flight delays, cancellations • Worker exposure to smoke from wildfires
Changes in freeze/thaw	• *Compounding*: Damage to runways, building foundations, underground utilities because of an increase in freeze/thaw cycles	• Disruption of services because of damage
Permafrost thaw	• Airport embankment failures • Thawing and subsidence of runway foundations • Airstrips, access roads built on permafrost may have to be relocated	• Increased road and runway maintenance costs. Repair and restore roads and runways

Bridges

This category includes bridge superstructure and substructure, bridge deck, and operator houses. There are more than 617,000 bridges across the United States. Forty-two percent have been in operation for more than 50 years, with the average age of 44 years. Seven and a half percent are structurally deficient. Officials are requiring that bridge design life be increased from 50 to 75 years.

(Continued)

(Continued)

	Effects	
Climate stressor	Physical	Operations and maintenance
Increased temperatures, extreme heat	• Thermal expansion of pavements can cause buckling • Thermal expansion may make bridge raising/lowering mechanisms inoperative • Electrical equipment overheating • Alaska: Increasing need for bridges as the climate warms. Rivers and lakes are freezing later and thawing earlier. Travel over ice is unpredictable and risky	• Greater chances of bridge operator, maintenance worker heat stress • Increase in bridge house cooling demands. Reduction in heating demands • Increased water demand for outdoor workers and site operations
Extreme cold	• Bridges freeze before roadways	• Increased use of de-icing salts contributes to corrosion of steel bridges and deterioration of concrete bridge decks • Ice buildup on bridge raising mechanisms may hinder operation
Extreme precipitation, inland flooding	• Heavy precipitation can increase stream or river flow and can affect local scour depth, the depth of erosion of bridge support • If water elevation reaches the low chord bridge elevation, scour depth can increase by 200% to 300% • Increased fog incidents can increase chances of boat collisions with bridge support structure • Erosion of bridge support structure	• Reduced operations: flooding, damage to bridge houses. Water damage to electrical components. • Winter storms can make driving on bridges more hazardous. Bridges freeze before roads

(Continued)

(Continued)

	Effects	
Climate stressor	**Physical**	**Operations and maintenance**
Sea level rise, extreme high tides, storm surge	• More than 1,000 bridges in coastal floodplains are vulnerable to extreme storms and hurricanes • *Accelerating*: Sea level rise may rise faster than anticipated because of ice losses in Antarctica and Greenland. Increasing vulnerability of bridges in coastal floodplains ○ Increase of baseline level of extreme storms can increase bridge scour ○ Waves from storm surge can damage bridge	• Loss of bridge clearance for ship passage under bridge • Loss of operation during storms. Flooding, damage to bridge houses. Water damage to electrical components • *Orphaned Resiliency*: Bridge may adapt to climate change but not adjacent properties. Items near bridge (e.g., barges, tugboats, oil rigs) can dislodge and damage bridge
Wind, dust storms	• Excessive wind loading on bridge structure • Debris on bridge deck	• High winds can limit bridge access
Drought	• Increased exposure of bridge support structure	
Wildfires	• Wildfires caused by dry conditions are hazardous to bridges constructed of timber • Post wildfire debris can damage bridge structure	• Cleanup of wildfire debris
Changes in freeze/thaw	• *Compounding*: Increase in freeze/thaw cycle causes damage to bridge pavements: concrete exposed to freeze/thaw conditions plus deicing salts	• More frequent repair of bridge pavements
Permafrost thaw	• Bridges with foundations on permafrost can be damaged or collapse. Cave-in of bridge supports • Rock slopes can become unstable and increase debris flow	• Additional structural support may be required

(Continued)

(Continued)

Broadband

Broadband is a critical element of this nation's infrastructure with Americans increasingly relying on the Internet for education, business, healthcare, and emergency notifications. Consists of network of lines, cell towers, power supplies, data transfer stations and other components. Push to expand broadband coverage to rural communities. Large portion installed 20 years ago haphazardly and opportunistically, and for climate and weather conditions that no longer exist. Private telecommunications companies own this infrastructure and keep their exact locations confidential. Buried cables are more reliable, especially in areas subject to high wind, wildfire, and extreme cold.

	Effects	
Climate stressor	**Physical**	**Operations and maintenance**
Increased temperatures, extreme heat	• Extreme heat events can degrade electrical systems	• Improved outdoor worker productivity with increasing warmer days. Reduced productivity on extremely hot days • Higher chances of outdoor worker heat stress • Increased water demand for outdoor workers and site operations
Extreme cold		• Decreased outdoor worker productivity on extremely cold days. Shortened work shifts
Extreme precipitation, inland flooding	• Flooding can inundate broadband infrastructure. Not all broadband infrastructure is made water resistant	• More frequent outages from water inundation
Sea level rise, extreme high tides, storm surge	• Sea level rise can inundate broadband infrastructure. Not all made water resistant • *Reinforcing*: Sea level rise combined with storm surge, high tide can magnify flooding	• Large portion of broadband infrastructure deployed near the coast
Wind, dust storms	• High winds can knock down trees and break tree limbs. Take down telecommunication lines	• More frequent repair and replacement of telecommunication lines

(Continued)

(Continued)

	Effects	
Climate stressor	**Physical**	**Operations and maintenance**
Drought		
Wildfires	• Wildfires can destroy telecommunication lines. Recent California wildfires knocked out Internet service in parts of the state	• More frequent repair and replacement of telecommunication lines
Changes in freeze/thaw		• Increase in maintenance requirements
Permafrost thaw	• Possible structural damage to broadband infrastructure components because of thawing and subsidence	• Increase in maintenance requirements

Buildings

There are 6 million commercial, industrial, residential, and school buildings in the United States with about 100 billion sq. ft. of floorspace. These buildings house a wide range of activities: lodging, health care, public order and safety, food service, mercantile, warehousing and storage, food sales, service, public assembly, office, education, and religious worship. In addition, there are about 84,000 public schools with nearly 100,000 buildings. US residential housing consists of 142 million units.

	Effects	
Climate stressor	**Physical**	**Operations and maintenance**
Increased temperatures, extreme heat	• Damage to electrical and mechanical equipment • Expansion of areas where damaging insects can live • *Reinforcing*: In Alaska coastal villages, loss of sea ice and extreme winter storms have caused high levels of coastal erosion. Requires additional shoreline protection or relocation	• Reduced building lifetimes • Improved outdoor worker productivity with increasing warmer days. Reduced productivity on extremely hot days • Higher chances of outdoor worker heat stress • Increased water demand for outdoor workers and site operations • Increased building cooling requirements, energy demand. Reduced heating requirements
Extreme cold	• Buildings in warmer climates may not have sufficient insulation, poor air sealing	• Decreased outdoor worker productivity on extremely cold days. Shortened work shifts

(Continued)

(Continued)

	Effects	
Climate stressor	Physical	Operations and maintenance
Extreme precipitation, inland flooding	• Flood damage from extreme events: rainstorms, storm surges • Water damage to structure and building components, contents • Water damage to electrical and mechanical equipment • Water leaks during intense rainstorms • 100-year flood estimates for the National Flood Insurance Program have influenced community planning. These estimates are no longer reliable	• *Adaptation Island*: Access roads flooded preventing users, operations and maintenance personnel from reaching building • Additional flood proofing needed for structures, electrical and mechanical equipment • Sealing of structures to prevent leaks
Sea level rise, extreme high tides, storm surge	• Flood damage: sea level rise, storm surges • *Reinforcing*: Sea level rise combined with storm surge, high tide can heighten flood levels • *Accelerating*: Sea levels may rise faster than anticipated because of ice losses in Antarctica and Greenland. Buildings located near the coast may have problems with erosion, shorter life spans • Salt spray can degrade material durability • Structural damage because of coastal erosion • Damage to electrical and mechanical equipment • Coastal exposure accelerates rusting from chloride in salt water	• Additional flood proofing needed for structures, electrical and mechanical equipment
Wind, dust storms	• Deterioration of siding, roofing. Increased leaking in rainstorms	• Additional siding and roofing maintenance. Sealing to prevent leaks
Drought	• Dry vegetation vulnerable to fire • Soil subsidence damage to foundations and wall structures	

(Continued)

(Continued)

	Effects	
Climate stressor	**Physical**	**Operations and maintenance**
Wildfires	• Damage from wildfires in urban forest interface	• Additional maintenance required to maintain defensible space around buildings
Changes in freeze/thaw	• *Compounding*: Increase in freeze/thaw cycles can cause damage to foundations, access roads, structures	• Increased road and building maintenance costs
Permafrost thaw	• Damage to foundations, structures • Melting permafrost causes infrastructure to sink into the earth. Alters the structural integrity of building foundations, walls • Alaska Native Communities: Federal government determined the location of communities, schools, associated infrastructure. Limited opportunity to move as the permafrost thaws and structures are damaged	• Increase road and building maintenance costs

Dams

Dams are structures that hold back water for the purpose of storage, flood control, or electric power generation. Dams can be earth filled or constructed from concrete. Earthen (erodible) dams are more vulnerable to climate change. They are more susceptible to erosion, high fluctuations in water levels, and long drying periods during hot weather. Nonerodible dams are less vulnerable to climate change but some climate variations can cause cracking or joint movement. The United States has more than 91,000 dams.

	Effects	
Climate stressor	**Physical**	**Operations and maintenance**
Increased temperatures, extreme heat	• Temperature fluctuations may cause additional stresses in concrete dams • Expansion, cracking of spillways	• Increase in vegetation growth—potential reduction in reservoir capacity and/or blocking of spillway • Increased water temperature could lead to increased duration and frequency of algal blooms

(Continued)

(Continued)

	Effects	
Climate stressor	**Physical**	**Operations and maintenance**
Extreme cold		• Gates can freeze in position reducing the ability to release water in a flood event • Ice runs can destroy gates, clog spillways, and lead to overtopping and dam failure
Extreme precipitation, inland flooding	• Floods can bring about excessive loads on dams • Catchments can become prone to landslides, erosion • Increased sedimentation from flood events could lead to a reduction in storage capacity • Changes in rainfall patterns (e.g., change from peak flow from short extreme rainfall event to storms that deliver larger water volumes over longer periods) • Dams can overflow in a less intense rainstorm if the ground is already saturated	• Increased flow to reservoirs increases flood risk • Additional maintenance of regulating gates and spillways is needed after being submerged in flood events • May need to reduce water levels at certain times during the year in anticipation of extreme storms • Greater sedimentation after storm events. Sedimentation made worse with drought periods when soils become dry and loose
Sea level rise, extreme high tides, storm surge	• *Reinforcing*: Sea level rise combined with storm surge, high tide can heighten flood levels	• Additional maintenance of regulating gates and spillways is needed after being submerged in flood events • May need to reduce water levels at certain times during the year in anticipation of extreme storms
Wind, dust storms	• *Reinforcing*: With high rainfall, increased wave action can damage liner and dam	• Additional maintenance on liner, dam

(Continued)

(Continued)

	Effects	
Climate stressor	**Physical**	**Operations and maintenance**
Drought	• Earthen dams can fail because of excessive cracking in dry conditions • Drier soils and water level fluctuations can increase internal erosion • Lower water levels can lead to increased concentrations of pollutants • Loss of vegetation cover on earthen dams increases the risk of cracking and surface erosion protection • Long periods of low or fluctuating water levels can lead to erosion, wave/sun damage, erosion of joint materials	• Additional maintenance during drought conditions. Dry, loose soils are easier to mobilize in rainstorm events • Increased maintenance and repair of vegetation cover • Increased monitoring of soil condition in earthen dams
Wildfires	• Wildfire can remove vegetative cover on catchment slopes • *Propagating*: Wildfire followed by storm events can increase erosion, risk of landslides • Grass fires on earthen dams can increase penetration of drying conditions into fill materials	• Increased maintenance and repair of vegetation cover, soils
Changes in freeze/thaw	• Change in the properties of soils • *Compounding*: Repeated freeze/thaw cycles causes change the engineering properties of soils and can affect dam stability and safety	• Increased monitoring of soil condition in earthen dams
Permafrost thaw	• Dam stability problems can arise with permafrost thaw	• Additional monitoring of permafrost conditions • Increased need for artificial refrigeration of the dam embankment

Drinking water

The United States' supply of drinking water is provided by water utilities that collect, treat, and distribute clean water to homes and businesses. There are more than 148,000 systems, more than 90% of which serve populations of fewer than 10,000 people. Water is distributed through about 2.2 million miles of underground piping.

(Continued)

(Continued)

	Effects	
Climate stressor	**Physical**	**Operations and maintenance**
Increased temperatures, extreme heat	• Reduced life spans of water treatment plant roofing systems because of higher temperatures • Reduced lifespan of motors in pump stations	• Increase in building cooling demands. Reduction in heating demands • Improved outdoor worker productivity with increasing warmer days. Reduced productivity on extremely hot days • Higher chances of outdoor worker heat stress • Increased water demand for outdoor workers and site operations • Additional outbreaks of harmful algal blooms. Diminish source water quality
Extreme cold	• Pipes can break and leak if they are not buried deep enough	• Increased maintenance requirements • Decreased outdoor worker productivity on extremely cold days. Shortened work shifts
Extreme precipitation, inland flooding	• Water utility infrastructure may become more vulnerable to storms. Can overwhelm or damage infrastructure, including treatment plants, intake facilities, water conveyance, and distribution systems	• Increased monitoring of source water quality. Intense storms can lead to greater pollutant runoff and adversely affect source water quality. Nutrient runoff into water bodies can feed algal blooms • *Adaptation Island*: Access roads flooded preventing operations and maintenance personnel from reaching facility. Alternative routes, modes of transportation required

(Continued)

(Continued)

	Effects	
Climate stressor	Physical	Operations and maintenance
Sea level rise, extreme high tides, storm surge	• Increased possibilities of saltwater intrusion into groundwater resources • *Accelerating*: Sea level rise may rise faster than anticipated because of ice losses in Antarctica and Greenland. Water treatment facilities located near the coast may have problems with erosion, shorter facility life spans • *Reinforcing*: Sea level rise combined with storm surge, high tide can magnify flooding • Can worsen the effects of storms, inundate and damage infrastructure • Desalination plants are vulnerable to flooding	• Increased monitoring of source water quality. Sea level rise would create more shallow and stable coastal water, conditions for algae growth in water sources
Wind, dust storms	• Deterioration of siding, roofing in buildings, facilities. Increased leaking in rainstorms	• Siding and roofing maintenance. Sealing to prevent leaks
Drought	• Reduction in water supplies • Loss of water pressure • Increased potential for salt water intrusion • New demand for water reuse. May need to use wastewater as a water supply source. Changes in water treatment processes	• Lower source water quality. Water treatment plants may have difficulty in handling diminished source water quality • Additional water treatment required to meet drinking water standards
Wildfires	• Damage to buildings, structures • *Propagating*: Wildfire followed by storm events in watershed can increase suspended solids, fire retardant contamination, pollutants into stormwater runoff and significantly affect source water quality	• Maintain a defensible space around buildings • Additional fireproofing
Changes in freeze/thaw	• Possible structural damage to components	• Increase in maintenance requirements

(Continued)

(Continued)

	Effects	
Climate stressor	**Physical**	**Operations and maintenance**
Permafrost thaw	• Possible structural stability problems with buildings, water treatment works	• Increased monitoring for structural stability

Energy

The energy infrastructure in the United States generates, transmits, and distributes electrical power to industrial, commercial, and residential users. Demands for more electricity to power computers, Internet, electric vehicle charging stations. Push to replace gas heating with heat pumps. Push to change to renewable energy sources.

	Effects	
Climate stressor	**Physical**	**Operations and maintenance**
Increased temperatures, extreme heat	• Demand for increased electrical power generation because of increased electrification: heating, electric vehicles • Electricity generation becomes less efficient as temperatures rise • At higher temperatures, thermal expansion of transmission and distribution power lines decreases the amount of power that can be transmitted	• Increase in building cooling demands. Reduction in heating demands • Improved outdoor worker productivity with increasing warmer days. Reduced productivity on extremely hot days • Higher chances of outdoor worker heat stress • Increased water demand for outdoor workers and site operations • Increased importance of maintaining the power grid • Warmer weather in cold regions reduces power demand for heating
Extreme cold	• Need for improved winterization in southern United States	• Many power plants in southern United States are not able to perform in extreme cold weather. Not winterized. Frequent power outages • Decreased outdoor worker productivity on extremely cold days. Shortened work shifts

(Continued)

(Continued)

	Effects	
Climate stressor	**Physical**	**Operations and maintenance**
Extreme precipitation, inland flooding	• Increased flooding of energy-generating facilities • Higher chances of freezing rain events coating tree limbs and power lines. Breaks in power lines causing power outages	• Increase in maintenance to clear tree limbs from power lines • Additional cable reinforcement of power lines • *Adaptation Island*: Access roads flooded preventing operations and maintenance personnel from reaching facility
Sea level rise, extreme high tides, storm surge	• Increased erosion and flooding of coastal oil and gas extraction, transport, storage and energy-generating facilities. Provisions for additional constructed flood protection • *Reinforcing*: Sea level rise combined with storm surge, high tide can magnify flooding • *Accelerating*: Sea level rise may rise faster than anticipated because of ice losses in Antarctica and Greenland. Thermal and nuclear power plants are generally located near the coast because of cooling needs. Result: power plant life spans shorter than their design life	• Requirements for additional flood and erosion protection • Requirements for increased shoreline protection
Wind, dust storms	• Increased blow down of tree limbs causing power lines to break • Changes in wind patterns can affect wind power generation • Deterioration of siding, roofing in buildings, facilities. Increased leaking in rainstorms	• Increase in maintenance to clear tree limbs from power lines • Additional cable reinforcement of power lines • Additional siding and roofing maintenance. Sealing to prevent leaks

(Continued)

(Continued)

	Effects	
Climate stressor	**Physical**	**Operations and maintenance**
Drought	• Problem for thermal, nuclear power plants located in water-stressed areas • Hydropower: less water available • *Propagating*: drought and high temperatures create conditions for wildfires	• Reductions of thermal, nuclear power generation because of reductions in cooling water supply • May need to shut off power to prevent wildfire propagation • Low water pressure for hydroelectric power generation • Reductions in biofuel feedstocks • Hydropower: possible construction of new water intakes if water levels in reservoirs drop
Wildfires	• Damage to energy infrastructure	• Maintain a defensible space around facilities • Fire-resistant upgrades are needed • May need to shut off power. Electricity arcs can cause fires
Changes in freeze/thaw	• Possible structural damage to components	• Increase in maintenance requirements
Permafrost thaw	• Risk of structural stability problems with buildings, structures, facilities	• Increased monitoring for structural stability

Hazardous waste

Consists of managed hazardous waste materials, as well as the hazardous waste sites containing uncontrolled hazardous materials. Federal data suggest about 60% of Superfund sites overseen by EPA are in areas that may be impacted by wildfires and flooding—natural hazards that may be exacerbated by climate change. Federal (EPA) and state agencies responsible for overseeing hazardous waste site cleanup are not sufficiently knowledgeable or equipped to address climate change risks.

(Continued)

(Continued)

	Effects	
Climate stressor	**Physical**	**Operations and maintenance**
Increased temperatures, extreme heat	• *Propagating*: Extreme heat events coupled with drought conditions are a contributor to wildfires	• Increase in cooling demands, reduction in heating demands at treatment works • Improved outdoor worker productivity with increasing warmer days. Reduced productivity on extremely hot days • Higher chances of outdoor worker heat stress • Increased water demand for outdoor workers and site operations
Extreme cold		• Extreme cold weather can hamper site operations, affect hazardous waste remediation activities, processes • Decreased outdoor worker productivity on extremely cold days. Shortened work shifts
Extreme precipitation, inland flooding	• If hazardous waste site is flooded, contaminants can be mobilized • Stormwater runoff can mobilize contaminants. *Reinforcing*: Can be made worse if storm follows a wildfire	• Need flood protection systems, stormwater runoff containment systems capable of handling intense storms, flooding • Flooding of a site with hazardous contaminants contained in place requires extra flood protection measures to avoid releases. Special attention needed for temporary containment and capping. As extreme precipitation and flooding events increase, removal actions become increasingly time critical • *Adaptation Island*: Access roads flooded preventing operations and maintenance personnel from reaching facility

(Continued)

(Continued)

	Effects	
Climate stressor	**Physical**	**Operations and maintenance**
Sea level rise, extreme high tides, storm surge	• If hazardous waste site is flooded, contaminants can be mobilized • *Reinforcing*: Sea level rise combined with storm surge, high tide can magnify flooding • *Accelerating*: Sea level rise may rise faster than anticipated because of ice losses in Antarctica and Greenland. Hazardous waste sites located near the coast may become inundated. May need to revise remediation plans, particularly those that contain the waste or stabilize in place	• Need flood protection systems, capable of handling storm surge, flooding
Wind, dust storms	• Strong wind events can mobilize hazardous pollutants mixed in soils	• Increased requirements to maintain ground covers over areas with contaminated soils
Drought	• Extreme heat events coupled with drought conditions are a contributor to wildfires	• Additional fire protection may be required
Wildfires	• Fire destruction of treatment facilities • Treatment facilities should be constructed of nonflammable materials. Fire protection systems should be incorporated	• Maintain a defensible space around the buildings, facilities, treatment works, sites • Wildfire on a hazardous waste site could result in mobilized hazardous pollutants, air, land and water • Fighting wildfires requires special firefighting equipment to handle hazardous contaminants safely • *Propagating*: Wildfire followed by storm events in watershed can increase stormwater runoff and mobilize hazardous pollutants
Changes in freeze/thaw	• Possible structural damage to components	• Increase in maintenance requirements

(Continued)

(Continued)

Climate stressor	Effects: Physical	Effects: Operations and maintenance
Permafrost thaw	• Possible structural stability problems with buildings, site remediation processes	• Increased monitoring for structural stability

Inland waterways

Consists of the 25,000 miles of inland navigable waterways and the associated locks and dams. About 12,000 miles of inland navigation channels are operated by the US Army Corps of Engineers. Inland waterways act as water highways for US freight.

Climate stressor	Effects: Physical	Effects: Operations and maintenance
Increased temperatures, extreme heat	• Higher temperatures contribute to lower water levels in waterways	• Low water levels restrict cargo carrying capacity • Increase in building cooling demands. Reduction in heating demands • Improved outdoor worker productivity with increasing warmer days. Reduced productivity on extremely hot days • Higher chances of outdoor worker heat stress • Increased water demand for outdoor workers and site operations
Extreme cold		• Ice formation, especially in low flowing rivers, can disrupt waterway operation • Decreased outdoor worker productivity on extremely cold days. Shortened work shifts
Extreme precipitation, inland flooding	• Heavy precipitation can change silt and debris buildup in waterways, reducing channel depths	• More frequent waterway closings from flooding, where water levels exceed maximum permitted levels

(Continued)

(Continued)

	Effects	
Climate stressor	Physical	Operations and maintenance
		• Navigation restrictions during extreme precipitation events • Increased dredging needs • Stronger currents increase the likelihood of accidents • More precipitation can raise water levels, resulting in fewer travel interruptions • Shallower channels are more likely to require dredging after storms because of debris
Sea level rise, extreme high tides, storm surge	• With sea level rise, some channels become more accessible • In some places, storm surge and sea level rise combinations can destroy barrier islands, eliminating waterway systems entirely • *Accelerating*: Sea level rise may rise faster than anticipated because of ice losses in Antarctica and Greenland. Damage to barrier islands and waterways may come sooner than anticipated • *Reinforcing*: Sea level rise combined with storm surge, high tide can magnify flooding	
Wind, dust storms	• Dust storms can reduce visibility	• Increase in dredging requirements from increased debris • Increase in waterway accidents because of high winds
Drought	• Lower water levels in waterways	• Low water levels restrict cargo carrying capacity, increase travel times • Increase in accidents, grounding in waterways

(Continued)

(Continued)

Climate stressor	Effects: Physical	Effects: Operations and maintenance
		• If low water levels persist, freight operators may have to switch to smaller, lower draft barges • Low water levels can force shippers to switch to more carbon-intensive modes (e.g., road travel) • Need to balance water for inland navigation with other water uses
Wildfires	• Debris from wildfires can disrupt traffic on waterways	
Changes in freeze/thaw	• River ice breakup in Alaska occurring earlier. Ice jam flooding risks	
Permafrost thaw	• Damage to infrastructure: coastal erosion affects road and rail links	

Levees

Levees are earthen embankments or concrete floodwalls built next to rivers to prevent river overflow or divert water flow from a defined area. Levee ownership and management ranges from those built and managed by the federal government to those built and managed by other entities. Levees protect critical infrastructure including 4,500 schools. There are 30,000 miles of levees across the United States that are part of the US Army Corps of Engineers (USACE) portfolio. An estimated 10,000 levees are not in the USACE portfolio. Their location and condition are unknown.

Climate stressor	Effects: Physical	Effects: Operations and maintenance
Increased temperatures, extreme heat	• High temperatures contribute to levee damage: desiccation, cracking	• Additional maintenance of vegetation • Inspection for ruts and depressions
Extreme cold	• In cold-region environments, ice-jam floods (IJFs) can result in high water levels in rivers to overtop levees, leading to devastating floods	• Increased level of inspection, maintenance, and repair requirements

(Continued)

(Continued)

	Effects	
Climate stressor	Physical	Operations and maintenance
Extreme precipitation, inland flooding	• Extreme rainfall events increase the probability of levee failure: slope stability, underseepage, uplift • Levee overtopping and breaching. Levee breaching with high water levels can cause a sudden flood rush with increased damage • *Compounding*: Repeated flooding increases risk of levee failure	• Increased level of inspection, maintenance, and repair requirements • Additional removal of debris • Additional riprap required for erosion control • Additional inspections, maintenance and repair for ruts and depressions
Sea level rise, extreme high tides, storm surge	• Reinforcing: Sea level rise combined with storm surge, high tide can magnify flooding • Levee overtopping and breaching. Levee breaching with high water levels cause a sudden flood rush with increased damage	• Increased level of inspection, maintenance, and repair requirements • Addition of material or bank protection from wave action • Increased requirements for removal of debris • Additional riprap for erosion control
Wind, dust storms	• Overtopping of levees caused by wind-generated waves • Soil erosion from wind and rain	• Increased level of inspection, maintenance, and repair requirements. Inspection for cracking
Drought	• Failure because of desiccation cracking, reductions in unit weight, shrinkage under drought conditions • *Propagating*: Extreme drought followed by heavy rainfall leading to failure of the levee. Drought weakens the levee structure. Subsequent heavy rainfall events can lead to failure	• Increased level of inspection, maintenance, and repair requirements. Inspection for cracking • Additional maintenance of vegetation • Additional inspections. Maintenance and repair for ruts and depressions

(Continued)

(Continued)

	Effects	
Climate stressor	Physical	Operations and maintenance
Wildfires	• *Propagating*: Stormwater runoff following a wildfire event is not absorbed because of the loss of ground cover, soils made hydrophobic. Flooding can overtop or breach levees. Additional debris flow	• Maintain vegetative cover
Changes in freeze/thaw	• Potential for weakening of earthen structures • Varying freeze/thaw cycles can affect levee designs that relied on permafrost	• Increased level of inspection, maintenance, and repair requirements. Inspection for cracking
Permafrost thaw	• Structural integrity of levee could be compromised	• Increased level of inspection, maintenance, and repair requirements. Inspection for cracking

Ports

The United States has more than 300 coastal and inland ports, consisting of piers, wharves and berths, channels, and terminal buildings. The operation of a port depends on the infrastructure (e.g., rail, roadways) that allows the transfer of cargo from the port to other destinations.

	Effects	
Climate stressor	Physical	Operations and maintenance
Increased temperatures, extreme heat	• Extreme heat can affect paved areas. Increased deterioration • Stresses, damage to motors because of high heat levels	• Increase in building cooling demands. Reduction in heating demands • Improved outdoor worker productivity with increasing warmer days. Reduced productivity on extremely hot days • Higher chances of outdoor worker heat stress

(Continued)

(Continued)

	Effects	
Climate stressor	**Physical**	**Operations and maintenance**
		• Increased water demand for outdoor workers and site operations • Higher energy consumption for refrigerated warehouses • Electrical outages during heat waves can spoil refrigerated goods • Higher temperatures can improve port operations in cold regions • Hotter and drier conditions increase risk of dust explosions in grain handling and storage
Extreme cold	• Freezing of water lines	• Decreased outdoor worker productivity on extremely cold days. Shortened work shifts
Extreme precipitation, inland flooding	• Flooding can damage port structures • Drainage systems can be overwhelmed	• Increased chances of water damage to electrical equipment • Water sensitive cargo can be damaged in flood events • Freezing rain can impair equipment operation. Safety risks • Safety risks in snow covered operations • *Adaptation Island*: Access roads flooded preventing operations and maintenance personnel from reaching facility. Need to increase diversity of land connections to port terminal • *Adaptation Island*: Port is protected from flooding but intermodal connections to move cargo are not protected

(Continued)

(Continued)

	Effects	
Climate stressor	Physical	Operations and maintenance
Sea level rise, extreme high tides, storm surge	• Effects on dock levels. May require retrofit, relocation • Increase in chronic flooding • Increase in coastal erosion at or adjacent to port • *Accelerating*: Sea level rise may rise faster than anticipated because of ice losses in Antarctica and Greenland. Ports may have problems with erosion, wave damage, sea water inundation, shorter life spans • Storm surges travel further inland. Need for additional flood protection	• Increase in metal corrosion of port equipment • Reduced clearance under bridges in ports reduces port access for large vessels • Deposition and sedimentation along port channels creates increased requirements for dredging
Wind, dust storms	• Damage of structures from high winds • Critical port facilities are classified as essential facilities and must be designed to withstand higher wind pressures	• Small boat and tug handling, ferry docking, and barge handling are restricted during high wind events. Increased port operation downtime because of high winds • Berthing of large vessels is limited during high wind events • Dust storms can reduce visibility and force ports to close. Navigation can be suspended. Increased risk of collisions
Drought	• Low river levels can reduce port and waterborne transportation operations • Reduction in cargo capacity • Possible closure because of low water levels	• Increased cargo transportation costs • Need to shift to alternate transport modes (e.g., rail or truck) • Increased risk of dust explosions in grain handling and storage
Wildfires	• Wildfires can damage port infrastructure	• Reduced visibility can force ports to close. Navigation can be suspended • Maintain defensible spaces around the buildings, facilities, wharves, transport linkages

(Continued)

(Continued)

Climate stressor	Effects: Physical	Effects: Operations and maintenance
Changes in freeze/thaw	• Freeze–thaw cycles can undermine infrastructure foundations	
Permafrost thaw	• Increased coastal erosion because of sea ice retreat and permafrost thaw. Shoreline increasingly exposed to direct wave action. Structure foundations at risk. Affects rail links to ports • Need to replace affected public port infrastructure • Damage to infrastructure: coastal erosion affects road and rail links to ports	• Disruption of port operations • Increased maintenance costs

Public parks

Parks are open, natural spaces provided for recreational use. Urban parks are located in urban settings as respite from city life. They are also an important line of defense in reducing the impact of urban heat islands on local climates. National parks are areas developed and preserved by the federal government for recreation and conservation purposes. Similarly, state parks are recreational and conservation areas protected by state governments. Linear parks, also called *greenways*, are corridors of land that run along canals, streams, power lines, or former railroads.

Climate stressor	Effects: Physical	Effects: Operations and maintenance
Increased temperatures, extreme heat	• Increased stress on electrical and mechanical systems	• Increase in building cooling demands. Reduction in heating demands • Improved outdoor worker productivity with increasing warmer days. Reduced productivity on extremely hot days • Higher chances of outdoor worker heat stress • Increased water demand for outdoor workers and site operations

(Continued)

(Continued)

	Effects	
Climate stressor	Physical	Operations and maintenance
		• Higher chances of user heat stress • Increased requirements for emergency services • Reduction in user fees collected
Extreme cold	• Freezing of water lines	• Decreased outdoor worker productivity on extremely cold days. Shortened work shifts
Extreme precipitation, inland flooding	• Flooding of parks	• Parks become inaccessible to users. Reduction in user fees collected
Sea level rise, extreme high tides, storm surge	• *Reinforcing*: Sea level rise combined with storm surge, high tide can magnify flooding • Flooding of parks, making them unusable • *Accelerating*: Sea level rise may rise faster than anticipated because of ice losses in Antarctica and Greenland. Parks located near the coast may have problems with erosion, shorter life spans	• Additional flooding. Reduction in user fees collected • Increased costs of cleanup after flooding
Wind, dust storms	• Deterioration of siding, roofing in buildings, facilities. Increased leaking of buildings in rainstorms	• Increased building siding and roofing maintenance. Sealing to prevent leaks
Drought	• Extreme heat events coupled with drought conditions are a contributor to wildfires	• Reductions in water and snow dependent recreational activities • Increased maintenance costs
Wildfires	• Damage to park infrastructure	• Increased number of fire bans preventing recreational activities • Maintain a defensible space around buildings, facilities
Changes in freeze/thaw	• Potential for weakening of foundations, roadways	• Increased level of inspection, maintenance, and repair requirements
Permafrost thaw	• Structural integrity could be compromised. Roadways can be damaged and made impassable	• Increased level of inspection, maintenance, and repair requirements

(Continued)

(Continued)

Rail

The US rail infrastructure consists of freight and passenger rail. Infrastructure and equipment include rail cars, railroad ties, tracks, rail switches, ballast, and electrical equipment (e.g., gates/flashers and signal bungalows). The US freight rail network consists of seven Class 1 railroads, 22 regional railroads, and 584 local or short line railroads. Together, they run on almost 140,000 route miles. Amtrak operates more than 21,400 route miles.

	Effects	
Climate stressor	**Physical**	**Operations and maintenance**
Increased temperatures, extreme heat	• Buckling of track. Sun or heat kinks. Buckled tracks must be taken out of service	• Reduced train speeds during high heat events • Increase in building cooling demands. Reduction in heating demands • Improved outdoor worker productivity with increasing warmer days. Reduced productivity on extremely hot days • Higher chances of outdoor worker heat stress • Increased water demand for outdoor workers and site operations
Extreme cold	• Increased risk of breakage of brittle rail track • Frozen switches may not fully close leading to derailment • Freezing of water lines	• Decreased outdoor worker productivity on extremely cold days. Shortened work shifts
Extreme precipitation, inland flooding	• Erosion of track supporting systems, embankments	• Damage to electrical equipment • Service disruptions because of flooding • Fewer days with snow and ice could mean lower removal costs • Winter storms increase icing on rails. Reduced speeds. Longer braking distances

(Continued)

(Continued)

	Effects	
Climate stressor	Physical	Operations and maintenance
Sea level rise, extreme high tides, storm surge	• Erosion of track supporting systems, embankments • Loss of rail bed from sea level rise, tides, storm surge combinations. Can scour rail bed • *Accelerating*: Sea level rise may rise faster than anticipated because of ice losses in Antarctica and Greenland. Railroad lines located near the coast may have problems with erosion, shorter life spans	• Scour of rail bed can lead to derailments, spills of hazardous cargo • Debris on rails can disrupt operations
Wind, dust storms	• High winds can damage platforms, stations, other rail structures	• Dust storms reduce visibility. Train speeds must be reduced
Drought	• For railway lines that pass through wetlands, extended drought can shrink the ground beneath the rail bed	• Increased inspection, maintenance and repair of track, rail beds
Wildfires	• Wooden bridge structures can be damaged by wildfires	• Wildfires can disrupt services in active wildfire areas
Changes in freeze/thaw	• Potential for ice jams flooding that can damage railroad bridges • Rail seat deterioration because of expansion of freezing water	• Increased rail maintenance costs
Permafrost thaw	• Risk of uneven settling and ground heaves can result in significant vertical movement of track. Rail track deformities • Subsidence of rail beds • Coastal erosion impacts railways	• Increased inspection, maintenance, and repair of track, rail beds

Roads

Consists of paved and unpaved roads. Stormwater drainage (culverts, side drains). Highway, road and street signs, traffic lights. Road work and maintenance. Driver safety, traffic, and service. The United States has over 4 million miles of public roadways which in 2021 handled more than 3.2 million vehicle miles. Because of the COVID-19 pandemic, that number dropped to 2.9 million. After the height of COVID-19, travel is returning to normal.

(Continued)

(Continued)

	Effects	
Climate stressor	**Physical**	**Operations and maintenance**
Increased temperatures, extreme heat	• Heat waves of long duration can cause concrete pavements to buckle • High temperatures can soften asphalt pavement, resulting in rutting • In Alaska, loss of sea ice causes loss of protection from winter storms resulting in coastal erosion. Loss of near shore ice pack. Shorter season of ice roads	• Increase in vehicle accidents in extreme heat conditions • Improved outdoor worker productivity with increasing warmer days. Reduced productivity on extremely hot days • Higher chances of outdoor worker heat stress. Limited hours for road maintenance and repair • Increased water demand for outdoor workers
Extreme cold	• Ice on roadways	• Decreased outdoor worker productivity on extremely cold days. Shortened work shifts
Extreme precipitation, inland flooding	• Water can enter through pavement cracks, and damage subgrade • Increases in road washout • Increase in landslides, mudslides that damage roads • Shoreline erosion can damage coastal roads	• Rain slows traffic decreasing road capacity • Road safety problems: low visibility, vehicle hydroplaning • Coastal roadways designated as evacuation routes could become flooded or damaged • Fewer days with snow and ice could mean lower removal costs • Less frequent but more intense winter storms mean more snow removal problems • Winter storms decrease visibility. Snow and ice on roads create hazardous driving conditions

(Continued)

(Continued)

	Effects	
Climate stressor	**Physical**	**Operations and maintenance**
Sea level rise, extreme high tides, storm surge	• 60,000 miles of US roads in coastal floodplains are vulnerable to extreme storms and hurricanes • *Accelerating*: Sea level rise may rise faster than anticipated because of ice losses in Antarctica and Greenland. Low-lying coastal roads and tunnels located near the coast may have problems with erosion, shorter life spans • Saltwater spray can cause concrete deterioration	• Road drainage systems become less effective • Barrier island communities can become inaccessible if access roads become flooded or damaged
Wind, dust storms		• Debris from windstorm or dust storm can disrupt traffic
Drought	• Drought contributes to cracking of pavements • With long heat waves, asphalt-concrete pavement can crack, allowing water to reach and damage subgrade • For roads that pass through wetlands, extended drought can shrink the ground beneath the roadway	• Increased sediment in culverts because of low flow • Dry conditions increase dusty conditions on gravel roads. Visibility problems. Health damage. Traffic safety hazard
Wildfires	• Debris flow in flooding following wildfires can damage roads	• Smoke from wildfires can disrupt traffic
Changes in freeze/thaw	• Increased wear and tear on pavement from freeze/thaw cycles	• Increased road maintenance costs
Permafrost thaw	• Permafrost thaw can cause road damage from loss of supporting ground structure	• Increased road maintenance costs

(Continued)

(Continued)

Solid waste

Solid waste management infrastructure includes solid waste transfer stations, recycling, incineration composting facilities, and landfills. Landfill sites consist of cells in which solid waste is placed and covered each day. Each cell has both an engineered lining system to collect and treat leachate and a landfill gas collection system. After the cell is filled, it is capped to prevent water infiltration. Gases produced by the decomposition of solid wastes, primarily methane and carbon dioxide, are captured and sent to energy utilities for use as fuel. Long-term management includes leachate treatment and gas collection. In addition to waste cells, a landfill site will have a leachate treatment facility, a landfill gas management system, waste handling equipment, a surface water management system, stockpiles of daily cell cover material, offices, workshops, and site security.

	Effects	
Climate stressor	**Physical**	**Operations and maintenance**
Increased temperatures, extreme heat	• Changes in waste decomposition, biological treatment processes • Changes in the generation of landfill gases • Changes in leachate production	• Increase in building cooling demands. Reduction in heating demands • Improved outdoor worker productivity with increasing warmer days. Reduced productivity on extremely hot days • Higher chances of outdoor worker heat stress • Increased water demand for outdoor workers and site operations • Increased leachate production in winter months • Changes in the distribution of vermin and pests • Increased odor problems • Stress on vegetation in landscaping and screening areas • Possible increase in rate of collection of mixed waste containing putrescibles and biodegradable wastes because of increased rate of decomposition. More odor and pest infiltration potential

(Continued)

(Continued)

	Effects	
Climate stressor	**Physical**	**Operations and maintenance**
Extreme cold	• Freezing of water lines	• Decreased outdoor worker productivity on extremely cold days. Shortened work shifts
Extreme precipitation, inland flooding	• Increased risk of flooding and inundation of landfill cells • Possible impact on slope stability • Erosion of capping layers • Increase in off-site pollution • Need for additional containers to keep waste dry	• Inundation of drainage systems • Disruption of road and rail transport • Increase in insurance costs. Limits on availability • Higher ground saturation may require changes in equipment • *Adaptation Island*: Facility is protected from flooding but flooded access roads prevent operations and maintenance personnel from reaching facility
Sea level rise, extreme high tides, storm surge	• Increased risk of flooding and inundation of landfill cells, waste management sites in coastal facilities • *Accelerating*: Sea level rise may rise faster than anticipated because of ice losses in Antarctica and Greenland. Solid waste facilities located near the coast may have problems with erosion, shorter life spans	• Increased erosion at coastal facilities • Disruption of water based transport
Wind, dust storms	• Storm damage to buildings and infrastructure	• Winds disrupting outdoor operations. More site closures • Increase in wind-blown debris • Increase in dust potential with high temperatures and drought
Drought	• Changes to surface water flows	• Reduced water availability
Wildfires	• Increased risk of wildfires. Damage to solid waste management facilities	• Additional maintenance required to maintain defensible space around buildings, infrastructure
Changes in freeze/thaw	• Increased wear and tear on solid waste facility pavement from freeze/thaw cycles	• Increased road maintenance costs

(Continued)

(Continued)

	Effects	
Climate stressor	**Physical**	**Operations and maintenance**
Permafrost thaw	• Permafrost thaw can cause road damage from loss of supporting ground structure	• Increased road maintenance costs

Stormwater

Stormwater infrastructure consists of a network of water retention and purification systems designed to slow and direct stormwater flow, prevent flooding, and reduce contamination of waterways. Range: large concrete storm sewers, roadside ditches, flood control reservoirs, rain gardens, and natural riverine systems. Stormwater runoff picks up pollutants as it flows over impervious or semipervious surfaces such as roadways, parking lots, and agricultural or landscaped areas. Stormwater runoff is increasing as more impervious areas are being created as urban and suburban areas expand. Climate change is expected to bring about larger and more intense rainfall events, increasing stormwater runoff.

	Effects	
Climate stressor	**Physical**	**Operations and maintenance**
Increased temperatures, extreme heat	• Creates conditions favorable to extreme precipitation events	• Improved outdoor worker productivity with increasing warmer days. Reduced productivity on extremely hot days • Higher chances of outdoor worker heat stress • Increased water demand for outdoor workers and site operations
Extreme cold		
Extreme precipitation, inland flooding	• Urbanization and associated increase in impervious surfaces increase stormwater flows • Stormwater management system overwhelmed with extreme flooding from heavy rainfall events	• Increased maintenance and repair costs • Increased cleanup costs following combined sewer overflows (CSO) event • CSOs into waterways. Release of untreated sewage into streams, rivers, and other water bodies

(Continued)

(Continued)

Climate stressor	Effects: Physical	Effects: Operations and maintenance
	• Release of stormwater on steep slopes can decrease slope stability. Increased erosion and the risk of landslides • Erosion can increase sediment loading in rivers and streams, resulting in decreased water quality • Increase in nonpoint source pollution	
Sea level rise, extreme high tides, storm surge	• Outfall pipes may be submerged for longer periods, thereby reducing hydraulic head. More difficult to move water continuously. Result: system backups, increasing ponding frequency and duration, reduced capacity to handle extreme events	• Combined sewer overflows (CSOs) into waterways • Increased corrosion in stormwater infrastructure • Additional maintenance and repair required
Wind, dust storms	• Increased sediment from dust storms • In coastal areas, increased exposure to wind, sand and salt spray and salt water, greater wave action	• Increase in blockages and breaks associated with rainfall events and storms • Accumulation of debris in pump stations
Drought		
Wildfires		
Changes in freeze/thaw	• Possible structural damage to stormwater management infrastructure	• Increase in maintenance and repair requirements
Permafrost thaw	• Possible structural damage to stormwater management infrastructure • Melting permafrost causes infrastructure, including water systems, to sink into the earth	• Increased monitoring for structural instability

(Continued)

(Continued)

Transit

The US transit infrastructure consists of urban and suburban rail and bus systems that provide local transport and access to jobs, schools, healthcare, shopping, and public/private services. Buses operate on more than 226,000 miles of streets and roads. Rail transit operates on more than 9,000 miles of rail.

	Effects	
Climate stressor	**Physical**	**Operations and maintenance**
Increased temperatures, extreme heat	• Buckling of track. Sun or heat kinks • Buckled tracks must be taken out of service	• Reduced train speeds during high heat events • Increase in building cooling demands. Reduction in heating demands • Improved outdoor worker productivity with increasing warmer days. Reduced productivity on extremely hot days • Higher chances of outdoor worker heat stress • Increased water demand for outdoor workers and site operations
Extreme cold	• Increased risk of breakage of brittle rail track • Frozen switches may not fully close leading to detailment • Freezing water lines	• Decreased outdoor worker productivity on extremely cold days. Shortened work shifts
Extreme precipitation, inland flooding	• Erosion of track supporting systems, embankments	• Underground transit systems may flood • Increase in damage to electrical equipment • More service disruptions because of flooding • Less frequent but more intense winter storms means more snow removal problems • Winter storms increase icing on rails. Reduced speeds. Longer braking distances
Sea level rise, extreme high tides, storm surge	• Erosion of track supporting systems, embankments • Scour or loss of rail bed from sea level rise, tides, storm surge combinations	• Scour of rail bed can lead to derailments, spills of hazardous cargo • Debris on rails can disrupt operations

(Continued)

(Continued)

	Effects	
Climate stressor	**Physical**	**Operations and maintenance**
Wind, dust storms	• High winds can damage platforms, stations, other rail structures	• Dust storms reduce visibility. Train speeds are reduced
Drought	• For railway lines that pass through wetlands, extended drought can shrink the ground beneath the rail bed	• Increased inspection, maintenance and repair of track, rail beds
Wildfires	• Wooden bridge structures can be damaged by wildfires	• Wildfires can disrupt services in active wildfire areas
Changes in freeze/thaw	• Potential for ice jams flooding that can damage transit rail bridges • Rail seat deterioration because of expansion of freezing water	• Increased maintenance costs
Permafrost thaw	• Risk of uneven settling and ground heaves can result in significant vertical movement of track	• Service disruptions. Railroad track becomes inoperable

Wastewater

Wastewater infrastructure consists of a network of pipes and treatment systems that collect, treat, and discharge effluents into the environment. The United States has more than 16,000 wastewater treatment plants operating on average, at 81% of design capacity. Fifteen percent have reached or exceeded their design capacity.

	Effects	
Climate stressor	**Physical**	**Operations and maintenance**
Increased temperatures, extreme heat	• Warmer temperatures can improve wastewater treatment process efficiencies • Increased odor problems • With warming temperatures, effluent will need to be cooled before discharging, increasing energy costs	• Increase in building cooling demands. Reduction in heating demands • Improved outdoor worker productivity with increasing warmer days. Reduced productivity on extremely hot days • Higher chances of outdoor worker heat stress • Increased water demand for outdoor workers and site operations

(Continued)

(Continued)

	Effects	
Climate stressor	**Physical**	**Operations and maintenance**
Extreme cold	• Colder temperatures can decrease wastewater treatment process efficiencies	
Extreme precipitation, inland flooding	• Risk of flood damage to the wastewater treatment plants • Increase in volume of wastewater. Extreme storms can overwhelm the design capacity of stormwater management systems resulting in surface flooding	• Increased corrosion and cracking in stormwater infrastructure (e.g., piping) • Increased corrosion in pump stations • Transport of larger loads of sediment and nutrients into waterways
	• Combined sewer overflows (CSOs) send untreated wastewater into waterways • Increased quantities of sediment, nutrients and bacteria in stormwater management system flowing to local waterways • Reductions in water quality. Threats to drinking water sources • More nonpoint source runoff in rural areas • Possible overflow of stormwater detention ponds • Deterioration of stormwater management systems because of frequent inundation: submerged outfalls • Rising downstream water levels can make effluent pumping a requirement, increasing energy needs • Increased water infiltration into wastewater networks	• More frequent clogging of inlets, outlets, and infiltration beds with debris • Greater requirements for sediment removal • *Adaptation Island*: Access roads flooded preventing operations and maintenance personnel from reaching facility

(Continued)

(Continued)

	Effects	
Climate stressor	**Physical**	**Operations and maintenance**
Sea level rise, extreme high tides, storm surge	• Combined sewer overflows into waterways • Saltwater intrusion, flooding, and inundation of low-lying stormwater system outfalls • Increasingly exposed to wave action can lead to deterioration • *Accelerating*: Sea level rise may rise faster than anticipated because of ice losses in Antarctica and Greenland. Wastewater treatment facilities located near the coast may have problems with erosion, shorter life spans • Sea level rise can increase groundwater levels, thus more water, saltwater infiltration into wastewater systems. Also, float and corrosion of pipes and pumping stations	• Increased operating costs of wastewater treatment
Wind, dust storms	• Increased sediment from dust storms • In coastal areas, increased exposure to wind, sand and salt spray and salt water	• Increase in blockages and breaks associated with rainfall events and storms • Accumulation of debris in pump stations
Drought	• Reduced water flows make it more difficult to meet effluent standards. Increase in pollutant concentration • Increased demand for recycled water from wastewater treatment plants. May conflict with use of treated wastewater discharge to support ecosystems, downstream users • Higher concentrations of wastewater, causing blockages and corrosion of pipes	• Higher maintenance and repair costs • Loss of revenue in areas with volumetric or variable rate structures

(Continued)

(Continued)

	Effects	
Climate stressor	**Physical**	**Operations and maintenance**
	• Ground settlement causing damage to pipes • Increase of odors: Low wastewater flow can cause more deposition of solids, more anaerobic decomposition • In low wastewater flow, possible cavitation of pumps and damage to piping system	
Wildfires	• Increased risk of wildfires. Damage to treatment facilities, buildings	• Additional maintenance required to maintain defensible space around buildings, infrastructure
Changes in freeze/thaw Permafrost thaw	• Possible structural damage to components • Possible structural stability problems with buildings, water treatment works • Melting permafrost causes infrastructure, including water and sewage systems, to sink into the earth and alters the structural integrity of these buildings	• Increase in maintenance and repair requirements • Increased monitoring for structural stability

References (Not Cited)

Alsharari, T. M. 2021. "Climate change impacts on slope stability of levees." Thesis. University of Illinois at Chicago. Dept. of Civil, Materials, and Environmental Engineering. https://indigo.uic.edu/articles/thesis/Climate_Change_Impacts_on_Slope_Stability_of_Levees/17026043.

Anastasios, Z., and A. Tolkou. 2015. "Effect of climate change in wastewater treatment plants: Reviewing the problems and solutions." In *Managing water resources under climate uncertainty*, edited by S. Shrestha, A. K. Anal, P. A. Salam, and M. van der Valk. Cham, Switzerland: Springer Publishing.

Armstrong, J. 2020. "Protecting bridges in extreme climates." Accessed January 1, 2020. https://csengineermag.com/protecting-bridges-in-extreme-climates/.

Baker, M. E. 2023. "Lesson learned: Dams in cold regions should account for ice." Accessed February 4, 2023. https://damfailures.org/lessons-learned/dams-in-cold-regions-should-account-for-ice/.

BBC News. 2020. "Climate change: Dams played key role in limiting sea level rise." Accessed August 19, 2020. https://www.bbc.com/news/science-environment-53836018.

Bebb, J., and J. Kersey. 2003. "Potential impacts of climate change on waste management." UK Environment Agency. https://www.nswai.org/docs/Potential%20Impacts%20of%20 Climate%20Change%20on%20Waste%20Management.pdf.

Berkowitz, B., A. Galocha, and J. Ledur. 2022. "We built a fake metropolis to show how extreme heat could wreck cities." *Washington Post,* Accessed August 11, 2022. https:// www.washingtonpost.com/climate-environment/interactive/2022/heat-infrastructure-transportation/.

Berkowitz, B., A. Galocha, and J. Ledur. 2023. "We built a fake metropolis to show how extreme cold could wreck cities." *Washington Post,* Accessed January 25, 2023. https:// www.washingtonpost.com/climate-environment/interactive/2023/cold-infrastructure-transportation/.

Bertanga, C. 2020. "Impacts and implications of climate change on wastewater treatment plants," with reference to Hughes, K. Cowper-Heays, E. Olesson, R. Bell, A. Stroombergen. Impacts and implications of climate change on wastewater systems: A New Zealand perspective. *Clim. Risk Manag.* **31** (2020).

Beuthe, M., B. Jourquin, N. Urbain, I. Lingemann, and B. Ubbels. 2014. "Climate change impacts on transport on the Rhine and Danube: A multimodal approach." *Transp. Res. D* **27** (2014): 6–11. https://www.sciencedirect.com/science/article/pii/S1361920913001442?via%3Dihub.

Bheemasetti, T. V., C. Tohm, and B. N. Lingwall. 2022. "Ice-water phase change studies in plastic and non-plastic silts." *Front. Built Environ.* **8**: Article 943614. https://doi.org/ 10.3389/fbuil.2022.943614.

Bierkandt, R., M. Auffhammer, and A. Levermann. 2015. "US power plant sites at risk of future sea-level rise." *Environ. Res. Lett.* **10**: 124022. https://iopscience.iop.org/article/10.1088/ 1748-9326/10/12/124022.

Boydell, R. 2021. "From melting steel to 'concrete cancer', our buildings aren't designed to withstand climate change." *Ideas.Ted.Com.* Accessed July 14, 2021. https://ideas.ted.com/ climate-change-extreme-weather-will-damage-buildings-and-infrastructure/.

Bronen, R. 2013. "Climate-induced displacement of alaska native communities," Brookings-LSE. Accessed January 30, 2013. https://www.brookings.edu/wp-content/uploads/2016/ 06/30-climate-alaska-bronen-paper.pdf.

Brunda, A. 2018. "The Internet is Drowning." National Geographic, Accessed July 16, 2018. https://www.nationalgeographic.com/science/article/news-internet-underwater-sea-level-rise.

Bureau of Transportation Statistics. 2023. "U.S. Highway Vehicle Miles." Accessed February 1, 2023. https://www.bts.gov/content/us-vehicle-miles.

Cai, M., E. Douglas, and M. Ferman. 2022. "How Texas' power grid failed in 2021—and who's responsible for preventing a repeat," *The Texas Tribune.* Accessed February 12, 2022. https://www.texastribune.org/2022/02/15/texas-power-grid-winter-storm-2021/.

Chappelle, C., H. McCann, D. Jassby, K. Schwabe and L. Szeptycki. 2019. "Managing wastewater in a changing climate." Public Policy Institute of California.

Chi, G., D. Holen, H. Randell, M. Mucioki, and R. Napolitano. 2021. "Rural Alaska has a bridge problem as permafrost thaws and crossing river ice gets riskier with climate change." United Nations Office for Risk Reduction, Prevention Web, October 12 2021. Accessed

January 30, 2023. https://www.preventionweb.net/news/rural-alaska-has-bridge-problem-permafrost-thaws-and-crossing-river-ice-gets-riskier-climate.

Christodoulou, A., and H. Demirel. 2018. "Impacts of climate change on transport—A focus on airports, seaports and inland waterways." EUR 28896 EN, Publications Office of the European Union, Luxembourg. ISBN 978-92-79-97039-9. JRC108865. https://doi.org/10.2760/378464.

Christodoulou, A., P. Christidis, and H. Demirel. 2019. "Sea-level rise in ports: A wider focus on impacts." *Marit. Econ. Logist.* **21**: 482–496. https://doi.org/10.1057/s41278-018-0114-z.

City Parks Forum. 2023. "How cities use parks for climate change management," Accessed January 29, 2023. https://planning-org-uploaded-media.s3.amazonaws.com/publication/download_pdf/Parks-for-Climate-Change.pdf.

Climate Change Post. 2014. "Extreme weather and climate change impacts on inland waterway transport." May 26, 2014. Accessed February 3, 2023. https://www.climatechangepost.com/news/2014/5/26/extreme-weather-and-climate-change-impacts-inland-/.

Coffey, M. J. 2021. "Arctic civil infrastructure and adaptation to climate change." AASHTO Climate Change Symposium 2021. Accessed February 6, 2023. https://environment.transportation.org/wp-content/uploads/2021/05/CC_Symposium_MichaelCoffey.pdf.

Das, A., P. Rokaya, and K.-E. Lindenschmidt. 2020. "Ice-Jam flood risk assessment and hazard mapping under future climate." *J. Water Resour. Plann. Manage.* **146** (6). https://doi.org/10.1061/(ASCE)WR.1943-5452.0001178.

EIA (US Energy Information Agency). 2020. "Commercial buildings have gotten larger in the United States, with implications for energy." Accessed January 31, 2023. https://www.eia.gov/todayinenergy/detail.php?id=46118#.

Ellfeldt, A. 2020. "Mounting Climate Impacts Threaten U.S. Nuclear Reactors," E&E News, Accessed August 20, 2020. https://www.scientificamerican.com/article/mounting-climate-impacts-threaten-u-s-nuclear-reactors/.

EPA (US Environmental Protection Agency). 2022. Climate change adaptation resource center (ARC-X). "Climate adaptation and stormwater runoff." Accessed January 29, 2023. https://www.epa.gov/arc-x/climate-adaptation-and-stormwater-runoff.

FHWA (Federal Highway Administration). 2008. *Highways in the coastal environment,* 2nd ed. Hydraulic Engineering Circular No. 25. FHWA-NHI-07-096. Department of Civil Engineering, University of South Alabama, Mobile, AL, 434 pp. https://www.fhwa.dot.gov/engineering/hydraulics/pubs/hif19059.pdf.

FHWA. 2014. Climate Change Sensitivity Matrix. https://toolkit.climate.gov/tool/transportation-climate-change-sensitivity-matrix.

FHWA. 2020. *Highways in the coastal environment,* 3rd ed. Hydraulic Engineering Circular No. 25. FHWA-NHI-19-059. Federal Highway Administration. Department of Civil Engineering, University of South Alabama, Mobile, AL, 250 pp. https://nca2018.globalchange.gov/chapter/8/.

Fluixá-Sanmartín, J., L. Altarejos-García, A. Morales-Torres, and I. Escuder-Bueno. 2018. "Review article: Climate change impacts on dam safety." *Nat. Hazards Earth Syst. Sci.* **18**: 2471–2488. https://doi.org/10.5194/nhess-18-2471-2018.

Fountain, H. 2020. "'Expect More': Climate change raises risk of dam failures," *New York Times,* May 21, 2020. Accessed January 29, 2023. https://www.nytimes.com/2020/05/21/climate/dam-failure-michigan-climate-change.html.

FRA (Federal Railroad Administration). 2020. "U.S. Department of Transportation, Freight Rail Overview." Accessed February 1, 2023. https://railroads.dot.gov/rail-network-development/freight-rail-overview.

GAO. 2021. Statement of J. Alfredo Gómez, Director, Natural Resources and Environment, "Superfund: EPA should take additional actions to manage risks from climate change effects." GAO-21-555T, Accessed May 13, 2021. https://www.gao.gov/assets/720/714315.pdf.

Gelman, B. 2021. "Can the internet be more resilient to climate change? Some are banking on it." WBUR, Accessed August 13, 2021. https://www.wbur.org/news/2021/08/13/internet-resilient-climate-change.

Gibbons, S. 2019. "As the climate crisis worsens, cities turn to parks," *National Geographic,* May 21, 2019. Accessed January 29, 2023. https://www.nationalgeographic.com/environment/article/climate-crisis-cities-turn-to-parks.

GOV.UK. 2013. "FD2628 impact of climate change on dams & reservoirs, final guidance report, May 2013. UK flood and coastal erosion risk management research and development programme." https://www.gov.uk/flood-and-coastal-erosion-risk-management-research-reports/review-of-the-impact-of-climate-change-on-dams-and-reservoirs.

Ingraham, C. 2017. "Houston is experiencing its third '500-year' flood in 3 years. How is that possible?" *Washington Post,* Accessed August 29, 2017. https://www.washingtonpost.com/news/wonk/wp/2017/08/29/houston-is-experiencing-its-third-500-year-flood-in-3-years-how-is-that-possible/.

Jin, J., S. Li, C. Song, et al. 2019. "Ageing deformation of tailings dams in seasonally frozen soil areas under freeze-thaw cycles." *Sci. Rep.* **9**: 15033. https://doi.org/10.1038/s41598-019-51449-6.

Kish, A., and Wesley Mui. 2003. "Track buckling research," Accessed July 9, 2003. U.S. Department of Transportation, Bureau of Transportation Statistics, National Transportation Library, ROSA P #11985. https://rosap.ntl.bts.gov/view/dot/11985.

Larsen, A., and B. Connor. 2021. "Safety on ice," U.S. Department of Transportation, FHWA, FHWA-HRT-22-001, Autumn 2021, Vol. **85** No. 3. https://highways.dot.gov/public-roads/autumn-2021/04.

Laskin, J. 2019. "How the severe cold affects airport ground operations," The Points Guy. Accessed February 4, 2023. https://thepointsguy.com/news/severe-cold-affects-airport-ground-operations/.

Liu, L., Y. Wen, Y. Liang, F. Zhang, and T. Yang. 2019. "ExtremeWeather impacts on inland waterways transport of Yangtze River." *Atmosphere* **2019** (10): 133. https://doi.org/10.3390/atmos10030133.

Major, D. C., A. Omojola, M. Dettinger, R. T. Hanson, and R. Sanchez-Rodriguez. 2011. *Climate change, water, and wastewater in cities. Climate change and cities: First assessment report of the urban climate change research network,* edited by C. Rosenzweig, W. D. Solecki, S. A. Hammer, and S. Mehrotra, 113–143, Cambridge, UK: Cambridge University Press.

MyMove. 2022. "Can climate change kill the internet?" March 10, 2022. Accessed January 29, 2023. https://www.mymove.com/internet/can-climate-change-kill-the-internet/.

Nasr, A., I. Björnsson, D. Honfi, O. L. Ivanov, J. Johansson, and E. Kjellström. 2021. "A review of the potential impacts of climate change on the safety and performance of bridges." *Sustain. Resilient Infrastruct.* **6** (3–4): 192–212. https://doi.org/10.1080/23789689.2019.1593003.

NC State University. 2020. College of Engineering news, "Study finds flooding damage to levees is cumulative—and often invisible," April 23, 2020. Accessed January 29, 2023. https://www.engr.ncsu.edu/news/2020/04/23/study-finds-flooding-damage-to-levees-is-cumulative-and-often-invisible-2/.

NOAA. 2019. National climate assessment, "Could climate change make Atlantic hurricanes worse?" Accessed May 29, 2019. https://www.climate.gov/news-features/climate-qa/could-climate-change-make-atlantic-hurricanes-worse.

NOAA. 2023. "Drought.gov. Water utilities. National integrated drought information system," Accessed January 29, 2023. https://www.drought.gov/sectors/water-utilities.

NRC (National Research Council). 2008. "Potential impacts of climate change on U.S. transportation," Transportation Research Board Special Report 290.

Olsen, R., B. Ayyub, D. Walker, A. Barros, et al. 2015. "Adapting infrastructure and civil engineering practice to a changing climate." https://doi.org/10.1061/9780784479193.

Resilient Analytics. 2020 "Heat impacts on infrastructure & personnel: An OKC case study," August 2020. Accessed January 29, 2023. https://www.wucaonline.org/assets/pdf/heat-impact-case-study-oklahoma1.pdf.

Resilient Analytics. 2020. "Heat impacts on infrastructure & personnel: A PWB case study," July 2020. Accessed January 29, 2023. https://www.wucaonline.org/assets/pdf/heat-impact-case-study-portland.pdf.

Rossetti, M. A. 2003. "Potential impacts of climate change on railroads," Paper presented at: *The Potential Impacts of Climate Change on Transportation: Workshop Summary and Proceedings*, September 2003. https://www.transportation.gov/sites/dot.gov/files/docs/rossetti_CC_Impact_Railroads.pdf.

Sasidharan, M., A. K. Parlikad, J. Schooling, G. M. Hadjidemetriou, M. Hamer, A. Kirwan, and S. Roffe. 2022. "A bridge scour risk management approach to deal with uncertain climate future," December 14 2022. *Transp. Res. D* **114** (2023): 103567. https://doi.org/10.1016/j.trd.2022.103567.

Sayles, F. H. 1987. "Embankment dams on permafrost," U.S. Army Corps of Engineers, Cold Regions Research and Engineering Laboratory, Special Report 87-11, July 1987.

Soroka, S. 2021. "Is climate change putting your county at risk from hazardous waste?" Conservation Law Foundation, October 27, 2021. Accessed January 29, 2023. https://www.clf.org/blog/hazardous-sites-at-risk-from-climate-change/.

Thaduri, A., A. Garmabaki, U. Kumar. 2021. "Impact of climate change on railway operation and maintenance in Sweden: A state-of-the-art review." August 3, 2021. Maintenance, Reliability and Construction Monitoring. https://doi.org/10.21595/mrcm.2021.22136.

Thompson, A. 2017. "The overlooked risk of levees as rainfall rises," Climate Central News, May 5, 2017. Accessed January 29, 2023. https://www.climatecentral.org/news/the-over-looked-risk-of-levees-as-rains-rise-21426.

Trekbaron. 2022. "10 types of parks," September 18, 2022. Accessed January 31, 2023. https://trekbaron.com/types-of-parks/.

UK Department of Environment, Food and Rural Affairs. 2009. "FD2628 impact of climate change on dams & reservoirs." https://assets.publishing.service.gov.uk/media/

6033dfa3e90e076605eab4de/Review_of_indirect_Impacts_of_climate_change_on_dams_and_reservoirs_final_report.pdf.

UNEP (United Nations Environment Programme). 2021. "A practical guide to climate-resilient buildings & communities." https://www.unep.org/resources/practical-guide-climate-resilient-buildings.

United States Census. 2023. "Quick facts." Accessed January 31, 2023. https://www.census.gov/quickfacts/fact/table/US/VET605221.

USACE (US Army Corps of Engineers). 2006. "Levee owner's manual for non-federal flood control works," March 2006. https://www.lrh.usace.army.mil/Portals/38/docs/civil%20works/Levee%20Owners%20Manual.pdf.

USACE. 2023. Portland District Website, "Levee and flood risk management," Accessed February 1, 2023. https://www.nwp.usace.army.mil/Missions/Flood-Risk-Management/Levees/.

USAID. 2015. "Bridges: A guide for USAID project managers," November 2015. https://www.climatelinks.org/sites/default/files/asset/document/BRIDGES_PRIMER_CCA_ENGINEERING_DESIGN.pdf.

USGCRP (US Global Change Research Program). 2017. Vol. 1 of *Climate Science Special Report: Fourth National Climate Assessment,* edited by Wuebbles, D. J., D. W. Fahey, K. A. Hibbard, D. J. Dokken, B. C. Stewart, and T. K. Maycock, Washington, DC: US Global Change Research Program. https://doi.org/10.7930/J0J964J6.

USGS (US Geological Survey). 2018. Water Science School, "Hydroelectric power water use," Accessed July 16, 2022. https://www.usgs.gov/special-topics/water-science-school/science/hydroelectric-power-water-use.

Vahedifard, F., F. H. Jasim, F. T. Tracy, M. Abdollahi, A. Alborzi, and A. AghaKouchak. 2020. "Levee fragility behavior under projected future flooding in a warming climate." *J. Geotech. Geoenviron. Eng.* **146** (12): 04020139. https://doi.org/10.1061/(asce)gt.1943-5606.0002399.

Vahedifard, F., J. D. Robinson, and A. AghaKouchak. 2016. "Can protracted drought undermine the structural integrity of California's Earthen Levees?" *J. Geotech. Geoenviron. Eng.* **142** (6): 02516001. https://amir.eng.uci.edu/publications/16_JGGE_Drought_Levees.pdf.

Rodrigues, V. S., S. Pettit, I. Harris, D. Zhang, and Y.-Y. Lau. 2017. "Risk and cost evaluation of port adaptation measures to climate change impacts." *Transp. Res. D* **61** (2018): 444–458. https://doi.org/10.1016/j.trd.2017.03.004.

Vidal, J. 2018. "Are coastal nuclear power plants ready for sea level rise?" *Hakai Magazine.* August 18, 2018. https://hakaimagazine.com/features/are-coastal-nuclear-power-plants-ready-for-sea-level-rise/.

WRI (World Resources Institute). 2018. *New Climate Economy, Unlocking the Inclusive Growth Story of the 21st Century: Accelerating Climate Action in Urgent Times,* Global Commission on the Economy and Climate, August 2018, p. 22. Accessed March 24, 2023. https://newclimateeconomy.report/2018/wp-content/uploads/sites/6/2018/09/NCE_2018_FULL-REPORT.pdf.

Yang, Z., A. K. Y. Ng, P. Tae-Woo, T. W. Lee, Z. Qu, Q. Yu, X.-L. Ren, P. Yue, G.-K. Zhang, J.-F. Wang, Y.-H. Mu, and E.-L. Liu. 2021. "Study on the freeze-thaw problems in the winter construction of the Lianghekou Earth-Core Rockfill Dam and the countermeasures for prevention," Hindawi, Vol. 2021of *Advances in Materials Science and Engineering,* Article ID **5539916**, 14 p. https://doi.org/10.1155/2021/5539916.

Yu, Y., E. Liu, Q. Yu, et al. 2022. "Effects of freeze-thaw cycles on the mechanical properties of the core-wall contact clay of a dam." *Int. J. Civ. Eng.* **20**: 779–791. https://doi.org/10.1007/s40999-022-00702-7.

Zhai, Z. J., J. M. Helman, "Implications of climate changes to building energy and design." *Sustain. Cities Soc.* https://doi.org/10.1016/j.scs.2018.10.043.

APPENDIX B

Envision Sustainable Infrastructure Rating System

Envision is a tool that enables engineers and other practitioners to benchmark civil infrastructure projects against measures of sustainable performance. The system comprises 64 sustainable performance credits organized into 5 sections and 14 subsections. Five of those credits are for innovation: exceptional performance relevant to each of the five sections. The Envision 3.0 Manual is available on the Institute for Sustainable Infrastructure website: https://sustainableinfrastructure.org/

Sustainable performance is measured as a relative departure from conventional performance, i.e., standard practices or regulatory requirements. As stated in the Envision manual, "[t]he Envision levels of achievement define the level and quality of project performance in each credit as follows:

- *Improved*: Performance that is above conventional. Slightly exceeds regulatory requirements.
- *Enhanced*: Sustainable performance that is on the right track. There are indications that superior performance is within reach.
- *Superior*: Sustainable performance at a very high level.
- *Conserving*: Performance that has achieved essentially zero negative impact.
- *Restorative*: Performance that restores natural or social systems. Such performance receives the highest award possible and is celebrated as such. The Restorative level does not apply to all performance objectives."

The Envision credits, their intent, the metrics used to measure performance, and the maximum points available are presented in the tables below. The total maximum point achievable through the credits is 1,000. Ten additional points can be achieved through demonstrated innovation in each of the five sections.

Quality of Life

Well-being				
Credit		Intent	Metric	Points
QL1.1	Improve community quality of life	Improve the net quality of life of all communities affected by the project and mitigate negative impacts to communities	Measures taken to assess community needs and improve quality of life while minimizing negative impacts	26
QL1.2	Enhance public health and safety	Protect and enhance community health and safety during operation	Measures taken to increase safety and provide health benefits on the project site, surrounding sites, and the broader community in a just and equitable manner	20
QL1.3	Improve construction safety	Enhance public and worker safety during construction	Commitments and measures to monitor safety, provide feedback mechanisms, train personnel, establish security plans, and make health programs available	14
QL1.4	Minimize noise and vibration	Minimize noise and vibrations during operations to maintain and improve community livability	The extent that operational noise and vibration is assessed and mitigated, and target levels achieved	12
QL1.5	Minimize light pollution	Reduce backlight, uplight, and glare without jeopardizing safety during operations	Lighting meets backlight, uplight, and glare requirements for lighting zones	12
QL1.6	Minimize construction impacts	Minimize or eliminate the temporary inconveniences associated with construction	Extent of issues addressed through construction management plans	8

Mobility				
Credit		Intent	Metric	Points
QL2.1	Improve community mobility and access	Plan the project as part of a connected network that supports all transportation modes for the efficient movement of people, goods, and services	The extent to which the project broadens mode choices, reduces commute times, reduces vehicle distance traveled, and improves levels of service	14
QL2.2	Encourage sustainable transportation	Expand accessibility to sustainable transportation choices including active, shared, and/or mass transportation	The extent to which active, shared, or mass transportation options are accessible, encouraged, and supported as part of a larger integrated transportation network	16
QL2.3	Improve access and wayfinding	Design the project to provide safe and appropriate access in and/or around the project in a way that integrates the project with the surrounding community	Incorporating and providing clear access, safety, and wayfinding measures to accommodate emergency services and regular vehicular or pedestrian traffic	14

Community				
Credit		Intent	Metric	Points
QL3.1	Advance equity and social justice	Ensure that equity and social justice are fundamental considerations within project processes and decision making	Degree to which equity and social justice are included in stakeholder engagement, project team commitments, and decision making	18
QL3.2	Preserve historic and cultural resources	Preserve or restore significant historical and cultural sites and related resources	Steps taken to identify, preserve, or restore cultural resources	18
QL3.3	Enhance views and local character	Preserve or enhance the physical, natural, and/or community character of the project site and its surroundings	Steps taken to assess valued community resources, implement preservation measures, and determine overall satisfaction	14
QL3.4	Enhance public space and amenities	Improve amenities and publicly accessible spaces to enhance community livability	Plans and commitments to preserve, conserve, enhance, and/or restore the defining elements of the amenity	14

Innovation				
Credit		Intent	Metric	Points
QL0.0	Innovate or exceed credit requirements	To reward exceptional performance beyond the expectations of the system and application of innovative methods that advance state-of-the-art sustainable infrastructure	Whether project sustainability performance qualifies as innovation, exceptional performance, or is not otherwise recognized in existing credits	+10

Leadership

Collaboration				
Credit		Intent	Metric	Points
LD1.1	Provide effective leadership and commitment	Provide effective leadership and commitment to achieve project sustainability goals	The degree to which the project owner and project team have made general, and project-specific, sustainability commitments and instituted sustainability management policies	18
LD1.2	Foster collaboration and teamwork	Enhance project sustainability through interdisciplinary collaboration and teamwork	The breadth and inclusivity of interdisciplinary and collaborative meetings and the resulting sustainability performance enhancements	18
LD1.3	Provide for stakeholder involvement	Early and sustained stakeholder engagement and involvement in project decision making	Establishment of sound and meaningful programs for stakeholder identification, early and sustained engagement, and involvement in project decision making	18
LD1.4	Pursue byproduct synergies	Critically reconsider whether traditional waste streams can be beneficially reused	The extent to which the project team works with external groups to find beneficial use of waste, excess resources, or capacity	18

Planning				
Credit		Intent	Metric	Points
LD2.1	Establish a sustainability management plan	Create a project sustainability management plan that can manage the scope, scale, and complexity of a project seeking to improve sustainable performance	Extent of organizational policies, authorities, mechanisms, education, and business processes put in place	18
LD2.2	Plan for sustainable communities	Incorporate sustainability principles into project selection/ identification to develop the most sustainable project for the community	The degree to which project selection/ identification includes sustainability performance assessments and is part of a larger sustainable development plan	16
LD2.3	Plan for long-term monitoring and maintenance	Put in place plans, processes, and personnel sufficient to ensure that long-term sustainable protection, mitigation, and enhancement measures are incorporated into the project	Comprehensiveness of long-term monitoring and maintenance plans, implementation goals, and commitment of resources to fund the activities	12
LD2.4	Plan for end-of-life	Ensure that the project team is informed by an understanding of the full impacts and costs of the project's end-of-life	The degree to which the project team analyzes, and communicates with stakeholders, the end-of-life impacts, cost, and value	14

Economy				
Credit		Intent	Metric	Points
LD3.1	Stimulate economic prosperity and development	Support economic prosperity and sustainable development, including job growth, capacity building, productivity, business attractiveness, and livability	The extent of job creation, increased operating capacity, access, quality, and/or improved socioeconomic conditions	20
LD3.2	Develop local skills and capabilities	Expand the knowledge, skills, and capacity of the community workforce to improve their ability to grow and develop	The inclusion of current and future training programs, informed by skill or capability gaps, and targeted to economically depressed or underemployed communities	16
LD3.3	Conduct a life-cycle economic evaluation	Utilize economic analyses to identify the full economic implications and the broader social and environmental benefits of the project	The comprehensiveness of the economic analyses used to determine the net impacts of the project, and their use in assessing alternatives to inform decision making	14

Innovation				
Credit		Intent	Metric	Points
LD0.0	Innovate or exceed credit requirements	To reward exceptional performance beyond the expectations of the system and application of innovative methods that advance state-of-the-art sustainable infrastructure	Whether project sustainability performance qualifies as innovation, exceptional performance, or is not otherwise recognized in existing credits	+10

Resource Allocation

Materials				
Credit		Intent	Metric	Points
RA1.1	Support sustainable procurement practices	Develop sustainable procurement policies and programs to source materials and equipment from manufacturers and suppliers that implement sustainable practices	The extent of sustainable procurement programs, and the percentage of materials sourced from manufacturers and/or suppliers that implement sustainable practices	12
RA1.2	Use recycled materials	Reduce the use of virgin natural resources and avoid sending useful materials to landfills by specifying reused materials, including structures, and material with recycled content	Percentage of project materials that are reused or recycled. Plants, soil, rock, and water are not included in this credit	16
RA1.3	Reduce operational waste	Reduce operational waste and divert waste streams from disposal to recycling and reuse	Percentage of total operational waste or byproducts diverted from disposal	14
RA1.4	Reduce construction waste	Divert construction and demolition waste streams from disposal to recycling and reuse	Percentage of total waste diverted from disposal	16
RA1.5	Balance earthwork on site	Minimize the movement of soils and other excavated materials off site to reduce transportation and environmental impacts	Percentage of excavated material retained on site or nearby	8

Energy				
Credit		Intent	Metric	Points
RA2.1	Reduce operational energy consumption	Conserve energy by reducing overall operational energy consumption throughout the project life	Percentage of operational energy reductions achieved	26
RA2.2	Reduce construction energy consumption	Conserve resources and reduce greenhouse gases and air pollutant emissions by reducing energy consumption during construction	The number of strategies implemented on the project during construction that reduce energy consumption and emissions	12
RA2.3	Use renewable energy	Meet operational energy needs through renewable energy sources	Extent to which renewable energy sources are incorporated	24
RA2.4	Commission and monitor energy systems	Ensure efficient functioning and extend useful life by specifying commissioning and monitoring of energy systems	The inclusion of monitoring equipment and software, the extent of commissioning, and the commissioning agent's independence from the project	14

Water				
Credit		Intent	Metric	Points
RA3.1	Preserve water resources	Assess and reduce the net negative impact on fresh water availability, quantity, and quality at a watershed scale to positively impact the region's water resources	The extent to which the project considers and contributes to positively addressing broader watershed issues	12
RA3.2	Reduce operational water consumption	Reduce overall water consumption while encouraging the use of greywater, recycled water, and stormwater to meet water needs	Percentage reduction in potable water use and overall water use	22
RA3.3	Reduce construction water consumption	Reduce potable water consumption during construction	The number of strategies implemented during construction that reduce potable water consumption	8
RA3.4	Monitor water systems	Improve operational performance by including monitoring capabilities	Extent and capability of water monitoring equipment and inclusion of response plans	12

Innovation				
Credit		Intent	Metric	Points
RA0.0	Innovate or exceed credit requirements	To reward exceptional performance beyond the expectations of the system and application of innovative methods that advance state-of-the-art sustainable infrastructure	Whether project sustainability performance qualifies as innovation, exceptional performance, or is not otherwise recognized in existing credits	+10

Natural World

Siting				
Credit		Intent	Metric	Points
NW1.1	Preserve sites of high ecological value	Avoid placing the project and temporary works on a site that has been identified as being of high ecological value	Avoidance of high ecological value sites and establishment of protective buffer zones	22
NW1.2	Provide wetland and surface water buffers	Protect, buffer, enhance, and restore wetlands, shorelines, and waterbodies by providing natural buffer zones, vegetation, and soil-protection zones	Type and quality of natural buffer zone established around all wetlands, shorelines, and waterbodies	20
NW1.3	Preserve prime farmland	Identify and protect soils designated as prime farmland, unique farmland, or farmland of importance	Percentage of farmland avoided or preserved during development	16
NW1.4	Preserve undeveloped land	Conserve undeveloped land by locating projects on previously developed land	Percentage of project development that is located on previously developed land	24

Conservation				
Credit		Intent	Metric	Points
NW2.1	Reclaim brownfields	Locate projects on sites classified as brownfields	The extent of remediation of the brownfield site	22
NW2.2	Manage stormwater	Minimize the impact of development on stormwater runoff quantity, rate, and quality	Degree to which the project infiltrates, evapotranspirates, reuses, and/or treats stormwater while not exceeding rate or quantity runoff targets	24
NW2.3	Reduce pesticide and fertilizer impacts	Reduce non-point-source pollution by reducing the quantity, toxicity, bioavailability, and persistence of pesticides and fertilizers	Reductions in quantity, toxicity, bioavailability, and persistence of pesticides and fertilizers used on site, selection of plant species, and use of integrated pest management techniques	12
NW2.4	Protect surface and groundwater quality	Preserve water resources by preventing pollutants from contaminating surface water and groundwater and monitoring impacts during construction and operations	Designs, plans, and programs instituted to prevent and monitor surface water and groundwater contamination during construction and operations	20

Ecology				
Credit		Intent	Metric	Points
NW3.1	Enhance functional habitats	Preserve and improve the functionality of terrestrial (land) habitats	The number of habitat functions addressed to preserve or enhance the net area and quality of functional habitat	18
NW3.2	Enhance wetland and surface water functions	Maintain and restore the ecosystem functions of streams, wetlands, waterbodies, and their riparian areas	Number of functions maintained and restored	20

(Continued)

(Continued)

Ecology				
Credit		Intent	Metric	Points
NW3.3	Maintain floodplain functions	Preserve floodplain functions by limiting development and impacts of development in the floodplain	Efforts to avoid floodplains or maintain natural-acting floodplain functions	14
NW3.4	Control invasive species	Use appropriate noninvasive species, and control or eliminate existing invasive species	Degree to which invasive species have been reduced or eliminated	12
NW3.5	Protect soil health	Preserve the composition, structure, and function of site soils	Degree to which the disruption of soil health has been minimized and restored	8

Innovation				
Credit		Intent	Metric	Points
NW0.0	Innovate or exceed credit requirements	To reward exceptional performance beyond the expectations of the system and application of innovative methods that advance state-of-the-art sustainable infrastructure	Whether project sustainability performance qualifies as innovation, exceptional performance, or is not otherwise recognized in existing credits	+10

Climate and Resilience

Emissions				
Credit		Intent	Metric	Points
CR1.1	Reduce net embodied carbon	Reduce the impacts of material extraction, refinement/manufacture, and transport over the project life	Percentage of reduction in net embodied carbon of materials	20
CR1.2	Reduce greenhouse gas emissions	Reduce greenhouse gas emissions during the operation of the project, reducing project contribution to climate change	Percentage of reduction in operational greenhouse gas emissions	26
CR1.3	Reduce air pollutant emissions	Reduce emissions of air pollutants: particulate matter (including dust), ground-level ozone, carbon monoxide, sulfur oxides, nitrogen oxides, lead, and volatile organic compounds	Reduction of air pollutants compared to baseline	18

Resilience				
Credit		Intent	Metric	Points
CR2.1	Avoid unsuitable development	Minimize or avoid development on sites prone to hazards	The degree to which the project is designed and/or sited to avoid or mitigate site-related risks	16
CR2.2	Assess climate change vulnerability	Develop a comprehensive climate change vulnerability assessment	Scope and comprehensiveness of climate change vulnerability assessment	20

(Continued)

(Continued)

Resilience				
Credit		Intent	Metric	Points
CR2.3	Evaluate risk and resilience	Conduct a comprehensive, multihazard risk and resilience evaluation	Scope and comprehensiveness of the multihazard risk and resilience evaluation	26
CR2.4	Establish resilience goals and strategies	To support increased project and community resilience through the establishment of clear objectives and goals	The degree to which resilience goals expand from initial commitments to quantifiable project objectives, long-term operating plans, and community-wide development plans	20
CR2.5	Maximize resilience	Increase resilience, life-cycle system performance, and the ability to withstand hazards by maximizing durability	The degree to which the project incorporates elements that increase durability, the ability to withstand hazards, and extend useful life	26
CR2.6	Improve infrastructure integration	Enhance the operational relationships and strengthen the functional integration of the project into connected, efficient, and diverse infrastructure systems	The degree to which the project is integrated into other connected systems, where beneficial and appropriate, to increase resilience and systems performance	18

Innovation				
Credit		Intent	Metric	Points
CR0.0	Innovate or exceed credit requirements	To reward exceptional performance beyond the expectations of the system and application of innovative methods that advance state-of-the-art sustainable infrastructure	Whether project sustainability performance qualifies as innovation, exceptional performance, or is not otherwise recognized in existing credits	+10

Robustness Matrix

The robustness matrix, adapted from Wall et al. (2015), is a risk management tool designed to identify a broad spectrum of risks across the project life cycle, including risks from climate change. The matrix is provided below as a generic example. Project managers can use this matrix to identify risks plus mitigating strategies and actions customized to their own projects. Strategies and their associated actions can be rated qualitatively regarding cost, robustness, and resilience achieved. They can be used to compare project alternatives and determine which alternative is most cost effective in reducing risk. Project managers are encouraged to identify additional vulnerabilities and opportunities beyond those shown in the example.

Legend

Conditions for Success: Conditions necessary for project success. What has to go right for the project to be successful? Conditions are divided into three phases: planning, project delivery, and operation, each with multiple subphases.

Vulnerability (V) or Opportunity (Op): Vulnerabilities or opportunities that could reasonably arise from deviations from conditions for success are included. The impacts, positive or negative, if the deviation from project success were to happen are also included:

Initial Risk: Risk of the effects and impacts of deviations from the conditions for success before the application of strategies and actions to reduce risk. Components of the initial risk (Lk = likelihood. Cn = Consequences) are rated qualitatively on a scale of 1 (lowest) to 9 (highest). Users who apply qualitative ratings can use the CAMPO risk rating matrix presented in Chapter 7. This simplified approach for visualizing risk and the numbers provided are for illustration purposes. In actual application, owners and project teams can apply more detailed approaches for risk assessment.

Strategies: Strategies to reduce vulnerabilities or take advantage of opportunities. Each strategy has a description. Users can also add a qualitative rating of relative cost ($$$). Strategy codes: (SM) Mitigation Strategies, (SH) Hedging Strategies, (SZ) Seizing Strategies, (SR) Resiliency Strategies.

Signposts, Triggers for t < 0: Signposts are the variables, trends, or events to be monitored before the operational phase. Predetermined actions are taken when they reach a predetermined threshold value or event (trigger).

Remedial, Adaptive, or Improvement Actions Taken at t < 0: Actions associated with the strategies, as described above. These actions are taken before the operational stage.

Signposts, Triggers for t > 0: Signposts, triggers for actions taken during the operational stage.

Remedial, Adaptive or Improvement Actions Taken at t > 0: These are actions associated with strategies taken during the operational phase. Action codes: (AD) Defensive Actions, (AC) Corrective Actions, (AP) Capitalizing Actions, (AS) Shaping Actions, (AR) Recovery Actions, (RE) Reassessment.

Residual Risk: Residual risk is the risk of the effects and impacts of deviations from the conditions for success after applying strategies and actions to reduce risk. In the robustness matrix, components of the residual risk (likelihood, consequences) are rated qualitatively on a scale of 1 (lowest) to 9 (highest). Like the initial risk ratings provided in this example matrix, the numbers provided are also arbitrary. Also, owners and project teams can apply more detailed risk assessment approaches.

Time: Time relative to the moment the project becomes operational. Planning (t < 0), Project Delivery (t < 0), Project Startup (t = 0), Operation (t > 0).

Robustness Matrix

This example matrix includes multiple examples of conditions for success, vulnerabilities, opportunities, strategies, and actions covering the full project life cycle.

Conditions for success	Vulnerability (V) or Opportunity (Op)	Initial risk Lk/Cn		Strategies	Signposts, triggers $t < 0$	Actions $t < 0$	Signposts, triggers $t > 0$	Actions $t > 0$	Residual risk Lk/Cn	
Strategies: (SM) Mitigation, (SH) Hedging, (SZ) Seizing, (SR) Resiliency. Actions: (AD) Defensive, (AC) Corrective, (AP) Capitalizing, (AS) Shaping, (AR) Recovery. (RE) Reassessment.										
Planning (time < 0) *Devising the best project alternative based on initial available information prior to project initiation*										
Location										
The property is free of any incumbrances that will affect the project or completed asset	**(V)** Deed restrictions on the property prevent intended usage. Restrictions can stall or stop project	4	8	**(SM)** Find and remove any significant title restrictions. Conduct title search and other investigations to determine if significant restrictions exist	*Signpost*: Site restrictions incorporated into the property title. *Trigger*: Restrictions that significantly limit proposed property use. **(AD)**	**(AD)** Work to remove restrictions or look for another site	Not applicable (n/a)	n/a	1	1
The property is free of significant quantities of regulated contaminants	**(V)** Regulated or hazardous contaminants in significant quantities are found on site during construction. Project stalled or stopped. Increased time and cost to remediate	4	7	**(SM)** Investigate the site to determine the existence of significant quantities of contaminants.	*Signpost*: Site use history. *Trigger-1*: Prior uses suggest a significant possibility of contamination. **(AD) (AC)** *Trigger-2*: Prior uses or soil sampling suggest a high level of contamination and significant cleanup costs. **(RE)**	**(AD)** Conduct site investigations to identify past site uses, possible presence of hazardous wastes. If found, change project location or cleanup site. **(AC)** Incorporate site sampling program during construction. **(RE)** If contamination is extensive, reassess the project site location, project viability	n/a	n/a	1	4

(Continued)

(Continued)

Conditions for success	Vulnerability (V) or Opportunity (Op)	Initial risk Lk/Cn		Strategies	Signposts, triggers $t < 0$	Actions $t < 0$	Signposts, triggers $t > 0$	Actions $t > 0$	Residual risk Lk/Cn	
Strategies: (SM) Mitigation, (SH) Hedging, (SZ) Seizing, (SR) Resiliency. Actions: (AD) Defensive, (AC) Corrective, (AP) Capitalizing, (AS) Shaping, (AR) Recovery. (RE) Reassessment.										
	(**Op**) The property could be classified as a brownfield site and eligible for redevelopment funding	—	—	(**SZ**) Investigate the possibility of brownfield redevelopment	*Signpost*: Former site use, level of contamination. *Trigger*: Low level of contamination (not a Superfund site). (**AP**)	(**AP**) Take advantage of brownfield redevelopment incentives.	n/a	n/a	—	—
Construction access and space is sufficient	(**V**) Possibility of space limitations during construction. Restricted construction access. Neighbors (stakeholders) upset. Increased time and costs	4	4	(**SM**) Determine construction access requirements	*Signpost*: Access and space availability vs. requirements for construction. *Trigger*: Access and space inadequate for construction needs. (**AD**)	(**AD**) Work with stakeholders. Establish sufficient construction access and space. If sufficient space is not available, change project location	n/a	n/a	2	2
Construction noise will not be a problem. Noise control is adequate	(**V**) Possibility of noise during construction reaching unacceptable levels	4	5	(**SM**) Design for noise control during construction	*Signpost*: Potential noise levels. *Trigger*: Potential noise levels would exceed local codes, acceptable levels. (**AD**)	(**AD**) Specify required noise reduction policies, practices in construction contracts	n/a	n/a	2	3

(Continued)

(Continued)

Conditions for success	Vulnerability (V) or Opportunity (Op)	Initial risk Lk/Cn		Strategies	Signposts, triggers $t<0$	Actions $t<0$	Signposts, triggers $t>0$	Actions $t>0$	Residual risk Lk/Cn	
Strategies: (SM) Mitigation, (SH) Hedging, (SZ) Seizing, (SR) Resiliency. Actions: (AD) Defensive, (AC) Corrective, (AP) Capitalizing, (AS) Shaping, (AR) Recovery. (RE) Reassessment.										
No significant risk of damage from adjacent structures, facilities	**(V)** *Orphaned Resiliency*: damage from nearby structures and facilities not made climate resilient. For example, barges upstream released from intense storm, wave action damaging infrastructure asset	6	5	**(SM)** Estimate risk of significant damage from nearby structures, facilities that are not climate resilient. Set up program to monitor adjacent facilities. Establish protection measures	*Signpost:* Risk of damage from nearby structures, facilities because of an extreme climate, weather event. *Trigger*: Significant risk. Owner, operator have no plans for improving climate resiliency. **(AS)** **(AD)**	**(AS)** Work with nearby property owners, operators to reduce hazards, exposures. **(AD)** Reinforce project to protect from damage from adjacent structures, facilities. Incorporate resiliency into the project to prevent harm and recover quickly and cost effectively in case of damage	*See Operation, Operations and Maintenance, Orphaned Resiliency*	*See Operation, Operations and Maintenance, Orphaned Resiliency*	3	3
Extreme climate and weather events are manageable at the project location	**(V)** Shoreline erosion and damage from wave action due to increased storm intensity, sea level rise	5	7	**(SM)** Reinforce shoreline using natural and engineered systems. **(SH)** Incorporate shoreline protection and monitoring programs to enable future shoreline reinforcement	*Signpost*: Potential for shoreline erosion. *Trigger*: Shoreline erosion is anticipated to be significant in the event of sea level rise with wave action. **(AD)** **(AC)**	**(AD)** Install riprap to an extent consistent with current and prevention of expected damage from wave action. **(AC)** Design in the capability to add riprap in case of inundation. Incorporate shore erosion monitoring into asset operations	*See Operation, Operations and Maintenance, shoreline erosion*	*See Operation, Operations and Maintenance, shoreline erosion*	3	3

(Continued)

(Continued)

Conditions for success	Vulnerability (V) or Opportunity (Op)	Initial risk Lk/Cn		Strategies	Signposts, triggers $t < 0$	Actions $t < 0$	Signposts, triggers $t > 0$	Actions $t > 0$	Residual risk Lk/Cn	
Strategies: (SM) Mitigation, (SH) Hedging, (SZ) Seizing, (SR) Resiliency. Actions: (AD) Defensive, (AC) Corrective, (AP) Capitalizing, (AS) Shaping, (AR) Recovery. (RE) Reassessment.										
	(V) Flood risk from sea level rise plus rainstorms of increasing duration, intensity and frequency	5	6	(**SM**) Incorporate flood walls to protect the project from flooding. (**SH**) Design flood walls to enable height adjustments	*Signpost:* Proximity to water bodies, flood zones. *Trigger:* Flooding is a significant risk to operations. (**AD**)	(**AD**) Design and construct flood walls at initial height XX. Incorporate ability to raise the height of flood walls incrementally to ZZ height	*See Operations, Operations and Maintenance, Flood risk*	*See Operations, Operations and Maintenance, Flood risk*	3	3
	(V) Heat waves of increasing frequency, intensity and duration	5	4	(**SM**) Incorporate the ability to handle a wide range of temperatures for extended periods of time. Backups for utilities in case of outages	*Signpost:* High temperatures and heat waves is a significant risk to operations. *Trigger:* Likelihood of temperatures, heat waves reaching XX levels. (**AD**)	(**AD**) Design project to operate in a wide range of ambient temperatures, extended heat waves	*See Operations, Operations and Maintenance, Heat waves*	*See Operations, Operations and Maintenance, Heat waves*	3	2
	(V) Wildfires due to extended drought plus extended heat wave	3	6	(**SM**) Reduce exposure to wildfires. Establish wildfire protection protocols	*Signpost:* Wildfires are a significant risk to operations. *Trigger:* Proximity to forests, woodlands. (**AD**)	(**AD**) Locate project away from urban-forest interfaces. Design the project with protection against wildfires	*See Operations, Operations and Maintenance, Wildfires*	*See Operations, Operations and Maintenance, Wildfires*	2	3

(Continued)

(Continued)

Conditions for success	Vulnerability (V) or Opportunity (Op)	Initial risk Lk/Cn		Strategies	Signposts, triggers $t < 0$	Actions $t < 0$	Signposts, triggers $t > 0$	Actions $t > 0$	Residual risk Lk/Cn	
Strategies: (SM) Mitigation, (SH) Hedging, (SZ) Seizing, (SR) Resiliency. Actions: (AD) Defensive, (AC) Corrective, (AP) Capitalizing, (AS) Shaping, (AR) Recovery. (RE) Reassessment.										
Users and workers can readily access the project site	(V) *Adaptation Island*: User and worker access is blocked. Transportation routes leading to the infrastructure asset become blocked because of flooding from severe storms	2	6	**(SM)** Assess flood risk of access roads, rail. **(SM)** Identify alternate means of access. **(SM)** Protect access routes from flooding	*Signpost*: Climate resiliency of transportation routes leading to the infrastructure asset. *Trigger*: Significant risk of flooding, other damage. **(AD)**	**(AD)** Devise plans for alternate worker access. **(AD)** Work with transportation agencies to improve flood protection of roads, rail	*See Operations, Access Adaptation Island: User and worker access*	*See Operations, Access Adaptation Island: User and worker access*	2	4
Supplies for asset operations can reach the project site	(V) *Adaptation Island*: Supplies are not available. Transportation routes leading to the infrastructure asset become blocked because of flooding from severe storms	2	6	**(SM)** Assess flood risk of access roads, rail. **(SM)** Provide room for storage of additional supplies. **(SM)** Identify alternate means of access. **(SM)** Protect access routes from flooding	*Signpost*: Climate resiliency of transportation routes leading to the infrastructure asset. *Trigger*: Significant risk of flooding, other damage. **(AD)**	**(AD)** Design in additional storage capacity for supplies. **(AD)** Devise plans for alternate access. **(AD)** Work with transportation agencies to improve flood protection of roads, rail	*See Operations, Supplies and Utilities Adaptation Island: Supplies*	*See Operations, Supplies and Utilities Adaptation Island: Supplies*	2	3
Utility services are sufficient and available	(V) *Adaptation Island*: Utility services, e.g., water, electrical, not made climate resilient. Asset loses services	2	6	**(SM)** Assess utility climate resilience. **(SM)** Provide emergency backups for water, electrical power	*Signpost*: Level of climate resiliency of utility services. *Trigger*: Utility services not made sufficiently climate resilient. **(AD)**	**(AD)** Design in backup systems for water, electrical power	*See Operations, Supplies, Utilities, Adaptation Island: Utility services*	*See Operations, Supplies, Utilities, Adaptation Island: Utility services*	2	3

(Continued)

(Continued)

Conditions for success	Vulnerability (V) or Opportunity (Op)	Initial risk Lk/Cn		Strategies	Signposts, triggers $t < 0$	Actions $t < 0$	Signposts, triggers $t > 0$	Actions $t > 0$	Residual risk Lk/Cn	
Strategies: (SM) Mitigation, (SH) Hedging, (SZ) Seizing, (SR) Resiliency. Actions: (AD) Defensive, (AC) Corrective, (AP) Capitalizing, (AS) Shaping, (AR) Recovery. (RE) Reassessment.										
Requirements										
No significant changes in applicable codes and regulations, permit requirements	(V) New relevant codes, regulations issued requiring changes in the scope of work. Potential increases in costs, delivery schedule	3	6	(**SM**) Understand and track possible changes in relevant codes and regulations. Incorporate changes as necessary into the project design	*Signpost*: Upcoming changes to applicable rules, codes, regulations. *Trigger*: Changes significantly affect the project. (**AS**) (**AD**)	(**AS**) Work with agencies to understand applicability of codes, regulations. (**AD**) Change plans, designs to comply with changes as required	n/a	n/a	3	3
Finance										
Project planning, design and construction budget is sufficient	(V) Tight municipal budgets force the project owner to cut project planning, design, and construction budget	4	6	(**SM**) Develop a sound project scope and budget. Develop a reasonable range of changes (scope, budget) that could be made and still substantially meet objectives. Solidify project funding, authorization	*Signpost*: Level of authorized funds for project. *Trigger-1*: Project funding is reduced to $XX. (**AD**) Trigger-2: Project funding is reduced to $YY. (**RE**)	(**AD**) Work with project owner to alter the project scope, schedule to accommodate changes in authorized funds. (**RE**) If authorized funds are too small, revise or abandon project	n/a	n/a	3	5
Sociopolitical										
Project is acceptable to stakeholders	(V) Loss of stakeholder support	5	6	(**SM**) Develop strong stakeholder relations program. Assess needs, concerns. Devise fixes	*Signpost*: Stakeholder acceptance and support. *Trigger*: Significant loss of support. (**AD**)	(**AD**) Set up information project programs to inform stakeholders. Meet with stakeholders regularly. Listen to their needs and concerns. Respond	n/a	n/a	3	4

(Continued)

(Continued)

Conditions for success	Vulnerability (V) or Opportunity (Op)	Initial risk Lk/Cn		Strategies	Signposts, triggers $t < 0$	Actions $t < 0$	Signposts, triggers $t > 0$	Actions $t > 0$	Residual risk Lk/Cn	
Strategies: (SM) Mitigation, (SH) Hedging, (SZ) Seizing, (SR) Resiliency. Actions: (AD) Defensive, (AC) Corrective, (AP) Capitalizing, (AS) Shaping, (AR) Recovery. (RE) Reassessment.										
No changes inadministration, elected officials	**(V)** Elections bring about changes in state, municipal governments. Changes in priorities	4	7	**(SM)** Develop relationships with administration and elected officials. Demonstrate strong stakeholder support	*Signpost:* Attitudes, priorities of administration, officials. *Trigger:* Significant changes in priorities that can significantly affect project. **(AD)**	**(AD)** Project owner and team meet with new administration and discuss project importance, criticality to stakeholders. Justify the project scope and budget	n/a	n/a	3	4
				Project delivery (time < 0) *Completing the project as specified on time and within budget*						
Scope										
Scope of work is sufficient for the owner's requirements and specifications	**(V)** Scope creep: additional work elements are added by project staff. Increased project schedule, cost	5	4	**(SM)** Incorporate scope creep guidance into the project change management plan	*Signpost:* project progress, costs, work tasks. *Trigger:* Project slippage > X weeks, or Costs > \$Y. New work elements added. **(AC)**	**(AC)** Educate project staff on managing work scope. Work with owner to make the appropriate scope and budget modifications	n/a	n/a	3	3
	(V) Scope creep: additional work elements are added by project owner. Increase in project schedule, cost	5	4	**(SM)** Institute procedures for handling client initiated scope of work changes	*Signpost:* Project progress, costs, work tasks. *Trigger:* Project slippage > X weeks. Costs > \$Y. New work elements added. **(AC)**	**(AC)** Educate project staff on managing owner additions to the scope of work. Work with owner to make the appropriate scope and budget modifications	n/a	n/a	3	3

(Continued)

(Continued)

Conditions for success	Vulnerability (V) or Opportunity (Op)	Initial risk Lk/Cn		Strategies	Signposts, triggers $t < 0$	Actions $t < 0$	Signposts, triggers $t > 0$	Actions $t > 0$	Residual risk Lk/Cn	
Strategies: (SM) Mitigation, (SH) Hedging, (SZ) Seizing, (SR) Resiliency. Actions: (AD) Defensive, (AC) Corrective, (AP) Capitalizing, (AS) Shaping, (AR) Recovery. (RE) Reassessment.										
The property remains free of significant quantities of regulated contaminants	(**V**) Hazardous contaminants in significant quantities are found on the site during construction. Project is stalled. Increased time and cost to remediate	4	7	(**SH**) Establish regular site sampling for contaminants throughout construction	*Signpost*: Composition of samples taken during construction. *Trigger-1*: Hazardous wastes detected in significant but at manageable concentrations. (**AC**) *Trigger-2*: Extensive contamination found requiring extensive cleanup. (**RE**)	(**AC**) Stop construction work. Conduct hazardous waste investigation and cleanup. (**RE**) Assess whether the site is too contaminated to proceed. If so, revise or abandon project	n/a	n/a	2	7
Schedule										
Project is delivered in accordance with agreed-on schedule	(**Op**) Opportunity to reduce project delivery time through improved scheduling, resulting in cost savings	—	—	(**SZ**) Work with owner and project team to find ways to shorten schedule	*Signpost*: Project progress as compared to established schedule. *Trigger*: Clear opportunities to improve project schedule and/or reduce project costs. (**AP**)	(**AP**) Convene meetings with project team, owner. Brainstorm ways to reduce cost and improve project delivery schedule	n/a	n/a	—	—

(Continued)

(Continued)

Conditions for success	Vulnerability (V) or Opportunity (Op)	Initial risk Lk/Cn		Strategies	Signposts, triggers $t < 0$	Actions $t < 0$	Signposts, triggers $t > 0$	Actions $t > 0$	Residual risk Lk/Cn	
Strategies: (SM) Mitigation, (SH) Hedging, (SZ) Seizing, (SR) Resiliency. Actions: (AD) Defensive, (AC) Corrective, (AP) Capitalizing, (AS) Shaping, (AR) Recovery. (RE) Reassessment.										
Cost										
Cost of labor, materials and equipment are the same as initially estimated	(**V**) Supply chain problems cause equipment delivery delays. Increased project costs. Schedule slippage	3	7	(**SH**) Schedule early delivery of equipment. (**SM**) Set up alternative sources for equipment. (**SH**) Build in slack time in project schedule to account for delays	*Signpost*: Promised vs. actual delivery dates. *Trigger-1*: Slippage in actual delivery dates exceeds XX weeks. (**AC-1**) *Trigger-2*: Slippage in actual delivery dates exceeds YY weeks. (**AC-2**)	(**AC-1**) Access alternate sources of equipment. Revise project schedule, using built in slack time. (**AC-2**) Negotiate with project owner for additional time for project completion			3	3
Quality										
Project quality is maintained in accordance with the established quality management plan	(**V**) Project quality differs due to lack of quality control. Increased project costs. Schedule slippage. Project owner dissatisfaction	5	5	(**SM**) Establish project quality standards and educate project team. Emphasize the importance of following the quality management plan to project staff. (**SH**) Assess quality of delivery	*Signpost*: Project quality as defined by owner, quality management plan. *Trigger*: Significant differences. (**AC**)	(**AC**) Conduct additional staff training on the project quality management plan, expectations	n/a	n/a	3	4

(Continued)

(Continued)

Conditions for success	Vulnerability (V) or Opportunity (Op)	Initial risk Lk/Cn		Strategies	Signposts, triggers $t<0$	Actions $t<0$	Signposts, triggers $t>0$	Actions $t>0$	Residual risk Lk/Cn	
Strategies: (SM) Mitigation, (SH) Hedging, (SZ) Seizing, (SR) Resiliency. Actions: (AD) Defensive, (AC) Corrective, (AP) Capitalizing, (AS) Shaping, (AR) Recovery. (RE) Reassessment.										
Staffing										
Qualified project staff is maintained throughout the project delivery process	(**V**) Loss of key project personnel resulting in increased project cost, schedule slippage	5	5	(**SM**) Improve project staff "bench strength." Establish programs for assessing, improving staff satisfaction	*Signpost*: Survey of project staff satisfaction. Staff voluntary turnover rate. *Trigger*: Significant unsatisfaction in key staff. Voluntary turnover rate greater than XX% per year. (**AD**) (**AC**)	(**AD**) Identify key personnel. Train staff for multiple roles. Able to sub for or take on other roles and responsibilities. (**AC**) Address causes of staff dissatisfaction	n/a	n/a	3	3
Technologies										
Applied technologies work as expected	(**V**) Unexpected problems with new and relatively untested technologies jeopardizes project schedule, cost	6	5	(**SM**) Establish an agreement with the project owner for risk and reward sharing on new and relatively untried technologies	*Signpost*: Technology performance across predetermined parameters. *Trigger*: Performance is not acceptable across one or more parameters. (**AD**)	(**AD**) Execute agreement terms for sharing costs of fixes	n/a	n/a	4	3

(Continued)

(Continued)

Conditions for success	Vulnerability (V) or Opportunity (Op)	Initial risk Lk/Cn		Strategies	Signposts, triggers $t < 0$	Actions $t < 0$	Signposts, triggers $t > 0$	Actions $t > 0$	Residual risk Lk/Cn	
Strategies: (SM) Mitigation, (SH) Hedging, (SZ) Seizing, (SR) Resiliency. Actions: (AD) Defensive, (AC) Corrective, (AP) Capitalizing, (AS) Shaping, (AR) Recovery. (RE) Reassessment.										
Acceptance										
Project is acceptable to the affected communities during construction	(**V**) Construction excessively disturbs adjacent neighborhoods. Many complaints	5	4	(**SH**) Establish local outreach programs to hear and address community concerns	*Signpost*: Extent of negative feedback. *Trigger*: Degree of negative feedback from community high enough to jeopardize project completion. (**AC**)	(**AC**) Alter construction practices to accommodate community concerns	n/a	n/a	4	3
Operation (time > 0) *Completed project delivers the expected services, levels of services while handling change*										
Access										
User and worker access to the infrastructure asset is maintained	(**V**) *Adaptation Island*: User and worker access is blocked. Transportation routes leading to the infrastructure asset become blocked because of flooding from severe storms	2	6	(**SH**) Use other means of transport to move essential staff to the asset. (**SM**) Improve climate resilience of transportation modes, routes	*See Planning, Location, Adaptation Island: Users and workers*	*See Planning, Location, Adaptation Island: Users and workers*	*Signpost*: Frequency, duration of infrastructure asset access restrictions. *Trigger*: Access blocked for XX days **YY** times per year. (**AC**) (**AS**)	(**AC**) Employ other infrastructure assets to provide service during asset downtime. Use alternate means of transport for essential workers. (**AS**) Continue push for improvements in climate resilience for transportation modes, access routes	2	4

(Continued)

(Continued)

Conditions for success	Vulnerability (V) or Opportunity (Op)	Initial risk Lk	Initial risk Cn	Strategies	Signposts, triggers $t<0$	Actions $t<0$	Signposts, triggers $t>0$	Actions $t>0$	Residual risk Lk	Residual risk Cn
Strategies: (SM) Mitigation, (SH) Hedging, (SZ) Seizing, (SR) Resiliency. Actions: (AD) Defensive, (AC) Corrective, (AP) Capitalizing, (AS) Shaping, (AR) Recovery. (RE) Reassessment.										
Operations and Maintenance (O&M)										
Maintenance is performed as required to maintain efficient operation	(**V**) Operating budgets are cut forcing maintenance to be deferred. Performance is compromised	3	6	(**SM**) Work with funding parties and communicate clearly the required operations and maintenance requirements and consequences if not provided. Clear documentation of O&M requirements, costs	n/a	n/a	*Signpost*: Authorized funds for O&M. *Trigger*: Funds available drops by XX%. (**AS**) (**AC**)	(**AS**) Convince funding parties to restore funding. (**AC**) Decrease service levels commensurate with safe operation. Work to locate other sources of O&M funding	3	4
Project operations are acceptable to stakeholders	(**V**) Loss of stakeholder support	3	5	(**SM**) Set up community outreach programs. Listen and respond to stakeholder needs, concerns	n/a	n/a	*Signpost*: Stakeholder acceptance and support based on user, stakeholder surveys. *Trigger*: Significant drop in support. (**AC**)	(**AC**) Schedule additional meetings with stakeholders to better understand needs and concerns. Make meaningful and timely responses	2	4
	(**V**) Noise during operations becomes excessive. Adjacent stakeholders disturbed	4	5	(**SH**) Maintain low levels of operational noise	n/a	n/a	*Signposts*: Noise level complaints by neighbors. *Triggers*: If levels exceed XX. (**AC**)	(**AC**) Install noise barriers. Replace noisy equipment. Alter operations	3	3

(Continued)

(Continued)

Conditions for success	Vulnerability (V) or Opportunity (Op)	Initial risk Lk	Initial risk Cn	Strategies	Signposts, triggers $t < 0$	Actions $t < 0$	Signposts, triggers $t > 0$	Actions $t > 0$	Residual risk Lk	Residual risk Cn
Strategies: (SM) Mitigation, (SH) Hedging, (SZ) Seizing, (SR) Resiliency. Actions: (AD) Defensive, (AC) Corrective, (AP) Capitalizing, (AS) Shaping, (AR) Recovery. (RE) Reassessment.										
Extreme climate and weather events are manageable at the project location	**(V)** Shoreline erosion and damage from wave action due to increased storm intensity, sea level rise	5	7	**(SH)** Track effectiveness of shoreline protection measures and instigate actions as needed	*See Planning, location, Shoreline erosion*	*See Planning, Location, Shoreline erosion*	*Signpost:* Shoreline erosion from wave action. *Trigger:* Shoreline erosion is significant. Storms inundate property XX times in YY time period. (**AC**)	(**AC**) Install additional riprap to enhance storm protection	3	3
	(V) Flood risk from sea level rise plus rainstorms of increasing duration, intensity, and frequency	5	6	**(SH)** Track sea level rise and adjust flood wall height in accordance with flood risk	*See Planning, Location, Flood risk*	*See Planning, Location, Flood risk*	*Signpost:* sea level rise. *Trigger-1:* greater than Y1. (**AC-1**) *Trigger-2:* Greater than Y2. (**AC-2**)	(**AC-1**) Raise the height of the flood wall to Z1. (**AC-2**) Raise the height of the flood wall to Z2	3	3
	(V) Heat waves of increasing duration, intensity, and frequency	5	4	**(SH)** Respond to extreme heat events	*See Planning, Location, Heat waves*	*See Planning, Location, Heat waves*	*Signpost:* Forecast temperatures, expected durations. *Trigger:* temperatures exceed XX for Y days. (**AC**)	(**AC**) Implement extreme heat protocols. Change equipment operations, work schedules to protect from high temperatures. Incorporate additional cooling capacity. Use utility backup.	3	2

(Continued)

(Continued)

Conditions for success	Vulnerability (V) or Opportunity (Op)	Initial risk Lk/Cn		Strategies	Signposts, triggers $t < 0$	Actions $t < 0$	Signposts, triggers $t > 0$	Actions $t > 0$	Residual risk Lk/Cn	
Strategies: (SM) Mitigation, (SH) Hedging, (SZ) Seizing, (SR) Resiliency. Actions: (AD) Defensive, (AC) Corrective, (AP) Capitalizing, (AS) Shaping, (AR) Recovery. (RE) Reassessment.										
	(**V**) Wildfires due to extended drought plus extended heat wave	3	6	(**SH**) Implement protocols for wildfire protection	*See Planning, Location, Wildfires*	*See Planning, Location, Wildfires*	*Signpost*: extended droughts and heat waves occur at the project location. *Trigger*: Fire danger rated as high. (**AC**)	(**AC**) Implement wildfire protection protocols. Shut down electrical power. Evacuate personnel	2	3
No significant risk of damage from adjacent structures, facilities	(**V**) *Orphaned Resiliency*: Damage from nearby structures and facilities not made climate resilient. For example, barges upstream released from intense storm, wave action	6	5	(**SH**) Monitor adjacent structures, facilities for vulnerabilities to extreme events	*See Planning, Location, Orphaned Resiliency*	*See Planning, Location, Orphaned Resiliency*	*Signpost*: Extreme storm events that could cause release. *Trigger*: Significant debris or objects released from nearby properties, structures, facilities. (**AC**) (**AR**)	(**AC**) Remove or reinforce vulnerable structures and facilities. (**AR**) Execute preplanned recovery actions as needed	3	3
Supplies and Utilities										
Supplies needed for asset operation are available as needed	(**V**) *Adaptation Island*: Supplies are not available. Transportation routes leading to the infrastructure asset become blocked because of flooding from severe storms	2	6	(**SM**) Increase supply storage capacity. (**SH**) Make use of alternative transport modes, routes to bring in supplies	*See Planning, Location, Adaptation Island: Supplies*	*See Planning, Location, Adaptation Island: Supplies*	*Signpost*: Infrastructure access. *Trigger*: Access blocked for YY days ZZ times per year. (**AC**)	(**AC**) Use stockpiled supplies for operation. Expand supply stockpile storage capacity	2	3

(Continued)

(Continued)

Conditions for success	Vulnerability (V) or Opportunity (Op)	Initial risk Lk	Initial risk Cn	Strategies	Signposts, triggers $t<0$	Actions $t<0$	Signposts, triggers $t>0$	Actions $t>0$	Residual risk Lk	Residual risk Cn
Strategies: (SM) Mitigation, (SH) Hedging, (SZ) Seizing, (SR) Resiliency. Actions: (AD) Defensive, (AC) Corrective, (AP) Capitalizing, (AS) Shaping, (AR) Recovery. (RE) Reassessment.										
Utility services are sufficient and available	**(V)** *Adaptation Island*: Utility services, e.g., water, electrical, not made climate resilient. Asset loses services	2	6	**(SM)** Provide emergency backups for water, electrical power	*See Planning, Location, Adaptation Island: Utility services*	*See Planning, Location, Adaptation Island: Utility services*	*Signpost*: Supplies of water and electrical power. *Trigger*: Loss of water, electrical power. **(AD)**	**(AD)** Shift to water and electrical backup supplies	2	3
Performance										
Project performs as required throughout its design life	**(V)** Performance is less than required	4	4	**(SH)** Devise performance metrics and monitor. Develop plans to improve performance	n/a	n/a	*Signpost*: Performance metrics. *Trigger*: Significant drop in performance measures. **(AC)**	**(AC)** Operations management and staff analyze and assess performance. Determine causes and fix	3	3
Service										
Demand for services develops as initially forecast	**(V)** Demand for services increases faster than expected. Additional stress on workers, equipment, facilities. Accelerated deterioration. Failure to meet expected quality of services	5	5	**(SH)** Develop and implement plans for handling increased demand	n/a	n/a	*Signpost*: Service demand. *Trigger*: Significant increase in demand for services. **(AC)** **(AD)** **(RE)**	**(AC)** Increase usage charges. Implement plans to manage increased usage. **(AD)** Construct new sections, components to add capacity. **(RE)** Reassess the overall plan	4	3

(Continued)

(Continued)

Conditions for success	Vulnerability (V) or Opportunity (Op)	Initial risk Lk/Cn		Strategies	Signposts, triggers $t < 0$	Actions $t < 0$	Signposts, triggers $t > 0$	Actions $t > 0$	Residual risk Lk/Cn	
Strategies: (SM) Mitigation, (SH) Hedging, (SZ) Seizing, (SR) Resiliency. Actions: (AD) Defensive, (AC) Corrective, (AP) Capitalizing, (AS) Shaping, (AR) Recovery. (RE) Reassessment.										
	(**V**) Demand for services is lower than expected. Loss of service revenue	5	5	(**SH**) Devise and implement options for increasing demand for services	n/a	n/a	*Signpost:* Service demand. *Trigger:* Significant decrease in demand for services. (**AC**)	(**AC**) Increase advertising to attract users. Change usage charges	4	3
	(**Op**) Demand for services is lower than expected. Possibility of alternate uses	5	5	(**SZ**) Conduct an analysis of potential for alternate uses if demand is below estimates	n/a	n/a	*Signpost:* Service demand. *Trigger:* Significant decrease in demand for services. (**AP**)	(**AP**) Implement alternate uses	4	4

Reference

Wall, T. A., W. E. Walker, V. A. W. J. Marchau, and L. Bertolini. 2015. "Dynamic adaptive approach to transportation-infrastructure planning for climate change: San-Francisco-bay-area case study." *J. Infrastruct. Syst.* **21** (4): 05015004. https://doi.org/10.1061/(ASCE)IS.1943-555X.0000257.

Index

Note: page numbers followed by *e*, *f*, or *t* indicate equations, figures, or tables.

About the Author

Bill Wallace has more than 40 years of professional experience, including 30 years in environmental engineering and management. The author is a recognized expert in sustainable development, specializing in work with the consulting engineering industry. His first book, *Becoming Part of the Solution: The Engineer's Guide to Sustainable Development*, has been a best seller for the American Council of Engineering Companies (ACEC). The author is also a member and elected fellow of ASCE.

Bill was the inventor and lead designer of the Envision sustainable infrastructure rating system. This system helps project owners and engineers determine the extent to which their infrastructure projects contribute to sustainability and climate resiliency. Envision is managed by the Institute for Sustainable Infrastructure (ISI) in Washington, DC. The Institute for Sustainable Infrastructure was founded by ASCE, ACEC, and the American Public Works Association. The Zofnass Program for Sustainable Infrastructure at the Harvard University Graduate School of Design was a partner in the development of Envision. They recently joined ISI and will be known as the Zofnass Research Program. There are more than 6,500 ENV SPs (Envision Sustainability Professionals) in the United States, Canada, and 19 other countries. More than 344 projects (constructed value of US$237 billion) have received Envision awards or are in the awards pipeline. For his work, Bill received *Engineering News-Record*'s 2013 Award of Excellence and the ASCE President's Medal in 2014.

Bill has spent 21 years at CH2M Hill (now part of the Jacobs Engineering Group) holding various senior management positions, including a 3-year term on the board of directors. He led the firm's hazardous waste management group and helped transform the company into a national leader in hazardous waste site remediation. On multiple occasions, he was invited to testify to US congressional committees on the use of innovative technologies in hazardous waste site cleanup. At CH2M Hill, Bill started the firm's sustainable engineering practice and led its foray into the Olympic Games. CH2M Hill provided environmental and engineering support services to the Olympic cities of Atlanta, Sydney, Salt Lake City, Beijing, and London.

Bill teaches a distance learning course at the University of Florida, "Principles of Sustainable Engineering Design." The author also developed and chaired ASCE's first International Conference on Sustainable Infrastructure (ICSI) in Long Beach, California, November 6–8, 2014.

Bill is a past member of the Board of GeoEngineers, Inc. and the ASCE Committee on Sustainability. The author also served as a liaison delegate to the World Business Council for Sustainable Development, an international organization of more than 200 multinational companies, all with a shared commitment to sustainability. The author served as the president and member of the Governing Board of Engineers Without Borders—USA.

Bill received a BS in chemical engineering from Clarkson University and an MS in management from Rensselaer Polytechnic Institute. He also attended the Advanced Management Program at the Harvard Business School.